Bertram

Numerical Solution of
Integral Equations

MATHEMATICAL CONCEPTS AND METHODS IN SCIENCE AND ENGINEERING

Series Editor: Angelo Miele
    Mechanical Engineering and Mathematical Sciences
    Rice University

Recent volumes in this series:

31 **NUMERICAL DERIVATIVES AND NONLINEAR ANALYSIS**
    • *Harriet Kagiwada, Robert Kalaba, Nima Rasakhoo, and Karl Spingarn*

32 **PRINCIPLES OF ENGINEERING MECHANICS**
    Volume 1: Kinematics—The Geometry of Motion • *Millard F. Beatty, Jr.*

33 **PRINCIPLES OF ENGINEERING MECHANICS**
    Volume 2: Dynamics—The Analysis of Motion • *Millard F. Beatty, Jr.*

34 **STRUCTURAL OPTIMIZATION**
    Volume 1: Optimality Criteria • *Edited by M. Save and W. Prager*

35 **OPTIMAL CONTROL APPLICATIONS IN ELECTRIC POWER SYSTEMS**
    G. S. Christensen, M. E. El-Hawary, and S. A. Soliman

36 **GENERALIZED CONCAVITY** • *Mordecai Avriel, Walter E. Diewert, Siegfried Schaible, and Israel Zang*

37 **MULTICRITERIA OPTIMIZATION IN ENGINEERING AND IN THE SCIENCES**
    • *Edited by Wolfram Stadler*

38 **OPTIMAL LONG-TERM OPERATION OF ELECTRIC POWER SYSTEMS**
    • *G. S. Christensen and S.A. Soliman*

39 **INTRODUCTION TO CONTINUUM MECHANICS FOR ENGINEERS**
    • *Ray M. Bowen*

40 **STRUCTURAL OPTIMIZATION**
    Volume 2: Mathematical Programming • *Edited by M. Save and W. Prager*

41 **OPTIMAL CONTROL OF DISTRIBUTED NUCLEAR REACTORS**
    • *G. S. Christensen, S. A. Soliman, and R. Nieva*

42 **NUMERICAL SOLUTION OF INTEGRAL EQUATIONS**
    • *Edited by Michael A. Golberg*

A Continuation Order Plan is available for this series. A continuation order will bring delivery of each new volume immediately upon publication. Volumes are billed only upon actual shipment. For further information please contact the publisher.

Numerical Solution of Integral Equations

Edited by
Michael A. Golberg
University of Nevada
Las Vegas, Nevada

Plenum Press • New York and London

Library of Congress Cataloging-in-Publication Data

Numerical solution of integral equations / edited by Michael A.
Golberg.
      p.   cm. -- (Mathematical concepts and methods in science and
engineering ; 42)
   Includes bibliographical references.
   ISBN 0-306-43262-5
   1. Integral equations--Numerical solutions.   I. Golberg, Michael
A. II. Series.
QA431.N852 1990
515'.45--dc20   90-7210
                                                                 CIP

© 1990 Plenum Press, New York
A Division of Plenum Publishing Corporation
233 Spring Street, New York, N.Y. 10013

All rights reserved

No part of this book may be reproduced, stored in a retrieval system, or transmitted
in any form or by any means, electronic, mechanical, photocopying, microfilming,
recording, or otherwise, without written permission from the Publisher

Printed in the United States of America

Contributors

R. S. Anderssen, Division of Mathematics and Statistics, Institute of Information and Communications Technologies, CSIRO, GPO Box 1965 Canberra, Australia 2601

K. E. Atkinson, Department of Mathematics, University of Iowa, Iowa City, Iowa 52242

F. R. de Hoog, Division of Mathematics and Statistics, Institute of Information and Communications Technologies, CSIRO, GPO Box 1965 Canberra, Australia 2601

D. Elliott, Department of Applied Mathematics, The University of Tasmania, Hobart, Tasmania, Australia 7001

M. A. Golberg, Department of Mathematical Sciences, University of Nevada, Las Vegas, Nevada 89154

G. Miel, Hughes Research Laboratories, Malibu, California 90265

I. H. Sloan, Department of Mathematics, The University of New South Wales, Kensington, New South Wales, Australia 2033

E. O. Tuck, Applied Mathematics Department, University of Adelaide, Adelaide, SA, Australia 5001

Preface

In 1979, I edited Volume 18 in this series: *Solution Methods for Integral Equations: Theory and Applications.* Since that time, there has been an explosive growth in all aspects of the numerical solution of integral equations. By my estimate over 2000 papers on this subject have been published in the last decade, and more than 60 books on theory and applications have appeared. In particular, as can be seen in many of the chapters in this book, integral equation techniques are playing an increasingly important role in the solution of many scientific and engineering problems. For instance, the boundary element method discussed by Atkinson in Chapter 1 is becoming an equal partner with finite element and finite difference techniques for solving many types of partial differential equations.

Obviously, in one volume it would be impossible to present a complete picture of what has taken place in this area during the past ten years. Consequently, we have chosen a number of subjects in which significant advances have been made that we feel have not been covered in depth in other books. For instance, ten years ago the theory of the numerical solution of Cauchy singular equations was in its infancy. Today, as shown by Golberg and Elliott in Chapters 5 and 6, the theory of polynomial approximations is essentially complete, although many details of practical implementation remain to be worked out.

Other topics treated here and largely unknown ten years ago are superconvergence, discussed by Sloan in Chapter 2 and Golberg in Chapter 3, and the implementation of numerical algorithms on multiprocessor computers, examined by Miel in Chapter 4.

The solution of ill-posed problems continues to present a major challenge to numerical analysts, and a number of important developments for the special case of Abel integral equations are surveyed by Anderssen and de Hoog in Chapter 8.

Although only a prophet could predict with certainty what directions this subject will take in the near future, we have no doubt that the topics presented here will continue to be actively investigated, and we hope that researchers and students will find this material useful in their work for years to come.

As with any multiply authored book, the efforts of many people have gone into producing the final product, and must be acknowledged. First there are the contributors, their colleagues, their secretaries, and no doubt their families, without whose outstanding efforts this work would not have come into being. Second are my own typists, Jennifer Lewis, Christine McKenna, and my wife, Gloria, who worked long and hard at deciphering my handwriting.

Thanks are also due to my children, Jonathan and Stefany, who helped with literature searches and proofreading, and to the interlibrary loan staff at the University of Nevada, Las Vegas, who were responsible for processing numerous requests for papers and books. I must also thank Dr. Sadanand Verma, the Chairman of the Mathematical Sciences Department at UNLV, for giving me the time to work on this project, and acknowledge the financial support of a Barrick Fellowship (a UNLV award), which provided funding for much of my own work. Finally, credit should be given to Professor Angelo Miele and Plenum Publishing Corporation for giving me the opportunity to get this project into print.

Michael A. Golberg

Las Vegas, Nevada

Contents

1. **A Survey of Boundary Integral Equation Methods for the Numerical Solution of Laplace's Equation in Three Dimensions** . . . 1
   K. E. Atkinson

   1. Introduction . . . 1
   2. Integral Equation Reformulations of Laplace's Equation . . . 2
      2.1. Direct BIE Methods . . . 4
      2.2. Indirect BIE Methods . . . 5
      2.3. The Smooth Boundary Case . . . 7
      2.4. Piecewise Smooth Boundaries . . . 8
      2.5. The Poisson Equation . . . 9
   3. Numerical Methods for Boundary Integral Equations . . . 9
      3.1. Global Methods . . . 11
   4. Boundary Element Methods . . . 14
      4.1. Strongly Elliptic Operator Framework . . . 14
      4.2. Collocation Methods . . . 18
   5. Numerical Integration . . . 21
      5.1. Boundary Element Integrations . . . 22
   6. Iterative Methods of Solution . . . 25
      6.1. A Two-Grid Method . . . 26
      References . . . 29

2. **Superconvergence** . . . 35
   I. H. Sloan

   1. Introduction . . . 35
   2. Galerkin and Collocation Methods . . . 38
   3. Iterated Projection Method . . . 40
      3.1. Iterated Kantorovich Method . . . 45
      3.2. An Example . . . 47
   4. Linear Functionals of the Galerkin Approximation . . . 48
      4.1. Approximation of (y, g) by (y_h, g) . . . 49
      4.2. The Iterated Galerkin Method Revisited . . . 51
      4.3. HMP Method for Approximating (y, g) . . . 53
      4.4. Discrete Galerkin and Iterated Galerkin Methods . . . 55
   5. Iterated Collocation Method . . . 56
      5.1. Iterated Collocation for Piecewise-Polynomial Spaces S_h . . . 58
      5.2. Collocation at the Gauss Points . . . 59

5.3. Iterated Collocation Method versus Iterated Galerkin Method		63
5.4. Discrete Collocation and Iterated Collocation Methods		64
6. Nonlinear Integral Equations		64
7. Concluding Remarks		67
Acknowledgments		67
References		67

3. Perturbed Projection Methods for Various Classes of Operator and Integral Equations — 71
M. A. Golberg

1. Introduction	71
2. Projection Methods and Their Variants	73
2.1. Notation	74
2.2. Projection Methods	75
2.3. Convergence Analysis of Projection Methods	81
2.4. Some Variants of the Projection Method	84
3. Perturbed Projection Methods	91
3.1. The Perturbed Sloan Iterate	92
3.2. The Perturbed Kantorovich Method	93
3.3. The Perturbed Kantorovich Iterate	94
3.4. The Perturbed Dellwo–Friedman Method	95
4. Galerkin's Method with Quadrature Errors	95
5. Direct Analysis of Galerkin's Method with Quadrature Errors	106
6. Collocation Methods with Quadrature Errors	110
6.1. Collocation Method	110
6.2. Direct Analysis of the Discrete Collocation Method	113
6.3. Superconvergence of Collocation for Volterra Equations	117
7. Galerkin's Method for Equations with Positive-Definite Dominant Parts	120
8. Conclusions	126
References	127

4. Numerical Solution of Parallel Processors of Two-Point Boundary-Value Problems of Astrodynamics — 131
G. Miel

1. Introduction	131
2. Mathematical Preliminaries	134
2.1. The Integral Equation	135
2.2. The Legendre Polynomials	136
2.3. The Hilbert Product Space	138
2.4. Operator Equations	140
2.5. The Newton–Kantorovich Method	141
3. Equations of Motion	144
3.1. Keplerian Motion	147
3.2. Aspherical Gravitational Potential	147
3.3. Earth Satellite Orbits	148
3.4. Perturbations of GPS Orbits	149

4. The Well-Posedness Issue	149
4.1. Example	150
4.2. Method of Patched Conics	151
4.3. Grand Tour of Voyager 2	152
4.4. Picard Iteration	154
4.5. Example: Perturbed Keplerian Motion	155
4.6. Newton–Kantorovich Iteration	157
5. Perturbed Galerkin Method	158
5.1. Equivalence	161
5.2. Analytic Principle	163
5.3. Convergence Result	164
6. Parallel Algorithms	165
6.1. Setup of the Matrix	167
6.2. Setup of the Constant Vector	168
6.3. Solution of the Linear System	170
6.4. Legendre Polynomial Expansion	171
6.5. Odds and Ends	172
7. Numerical Examples	174
7.1. Earth–Mars Trajectories	174
7.2. Trajectory Optimization	177
8. Conclusion	179
References	179

5. Introduction to the Numerical Solution of Cauchy Singular Integral Equations 183

M. A. Golberg

1. Introduction	183
2. Analytical Solution of the Airfoil Equation	185
3. Determining c	190
3.1. The Kutta Condition	191
3.2. $l(v) = \int_{-1}^{1} v(t)\, dt = M$	192
3.3. $v(1) = v(-1) = 0$	192
4. Numerical Methods for the Generalized Airfoil Equation	193
5. Indirect Methods	194
6. Direct Methods	197
7. Some Mapping Properties of the Airfoil Operator	198
8. Operator Formulation of the Generalized Airfoil Equation	204
9. Degenerate Kernel Methods	207
10. Galerkin's Method	209
10.1. Galerkin's Method: $\nu = 0$	210
10.2. A Superconvergence Result for $\nu = 0$	214
10.3. Galerkin's Method: $\nu = 1$	218
10.4. Galerkin's Method: $\nu = 1$	220
11. Collocation	220
11.1. $\nu = 0$: Continuous Data	221
11.2. $\nu = 0$: $k(x, t) = a(x, t)\log(|x - t|) + b(x, t)$	222
11.3. Polynomial Collocation: $\nu = \pm 1$	228

12. Quadrature Methods	230
12.1. Quadrature Rules for Cauchy Principal-Value Integrals	230
12.2. The Gaussian Quadrature Method	233
12.3. Lobatto Quadrature: $\nu = 1$	236
12.4. Endpoint Convergence of the Lobatto Quadrature Method	238
12.5. Lobatto Quadrature: $\nu = 0$	240
13. Kantorovich Regularization	241
14. Product Quadrature	246
15. Kalandiya's Method	247
16. Some Other Numerical Methods for the GAE	249
16.1. Conversion to a Logarithmic Equation	250
16.2. Some Other Polynomial Methods for Solving the GAE	255
17. Numerical Solution of CSIES of the Second Kind with Constant Coefficients	257
17.1. The Standard Polynomial Algorithms	258
18. Other Polynomial Approximation Methods	268
18.1. Cohen's Method	269
18.2. The Method of Chawla and Kumar	270
18.3. Hashmi and Delves' Algorithm	272
18.4. Piecewise Polynomial Methods	272
19. Convergence	272
19.1. Mean-Square Convergence of Galerkin's Method: $\nu = 0$	273
19.2. Mean-Square Convergence of Galerkin's Method: $\nu = 1$	278
19.3. Uniform Convergence of Galerkin's Method: $\nu = 0, 1$	281
19.4. The Sloan Iterate	283
19.5. Convergence of the Collocation Method	284
19.6. Convergence of the Discrete Galerkin Method	287
19.7. Convergence of the Gaussian Quadrature Method	293
20. Conclusions	296
References	296

6. Convergence Theorems for Singular Integral Equations 309

D. Elliott

1. Introduction	309
2. A Theory for the Singular Integral Equation	310
3. The Approximate Solution of Singular Integral Equations	322
3.1. An Indirect Method	323
4. Direct Methods and Analysis of their Convergence	328
5. Some Examples of Direct Methods	342
5.1. The Galerkin-Petrov Method	342
5.2. A Collocation Method	348
5.3. A Discrete Galerkin Method	356
6. Conclusions	359
Acknowledgment	359
References	359

7. Planing Surfaces 363

E. O. Tuck

1. Introduction	363
2. The Planing Equation	364

3. Generalizations 368
   References 371

8. Abel Integral Equations 373
R. S. Anderssen and F. R. de Hoog

1. Introduction 373
2. Abel Integral Equations in Applications 376
   2.1. First-Kind Abel Integral Equations in Geometric Probability 377
   2.2. First-Kind Abel Integral Equations in Interferometry 377
   2.3. First-Kind Abel Integral Equations in Stereology 378
   2.4. Second-Kind Abel Integral Equations in Stereology 379
   2.5. The Abel Integral Equation of Seismology 379
   2.6. Abel Integral Equations in Tomography 380
3. The Numerical Analysis of Abel Integral Equations 382
4. Pseudoanalytic Methods 387
5. Wiener Filtering 394
6. Stabilized Evaluation of Inversion Formulas 400
7. The Data Functional Strategy 402
8. Choice of Algorithm 403
   References 406

Index 411

1

A Survey of Boundary Integral Equation Methods for the Numerical Solution of Laplace's Equation in Three Dimensions

K. E. ATKINSON

Abstract. The principal reformulations of Laplace's equation as boundary integral equations (BIEs) are described, together with results on their solvability and the regularity of their solutions. The numerical methods for solving BIEs are categorized, based on whether the method uses local or global approximating functions, whether the method is of collocation or Galerkin type, and whether the equation being solved is defined on a region whose boundary is smooth or only piecewise smooth. Some of the major ideas in the mathematical analysis of these numerical methods are outlined. Certain problems are associated with all numerical methods for boundary integral equations. Principal among these are numerical integration and the iterative solution of linear systems of equations. The research literature for these topics as they arise in solving BIEs is reviewed, and some of the major ideas are discussed.

1. Introduction

Consider boundary integral equation (BIE) methods for solving Laplace's equation in three dimensions. During the last 10–20 years, there has been a large increase in the application of such methods in the engineering literature. For example, see Refs. 1–16. In this survey, in contrast, we focus on the mathematical analysis of BIE methods, describing what is known and suggesting areas where further research is needed. In general, all aspects

---

K. E. ATKINSON • Department of Mathematics, University of Iowa, Iowa City, Iowa 52242. This chapter was written while the author was a Visiting Professor in the School of Mathematics of the University of New South Wales, Sydney, Australia. The author was supported by the University of New South Wales ARC Program Grant on "Numerical Analysis for Integrals, Integral Equations, and Boundary-Value Problems."

of the three-dimensional problem are far less developed than for the two-dimensional problem.

The equation being solved is

$$\Delta u(P) = 0, \qquad P \in D \subset \mathbb{R}^3, \tag{1}$$

with D an open set, possibly unbounded. The boundary of D, call it S, is assumed to be piecewise smooth and to be such that the divergence theorem is valid on $\bar{D}$. It is assumed that

$$S = S_1 \cup \cdots \cup S_J, \tag{2}$$

with each S_j of class C^k, $k \geq 2$. Boundaries that have a cuspoidal type of behavior at some point are excluded. For a further discussion of these requirements, see Ref. 17, Section 1.2, Ref. 18, p. 85, and Ref. 19.

In Section 2, various standard BIE formulations of boundary value problems for Eq. (1) are presented, together with references to the existence and uniqueness theory for them. Numerical methods of discretizing these integral equations are discussed in Sections 3 and 4, including both global and local methods, and both collocation and Galerkin methods. Numerical integration problems and iterative methods for solving the approximating linear systems are discussed in Sections 5 and 6, respectively.

2. Integral Equation Reformulations of Laplace's Equation

We first state the basis of the most popular boundary integral equation reformulations of boundary value problems for Eq. (1). By using Green's second identity, we obtain the following classical formulas.

Initially let D be a bounded simply connected region, with its boundary S described as earlier following Eq. (1); let $\bar{D}_e$ denote the complement of D, $\bar{D}_e = \mathbb{R}^3 \setminus D$.

(F1) Assume that $u \in C^2(D) \cap C^1(\bar{D})$, and let $\Delta u(P) = 0$ for all $P \in D$. Then,

$$\int_S \frac{\partial u(Q)}{\partial \mathbf{n}_Q} \frac{dS(Q)}{|P-Q|} - \int_S u(Q) \frac{\partial}{\partial \mathbf{n}_Q}\left[\frac{1}{|P-Q|}\right] dS(Q)$$

$$= \begin{cases} 4\pi u(P), & P \in D, \\ \mathscr{I}(P)u(P), & P \in S, \\ 0, & P \in D'; \end{cases} \tag{3}$$

here, $\mathbf{n}_Q$ denotes the normal to S at Q, directed outward from D; $\mathscr{I}(P)$ denotes the interior solid angle at $P \in S$; and, for S possessing a unique tangent plane at P, $\mathscr{I}(P) = 2\pi$. See Ref. 20, Section 18.2 for more information on the concept of solid angle.

(F2) Let D be bounded as above, with $\mathbf{n}_Q$ and $\mathscr{I}(P)$ defined exactly as in (F1). Assume that $u \in C^2(D_e) \cap C^1(\bar{D}_e)$, let it satisfy $\Delta u(P) = 0$ at all $P \in D_e$, $D_e = \mathbb{R}^3 \setminus \bar{D}$, and assume that

$$u(P) = O(|P|^{-1}), \tag{4a}$$

$$|\nabla u(P)| = O(|P|^{-2}), \tag{4b}$$

as $|P| \to \infty$. Then

$$\int_S \frac{\partial u(Q)}{\partial \mathbf{n}_Q} \frac{dS(Q)}{|P-Q|} - \int_S u(Q) \frac{\partial}{\partial \mathbf{n}_Q}\left[\frac{1}{|P-Q|}\right] dS(Q)$$

$$= \begin{cases} 0, & P \in D, \\ -[4\pi - \mathscr{I}(P)]u(P), & P \in S, \\ -4\pi u(P), & P \in D_e. \end{cases} \tag{5}$$

(F3) The two preceding formulas are now combined. Let u be harmonic in D and D_e, while satisfying the growth conditions (4). Also, let $u \in C^2(D) \cap C^1(\bar{D})$ and $u \in C^2(D_e) \cap C^1(\bar{D}_e)$. Then,

$$\int_S \left[\frac{\partial u}{\partial \mathbf{n}_Q}\right] \frac{dS(Q)}{|P-Q|} - \int_S [u(Q)] \frac{\partial}{\partial \mathbf{n}_Q}\left[\frac{1}{|P-Q|}\right] dS(Q)$$

$$= \begin{cases} 4\pi u(P), & P \in D \cup D_e, \\ \mathscr{I}(P)u^i(P) + [4\pi - \mathscr{I}(P)]u^e(P), & P \in S. \end{cases} \tag{6}$$

We define

$$u^i(P) = \lim_{Q \to P} u(Q), \quad Q \in D, \quad P \in S, \tag{7a}$$

$$u^e(P) = \lim_{Q \to P} u(P), \quad Q \in D_e, \quad P \in S, \tag{7b}$$

$$[u(P)] = u^i(P) - u^e(P). \tag{7c}$$

The quantities $\partial u^i/\partial \mathbf{n}_Q$, $\partial u^e/\partial \mathbf{n}_Q$, and $[\partial u/\partial \mathbf{n}_Q]$ are defined analogously.

2.1. Direct BIE Methods. Methods based on (F1) and (F2) are generally referred to as direct methods. The unknown function being sought on S is either u or $\partial u/\partial n$, both usually of direct interest to the person solving Laplace's equation.

For example, consider solving the interior Dirichlet problem

$$\Delta u(P) = 0, \qquad P \in D, \tag{8a}$$

$$u(P) = f(P), \qquad P \in S, \tag{8b}$$

with f given. Using Eq. (3), we have

$$u(P) = \frac{1}{4\pi} \int_S \frac{\partial u(Q)}{\partial \mathbf{n}_Q} \frac{dS(Q)}{|P-Q|}$$

$$- \frac{1}{4\pi} \int_S u(Q) \frac{\partial}{\partial \mathbf{n}_Q} \left[\frac{1}{|P-Q|}\right] dS(Q), \qquad P \in D. \tag{9}$$

We need to know $\partial u/\partial \mathbf{n}_Q$ on S to be able to use this formula. To find $\rho \equiv \partial u/\partial \mathbf{n}_Q$, use (3) to write

$$\int_S \frac{\rho(Q)}{|P-Q|} ds(Q) = \mathscr{I}(P)f(P) + \int_S f(Q) \frac{\partial}{\partial \mathbf{n}_Q} \left[\frac{1}{|P-Q|}\right] dS(Q)$$

$$\equiv \hat{f}(P), \qquad P \in S. \tag{10}$$

This is an integral equation of the first kind. For right-hand sides $\hat{f}$ of the given form, the equation will always have a solution, based on the solution of (8). The solvability for general $\hat{f}$ and the uniqueness of the solution ρ must be examined separately, as these questions do not follow from (3).

The Neumann problem

$$\Delta u(P) = 0, \qquad P \in D, \tag{11a}$$

$$\frac{\partial u(P)}{\partial \mathbf{n}_Q} = g(P), \qquad P \in S, \tag{11b}$$

has the solution (9), but now we need to know $u(Q)$ on S. Using (3), we obtain the integral equation of the second kind

$$\mathscr{I}(P)u(P) + \int_S u(Q) \frac{\partial}{\partial \mathbf{n}_Q}\left[\frac{1}{|P-Q|}\right] dS(Q) = \int_S \frac{g(Q)}{|P-Q|} dS(Q)$$

$$\equiv \hat{g}(P). \qquad (12)$$

For functions g satisfying the standard compatibility condition

$$\int_S g(Q)\, ds(Q) = 0, \qquad (13)$$

the integral equation (12) is solvable. However, it is not solvable for those right-hand sides $\hat{g}$ where (13) is not satisfied; and the equation is not uniquely solvable, since $u(Q) \equiv 1$ is a solution of the homogeneous equation. The lack of uniqueness is easily fixed by specifying $u(P)$ at some particular point P.

A more serious concern with both Eqs. (10) and (12) is that the right-hand functions must be evaluated by numerical integration of a singular surface integral. We comment further on this in Section 5. Exterior problems for Laplace's equation can be handled similarly, using (F2). We omit the details; for example, see Ref. 21.

An important practical aspect of direct methods is that the quantity being found on S, either u or $\partial u/\partial \mathbf{n}$, may be all that is needed by the user. In such a case, it will not be necessary to consider the numerical integration of the quantities in Eq. (9), for $P \in D$.

2.2. Indirect BIE Methods. These methods are based on (F3). The solution of the integral equation is usually not of immediate physical interest, but is merely an intermediate step in obtaining the harmonic function u.

As an example, consider again the interior Dirichlet problem (8). To use (F3), also consider the associated exterior Dirichlet problem

$$\Delta u(P) = 0, \qquad P \in D_e, \qquad (14a)$$

$$u(P) = f(P), \qquad P \in S, \qquad (14b)$$

with u also satisfying the growth conditions (4). This problem is known to have a unique solution.

Applying Eq. (6) to problems (8) and (14), we have

$$u(P) = \frac{1}{4\pi} \int_S \frac{\rho(Q)}{|P-Q|} dS(Q), \qquad P \in D \cup D_e, \qquad (15)$$

with $\rho = [\partial u/\partial n]$, the difference of the normal derivatives of u on D and D_e. An equation for the function ρ can be found by letting P tend to a boundary point in (15), or (6) can be applied directly, obtaining

$$\frac{1}{4\pi} \int_S \frac{\rho(Q)}{|P-Q|} dS(Q) = f(P), \qquad P \in S. \tag{16}$$

This equation is of the same form as (9), but now the right side is given explicitly rather than as an integral that must be evaluated numerically.

Another reformulation of the interior Dirichlet problem (8) can be found by considering an exterior problem different than that in (14). Solve the Neumann problem

$$\Delta u(P) = 0, \qquad P \in D_e, \tag{17a}$$

$$\frac{\partial u(P)}{\partial \mathbf{n}_P} = \frac{\partial u^i(P)}{\partial \mathbf{n}_P}, \qquad P \in S, \tag{17b}$$

with the growth conditions (4). The function $u^i(P)$ refers to the solution of problem (8). This problem is known to have a unique solution.

Using the formula (6) with function u satisfying (8) and (17), we have

$$u(P) = \frac{1}{4\pi} \int_S \rho(Q) \frac{\partial}{\partial \mathbf{n}_Q} \left[\frac{1}{|P-Q|} \right] dS(Q), \qquad P \in D \cup D_e. \tag{18}$$

Generally we are interested only in the case $P \in D$. The unknown ρ represents the jump in the values for $u(P)$, for problems (8) and (17), as P passes through S. To find ρ, use $u^e = u^i - \rho$ and Eq. (6) to write

$$[4\pi - \mathscr{I}(P)]\rho(P) - \int_S \rho(Q) \frac{\partial}{\partial \mathbf{n}_Q} \left[\frac{1}{|P-Q|} \right] dS(Q) = 4\pi f(P), \qquad P \in S. \tag{19}$$

This is an integral equation of the second kind. For the case of a smooth boundary, Eqs. (18) and (19) give the classical approach used by Fredholm in examining the solvability of the interior Dirichlet problem for Laplace's equation. By a "smooth boundary," we mean one that has a C^k parameterization at all points of the boundary, with $k \geq 2$.

For a smooth boundary S, Eq. (19) can be shown to be uniquely solvable for all $f \in C(S)$ or $L^2(S)$, with the solution ρ in these respective spaces. The advantage of (19) over an integral equation of the first kind is that both the solvability theory of the equation and its numerical analysis are better understood for (19). Equation (19) is also slightly better conditioned than the first-kind equations (10) and (16), but this is not very significant in practice.

The choice of an indirect vs. a direct method will depend on a number of factors. The direct methods lead to unknowns that are more meaningful physically. The indirect methods often lead to equations that are slightly easier to deal with numerically. In many cases, personal preference is probably the determining factor.

2.3. The Smooth Boundary Case. Although most boundaries S are only piecewise smooth, it is only the smooth boundary case that is even moderately well understood for solving Laplace's equation in three dimensions.

For S a smooth boundary, the integral operators

$$\mathcal{S}\rho(P) = \int_S \frac{\rho(Q)}{|P - Q|} \, dS(Q), \qquad P \in S, \tag{20}$$

and

$$\mathcal{D}\rho(P) = \int_S \rho(Q) \frac{\partial}{\partial \mathbf{n}_Q}\left[\frac{1}{|P - Q|}\right] dS(Q), \qquad P \in S, \tag{21}$$

are compact operators from $\mathcal{X}$ to $\mathcal{X}$, for $\mathcal{X} = C(S)$ or $L^2(S)$. Quantities $\mathcal{S}$ and $\mathcal{D}$ are called, respectively, *single layer* and *double layer* integral operators; the function ρ is called a density function. For integral equations of the second kind, such as (19), the compactness of $\mathcal{S}$ and $\mathcal{D}$ leads to a complete solvability theory for the associated integral equation. For example, see Refs. 21–22, Ref. 20, Chapter 18, and Ref. 23, Chapter 12. The compactness also leads to a fairly well developed numerical analysis for these integral equations.

The compactness is of less value for formulations leading to integral equations of the first kind, such as those in (10) and (16). Based on the derivation of (16) and associated results in potential theory, Eq. (16) can be shown to have a unique solution $\rho \in L^2(S)$ for all given functions f in the Sobolev space $H^1(S)$. For a discussion of the solvability of (16), see Ref. 21, Section 2.5.

Another framework for analyzing the boundary integral equation reformulations, particularly first-kind integral equations, is to regard them as *strongly elliptic* operator equations on appropriate Sobolev spaces. This approach has been used (very successfully) in the work of Nedelec, Wendland, and others. It makes possible a fairly straightforward generalization of finite element methods to the solution of BIEs. An introduction to this approach is given in Ref. 24 in the context of finite element methods for partial differential equations. A more general presentation is given in Wendland (Ref. 25), for BIEs for more general partial differential equations.

A discussion and numerical analysis of the particular equation (16) is given in the paper of Nedelec (Ref. 26). He introduces the bilinear functional

$$a(\rho, \psi) = \frac{1}{4\pi} \int_S \int_S \frac{\rho(P)\psi(Q)}{|P - Q|} \, dS(P) \, dS(Q). \tag{22}$$

He quotes results from an earlier paper, showing that $\mathscr{S}$, from (16) or (20), is a one-to-one mapping of the Sobolev space $H^{-1/2}(S)$ onto $H^{1/2}(S)$. Moreover, the bilinear functional a satisfies

$$a(\rho, \rho) \geq \alpha \|\rho\|^2_{H^{-1/2}(S)}, \qquad \rho \in H^{-1/2}(S), \tag{23}$$

for some $\alpha > 0$. This says $\mathscr{S}$ is a strongly elliptic operator, and the standard finite element method and theory can then be applied to the solution of (16). We return to this and Nedelec's analysis in Section 4.

2.4. Piecewise Smooth Boundaries. In the past few years, much progress has been made on the numerical solution of BIEs for two-dimensional problems on regions with only a piecewise smooth boundary. For example, see Refs. 27–31. In contrast, very little is understood about the corresponding three-dimensional problems.

From results for the two-dimensional problem, we know that the double-layer operator $\mathscr{D}$ of (21) is not compact nor is any power of it compact. This removes an important tool used in the analysis of the second-kind equations involving $\mathscr{D}$, such as (19). In addition, in the two-dimensional case, the solution function $\rho(P)$ is usually poorly behaved as P approaches a vertex of the boundary. To obtain rapid convergence of the numerical methods that solve for ρ, the mesh points of the method must be distributed more densely near the vertices of the boundary. This is referred to as using a *graded mesh*, and details of it are given in the papers of Chandler and Graham (Refs. 28, 29, and 31).

For the three-dimensional Laplace equation on a piecewise smooth boundary, the book of Grisvard (Ref. 17) gives a current account of the solvability and regularity theory for two- and three-dimensional boundary value problems. Grisvard deals with the following Dirichlet problem for the Poisson equation:

$$\Delta u(P) = F(P), \qquad P \in D, \tag{24a}$$

$$u(P) = 0, \qquad P \in S. \tag{24b}$$

Thus his results for the behavior of u must be modified in order to apply them to Laplace's equation. The results can then be used with (F1)–(F3) to give information on the regularity of the density functions for which one is solving.

In analyzing the behavior of system (24), Grisvard considers separately the behavior of $u(P)$ near points P of an edge of S and points P that are vertices. In the first case, the behavior is linked closely to that of similar problems for two-dimensional problems on polygonal regions; see Ref. 17, Section 8.2.1. For the second case, with P a vertex, we must consider the set G formed by intersecting D with the surface of a small sphere centered at P. Assuming for simplicity that D is truly conical near P, consider G and its boundary. The singular solutions are connected to the eigenfunctions of the Laplace–Beltrami equation on G, with zero boundary conditions. If the boundary of G is smooth, then there is a relatively well understood class of singular solutions, and asymptotic expansions of u about P that incorporate the singular behavior are possible. But if the boundary of G is only piecewise smooth, then an additional category of singular solutions is possible, one that is much less understood. For details of what we have described, see Ref. 17, Section 8.2.2. Also see Add. Ref. 17.

2.5. The Poisson Equation. To use BIE methods for solving boundary value problems for the Poisson equation

$$\Delta u(P) = F(P), \qquad P \in D, \tag{25}$$

the problem must be converted (explicitly or implicitly) to an equivalent problem for Laplace's equation. An implicit method is described in Ref. 6, Section 2.7, and some explicit methods are given in Ref. 32. In the latter paper, one method considered is the efficient numerical integration of the Poisson integral formula

$$u_0(P) = \frac{1}{4\pi} \int_{\hat{D}} \frac{F(Q)\, dQ}{|P - Q|}, \qquad P \in \bar{D}, \tag{26}$$

where $\hat{D}$ is a larger (usually ellipsoidal) region containing $\bar{D}$. Then equation (25) can be made homogeneous by subtracting u_0 from u, obtaining $\Delta v = 0$, $v = u - u_0$; the boundary values for v are found similarly from u_0 and those given for u. This leads to an effective numerical procedure; an example is given in Ref. 32.

3. Numerical Methods for Boundary Integral Equations

There are a wide variety of numerical methods that have been used for solving three-dimensional BIEs. This is partly a reflection of the wide variety of regions and problems, where one wants to take into account the special nature of the problem at hand. It also reflects the limited general knowledge concerning numerical methods for BIEs for the three-dimensional Laplace equation. Examples of the variety of approaches can be seen in the various proceedings of engineering boundary element conferences given in the references at the end of this paper.

Even so, most numerical methods can be considered to be of *collocation* or *Galerkin* type, with special adaptions having been made for the particular problem at hand. We describe these methods and look at their analysis.

Let the boundary integral equation being solved be denoted by

$$\mathscr{L}\rho = f. \tag{27}$$

Both Galerkin and collocation methods (and all other projection methods) approximate ρ with functions from a finite-dimensional family, denoted here by $\mathscr{X}_n$. The methods attempt to minimize

$$r_n = \mathscr{L}\rho_n - f,$$

as ρ_n ranges over $\mathscr{X}_n$, in the hope that the minimizer ρ_n will also be close to ρ. The methods differ in the criterion for making r_n small.

Galerkin methods minimize r_n by forcing the Fourier coefficients of r_n to be zero for all elements of some basis of $\mathscr{X}_n$. If $\{\varphi_1, \ldots, \varphi_n\}$ is such a basis, then we choose $\rho_n \in \mathscr{X}_n$ such that

$$(r_n, \varphi_i) = 0, \qquad i = 1, \ldots n, \tag{28}$$

where $(\cdot, \cdot)$ denotes some type of inner product for the functions under consideration. Writing

$$\rho_n = \sum_{j=1}^{n} \alpha_j \varphi_j, \tag{29}$$

the coefficients $\{\alpha_j\}$ are determined from

$$\sum_{j=1}^{n} \alpha_j (\mathscr{L}\varphi_j, \varphi_i) = (f, \varphi_i), \qquad i = 1, \ldots, n. \tag{30}$$

The inner products are integrals which must be evaluated numerically in most cases, and this will be discussed later for some particular cases.

Collocation methods make r_n small by forcing $r_n(P)$ to be zero at a preselected set of points P in the integration domain S. Letting $\{P_1, \ldots, P_N\}$

be points in S, we require

$$r_n(P_i) = 0, \qquad i = 1, \ldots, n. \tag{31}$$

If $\{\varphi_1, \ldots, \varphi_n\}$ is a basis of $\mathcal{X}_n$, then we again seek the solution ρ_n in the form (29). The coefficients are now determined from

$$\sum_{j=1}^{n} \alpha_j \mathcal{L}\varphi_j(P_i) = f(P_i), \qquad i = 1, \ldots, n. \tag{32}$$

For integral equations of the second kind, an introduction to the error analysis and the implementation of Galerkin and collocation methods is given in Ref. 33, pp. 50-84. We give some error results and comments on implementing such methods below in more specific contexts.

When the functions of $\mathcal{X}_n$ are defined globally, in some sense, as smooth functions on the integration domain S, we say the numerical method is a *global method*. In contrast, if the region is divided into small *elements* Δ_i,

$$S = \bigcup_{j=1}^{N} \Delta_i, \tag{33}$$

and if the elements of $\mathcal{X}_n$ are piecewise polynomial functions of a fixed degree on the elements of this subdivision, we say the method is a *local method*. Most of the numerical methods used to solve BIEs in the engineering literature have been local methods. They are generally referred to as *boundary element methods* (or BEMs) in the engineering literature, and therefore we will also use that name.

The boundary element methods are very flexible in handling a variety of surfaces S, boundary conditions, and changes in the smoothness of the unknown density function ρ. For surfaces that are only piecewise smooth, BEMs are almost mandatory, since the elements can be graded more finely near edges and vertices to compensate for the lack of smoothness in the density function ρ. In contrast, for surfaces that are smooth, the global methods can be much more efficient as regards computing time and storage, when compared to the BEMs. We look at examples of global methods in the remainder of this section, and we consider BEMs in the following section.

3.1. Global Methods. If the unknown density function $\rho(P)$ is smooth and if the boundary S is smooth, then one can consider approximating ρ by combinations of functions from some complete family of globally defined smooth functions. For regions in $\mathbb{R}^2$ with a smooth boundary, this leads to approximations by trigonometric functions. For three dimensions, there are more possibilities, depending on the connectivity of the region D.

Let D be simply-connected, and let S be smooth. We summarize the method from Ref. 34 which uses the double-layer representation (18) to solve the interior Dirichlet problem for Laplace's equation.

The integral equation to be solved is (19) [with $\mathscr{I}(P) = 2\pi$]. From results in Ref. 22, pp. 49, 106, we have results on the differentiability of ρ on S. Assume that S is C^{k+1}, with all $(k+1)$st derivatives satisfying a Hölder condition with exponent λ, some $0 < \lambda \leq 1$ and $k \geq 0$. Also assume the given boundary value $f \in C^k(S)$ with all kth-order derivatives satisfying a Hölder condition with exponent λ. Then the density $\rho \in C^k(S)$, with all kth-order derivatives Hölder continuous with exponent λ', any $\lambda' < \lambda$.

The spherical polynomials on the unit sphere U are the generalization of the trigonometric polynomials on the unit circle. Convergence results concerning approximation of functions on U by spherical polynomials are given in Ref. 35. To use spherical polynomials in solving Eq. (19), we first convert the integral equation and all functions in it to an equation and functions defined on U. With common boundaries like an ellipsoid, this is a simple and straightforward change of variables. For more discussion of this, see Ref. 36, pp. 86–88. We write the new integral equation on U as

$$2\pi\hat{\rho}(P_u) - \int_U \hat{\rho}(Q_u)\hat{K}(P_u, Q_u)\, dS(Q_u) = \hat{f}(P_u), \qquad P_u \in U, \quad (34)$$

with $\hat{\rho}$ and $\hat{f}$ equivalent to ρ and f under the mapping between U and S. The kernel function $\hat{K}$ incorporates the double-layer kernel from (19) and the Jacobian for the change of integration variable.

Symbolically we write Eq. (34) as

$$(2\pi - \mathscr{K})\hat{\rho} = \hat{f}. \qquad (35)$$

The operator $\mathscr{K}$ is compact on $\mathscr{X}$ to $\mathscr{X}$, with $\mathscr{X} = C(U)$ and $L^2(U)$. We apply the Galerkin method to the solution of (35), with the approximating subspace $\mathscr{X}_n$ defined as all spherical polynomials of degree $\leq n$. The dimension of $\mathscr{X}_n$ is $d_n = (n+1)^2$, and a standard orthogonal basis of spherical harmonics is available, denoted here by $\{\varphi_1, \ldots, \varphi_d\}$. Thus we write

$$\hat{\rho}(P_u) = \sum_{j=1}^{d_n} \alpha_j \varphi_j(P_u), \qquad (36)$$

with $\{\alpha_j\}$ determined from

$$2\pi(\varphi_i, \varphi_i)\alpha_i - \sum_{j=1}^{d_n} \alpha_j(\mathscr{K}\varphi_j, \varphi_i) = (\hat{f}, \varphi_i), \qquad i = 1, \cdots, d_n. \quad (37)$$

The inner product is that of $L^2(U)$.

For convergence, assume f and S satisfy the earlier differentiability assumptions, so that $\rho \in C^k(S)$ with Hölder continuous kth derivatives. Then the solution of (36)-(37) exists for sufficiently large n, and it satisfies

$$\|\rho - \rho_n\|_\infty \le \frac{c_k}{n^{k+\lambda}}, \qquad n \ge n_0. \tag{38}$$

For a C^∞ boundary S and C^∞ boundary function f, the speed of convergence is faster than any power of $1/n$.

Details of the practical implementation of (36)-(37) and extensions to other boundary value problems are given in Refs. 37, 34 and 36. Extensions to solving the Helmholtz equation in three dimensions is given in Ref. 38.

A noteworthy feature of the method is that the order of the linear system in (37) is reasonably small in general, for example, $d_n \le 100$. Thus iterative methods are less likely to be needed, unlike the usual case for BEMs in three dimensions. A major difficulty of the method is the calculation of the integrals $(\hat{f}, \varphi_i)$ and $(\mathcal{K}\varphi_j, \varphi_i)$. The first is a surface integral involving smooth integrands. The second is a double surface integral, with $\mathcal{K}\varphi_j$ having a singular kernel function. A discussion of the efficient numerical evaluation of these integrals is given in Ref. 36.

To reduce the preceding problems of numerical integration, it would be desirable to have a collocation method involving the spherical polynomials as approximating functions. Unfortunately, it appears to be unknown as to how to choose the collocation points. It is necessary that the collocation points $\{P_i\}$ on U satisfy

$$\det[\varphi_j(P_i)] \ne 0, \qquad i, j = 1, \ldots, d_n.$$

This guarantees that the following interpolation problem has a unique solution: Find $\psi_n \in \mathcal{X}_n$ such that

$$\psi_n(P_i) = \rho(P_i), \qquad i = 1, \ldots, d_n,$$

where ρ is any given function on U. In addition, we require that the interpolation functions ψ_n converge to ρ as $n \to \infty$. Virtually nothing is known of any of this needed theory, and thus global collocation methods using spherical polynomials are not yet available for solving boundary integral equations.

Another example of a global method arises for D a toroidal region. In this case, S is homeomorphic to the cross-product of the unit circle with itself. A natural way to approximate ρ is to use tensor products of trigonometric polynomials. This yields both Galerkin and collocation methods for solving the double-layer equation (19). For more discussion of this, see Ref. 39.

In the next section, we describe the general framework that regards BIEs as strongly elliptic pseudodifferential operator equations on suitable Sobolev spaces. That theory has been applied mainly to the analysis of BEMs, but much of it applies equally well to the analysis of global methods. An important theorem for the error analysis is in Ref. 40, Theorem 3.1 of W. Wendland. Preceding it, Wendland notes that spherical polynomials can be used as approximating functions within his general framework.

4. Boundary Element Methods

Let the surface S be subdivided into a fixed number of subregions,

$$S = S_1 \cup \cdots \cup S_J,$$

with each S_j closed and the interiors of the S_js pairwise disjoint. We assume each S_j can be parameterized by a mapping of a polygonal region R_j:

$$F_j : R_j \to S_j, \qquad j = 1, \ldots, J. \tag{39}$$

To triangulate each S_j, first triangulate each R_j, and then use the image under F_j of this triangulation. Denote the collective triangulation of S by $\{\Delta_i \mid i = 1, \ldots, N\}$. Much of the approximation of functions on S is carried out by using approximations of functions on the regions R_j. In particular, if $f(x, y)$ is a polynomial or piecewise polynomial function on R_j, then we regard

$$g(P) \equiv f(F_j^{-1}(P))$$

as a polynomial or piecewise polynomial function on the surface S_j. We will look at two approaches to defining and analyzing boundary element methods for solving BIEs defined on S.

4.1. Strongly Elliptic Operator Framework. Denote the BIE being solved as

$$\mathscr{L}\rho = f. \tag{40}$$

Assume that

$$\mathscr{L} : H^s(S) \xrightarrow[\text{onto}]{1\text{-}1} H^{s-2\alpha}(S), \tag{41}$$

for $s \geq \alpha$. The surface S must be sufficiently smooth, with no vertices or edges; and for simplicity, we assume S is C^∞. The number 2α is called the index of the pseudodifferential operator $\mathscr{L}$. For the single-layer integral operator in (20), $\alpha = -1/2$.

In addition, we assume that

$$(\mathscr{L}\varphi, \varphi) \geq \gamma \|\varphi\|^2_{H^\alpha(S)}, \qquad \varphi \in H^\alpha(S). \tag{42}$$

This makes $\mathscr{L}$ a strongly elliptic operator. For a more general treatment of this framework for analyzing BIE methods, see Refs. 25, 40-41.

We assume the existence of a sequence of approximating subspaces $\mathscr{X}_n$, with $\mathscr{X}_n \subset H^\alpha(S) \cap L^2(S)$. As before, let $\{\varphi_1, \ldots, \varphi_n\}$ be a basis for $\mathscr{X}_n$ and consider solving $\mathscr{L}\rho = f$ with the Galerkin method, given in Eqs. (29)-(30). Let P_n be the orthogonal projection operator of $L^2(S)$ onto $\mathscr{X}_n$. Assume that the extension of P_n to $H^\alpha(S)$ satisfies

$$\|P_n\rho - \rho\|_{H^\alpha(S)} \to 0, \quad \text{as } n \to \infty, \tag{43}$$

for all $\rho \in H^\alpha(S)$. Then we have the following convergence result of Wendland from Ref. 40, Theorem 3.1.

Theorem 4.1. Let $\mathscr{L}\rho = f$ be a strongly elliptic equation of order 2α with a unique solution $\rho \in H^\alpha(S)$ for all $f \in H^{-\alpha}(S)$. Then there exists n_0 such that the Galerkin system (30) is uniquely solvable for all $n \geq n_0$. In addition, the Galerkin solution ρ_n satisfies

$$\|\rho - \rho_n\|_{H^\alpha(S)} \leq c \min_{\psi \in \mathscr{X}_n} \|\rho - \psi\|_{H^\alpha(S)}, \tag{44}$$

with c a constant independent of f and n.

This result has been applied by Wendland to a variety of two- and three-dimensional boundary problems; for example, see Refs. 25, 40-43. This type of analysis has also been applied to three-dimensional BEMs by Nedelec and his colleagues; see Refs. 26 and 44. The above framework and convergence theorem can also be applied to the analysis of global methods using spherical polynomials (as noted by Wendland in Ref. 25, p. 295), but this does not seem to have been done. The use of BEMs within a Galerkin framework is also discussed in Ref. 45, where the practical questions of implementation are examined, although no convergence analysis is provided.

We consider the work of Nedelec (Ref. 26), as it gives a very careful and detailed error analysis in the special case of the first-kind equation (16). It also considers the important case in which the boundary S is approximated by a simpler boundary, one for which integrals and derivatives on the boundary are more easily computed.

Recall the definition of the triangulation $\{\Delta_i\}$ of S, following (39). To define interpolation and numerical integration on Δ_i, we first consider Δ_i as the image of the standard simplex

$$\sigma = \{(s, t) | 0 \le s, t, s + t \le 1\}. \qquad (45)$$

If $\hat{\Delta}_i$ is a triangle in the triangulation of one of the subregions of S, say S_k, and $\Delta_i = F_k(\hat{\Delta}_i)$, then we introduce the mapping

$$m_i : \sigma \xrightarrow[\text{onto}]{1\text{-}1} \Delta_i,$$

$$m_i(s, t) = F_k(u\hat{v}_1 + s\hat{v}_2 + t\hat{v}_3), \qquad u = 1 - s - t, \qquad (s, t) \in \sigma, \qquad (46)$$

with $\hat{v}_1$, $\hat{v}_2$, $\hat{v}_3$ the vertices of $\hat{\Delta}_i$ in $\mathbb{R}^2$. The vertices of σ, namely $\{(0, 0), (1, 0), (0, 1)\}$, map to the vertices of Δ_i, namely $v_j \equiv F_k(\hat{v}_j)$, $j = 1$, 2, 3.

To define interpolation polynomials of degree d on σ, let $n_d \equiv n = (d + 1)(d + 2)/2$ points in σ be given, with respect to which the polynomials in (s, t) of degree $\le d$ are unisolvent. For example, consider the following choices:

$$d = 0, \qquad (\tfrac{1}{3}, \tfrac{1}{3});$$

$$d = 1, \qquad (0, 0), (1, 0), (0, 1);$$

$$d = 2, \qquad (0, 0), (1, 0), (0, 1), (\tfrac{1}{2}, 0), (\tfrac{1}{2}, \tfrac{1}{2}), (0, \tfrac{1}{2}).$$

Denote the n points in σ by $q_1, \ldots, q_n$. Then we can define the polynomial of degree $\le d$ that interpolates a given $g \in C(\sigma)$ by

$$L_d g(s, t) \equiv \sum_{j=1}^{n} g(q_j) l_j(s, t) \doteq g(s, t), \qquad (47)$$

where $l_j(q_i) = \delta_{ij}$ and degree$(l_j) \le d$.

To interpolate over a triangle Δ_i on S, let Δ_i be obtained as the image of m_i. For f defined on Δ_i, let its interpolant of degree $\le d$ be defined by

$$L_d f(P) \equiv \sum_{j=1}^{n} f(m_i(q_j)) l_j(s, t), \qquad P = m_i(s, t). \qquad (48)$$

It is relatively straightforward to show that if $f \in C^{d+1}(S)$, then

$$\|f - L_d f\|_{C^\infty(S)} \le ch^{d+1}, \tag{49}$$

with

$$h = \max_{1 \le i \le N} \text{diameter}(\Delta_i).$$

In many numerical integrations that arise with both collocation and Galerkin methods, the Jacobian of the transformation (46) must be evaluated; and it is also needed in evaluating the normals to S. Thus the BEM requires knowledge of the derivatives of each m_i, or equivalently, of the derivatives of the mappings F_k in (39). To simplify these calculations, we introduce an approximating surface S_h. Its derivatives can be calculated without a need for the derivatives of the original surface S. To approximate $\Delta_i \subset S$, we define

$$m_{h,i}(s, t) = \sum_{j=1}^{n} m_i(q_j) l_j(s, t). \tag{50}$$

Then define $\Delta_{h,i}$ approximating Δ_i by

$$\Delta_{h,i} = m_{h,i}(\sigma). \tag{51}$$

We require that if Δ_i and Δ_k have a common edge, then $\Delta_{h,i}$ and $\Delta_{h,k}$ must join continuously along their corresponding edges. Define

$$S_h = \bigcup_{i=1}^{N} \Delta_{h,i}. \tag{52}$$

We define function f to be piecewise polynomial on S [on S_h] by saying if $P \in \Delta_i$ [$P \in \Delta_{h,i}$], then

$$f(P) = g(s, t), \tag{53}$$

with $P = m_i(s, t)$ [$P = m_{h,i}(s, t)$] and g a polynomial in (s, t). For the approximating space in Nedelec's method, let $\mathcal{X}_h$ be the set of all piecewise polynomial functions on S_h of degree $\le r$, and let the surface S_h be defined using interpolation of degree $\le k$. Because Eq. (16) uses $f \in C(S)$, we must extend it to S_h. Let $f_h \in C(S_h)$ and let it approximate f in some sense. For example, define

$$f_h(m_{h,i}(s, t)) = f(m_i(s, t)).$$

Nedelec introduces the approximating bilinear functional

$$a_h(\rho, \psi) = \frac{1}{4\pi} \int_{S_h} \int_{S_h} \frac{\rho(P)\psi(Q)}{|P - Q|} dS(P) \, dS(Q). \tag{54}$$

He then seeks the Galerkin solution $\rho_h \in \mathcal{X}_h$ that satisfies the variational equation

$$a_h(\rho_h, \psi) = (f_h, \psi), \quad \text{all } \psi \in \mathcal{X}_h, \tag{55}$$

with the right-hand inner product using integration over S_h. The function f_h is defined on S_h and approximates f. This yields the earlier Galerkin system (30), with $\mathcal{L}$ replaced by the single-layer integral operator over the approximate surface S_h. The following convergence theorem is given in Ref. 26, p. 65.

Theorem 4.2. Let ρ be the solution of (16), and let ρ_h be the solution of Eq. (55). Also assume $k \geq r$, where S_h is defined using piecewise polynomial interpolation of degree k and $\mathcal{X}_h$ contains piecewise polynomials of degree $\leq r$. Then

$$\|\rho - \rho_h \circ \Psi_h^{-1}\|_{L^2(S)}$$
$$\leq c \left[\frac{1}{\sqrt{h}} \|f - f_h \circ \Psi_h^{-1}\|_{H^{1/2}(S)} + h^{r+1} \|\rho\|_{H^{r+1}(S)} + h^k \|\rho\|_{L^2(S)} \right]. \tag{56}$$

The function Ψ_h is a 1-1 mapping of S onto S_h. For $P \in S$, let $\Psi_h^{-1}(P)$ be the point on S_h obtained as follows: construct the normal to S at P, and then let that normal intersect S_h, with $\Psi_h^{-1}(P)$ the point of intersection. For h sufficiently small, the mapping Ψ_h is well-defined.

Direct extensions of these ideas to BIEs for the Helmholtz equation are given in Refs. 44 and 46. As mentioned earlier, Wendland in Ref. 25 gives a general framework for analyzing Galerkin's method for solving all known BIEs.

4.2. Collocation Methods. The general idea of collocation was defined earlier, in (31)-(32). When combined with the decomposition of S into the union of many small elements $\{\Delta_i\}$ and the use of low-degree piecewise polynomial approximations, collocation has been the preferred method in the engineering literature for solving BIEs for three-dimensional problems.

Nonetheless, much less is known about the mathematical analysis of such methods as compared to Galerkin's method. To date, all analyses have

A Survey of Boundary Integral Equation Methods

been for integral equations of the second kind, often making use of compactness of the associated integral operator. We describe some of those results below. We know of no analysis of collocation methods for BIEs of the first kind for the three-dimensional Laplace equation.

The most general convergence results for collocation methods are those given by Wendland in Ref. 47, and we briefly summarize his results. Recall the notation on the triangulation of S, given earlier in this section. Let the triangulation of S be denoted by $\{\Delta_i | i = 1, \ldots, N\}$, and recall the interpolation polynomial defined in (48). For a particular triangle Δ_i, let the interpolation nodes contained in Δ_i be denoted by $\{v_{i,k} | k = 1, \ldots, n_d\}$, where the interpolation is of degree d and $n = n_d$ is the number of nodes. We also refer to the interpolation nodes collectively by $\{v_i | i = 1, \ldots, N_v\}$. For f defined on Δ_i, the function of degree d that interpolates f on Δ_i is given by

$$f(m_i(s, t)) \doteq \sum_{k=1}^{n_d} f(v_{i,k}) l_k(s, t), \qquad (s, t) \in \sigma. \tag{57}$$

To be clearer in our description of the collocation method, we consider the specific second-kind equation (19) arising from solving the interior Dirichlet problem using a double-layer potential representation of the solution. Also, assume the boundary is smooth at all points, so that $\mathcal{I}(P) = 2\pi$. This is critical for the convergence analysis in Ref. 47.

Write the approximate solution in the form

$$\rho_h(m_i(s, t)) = \sum_{k=1}^{n_d} \alpha_{k,i} l_k(s, t), \qquad (s, t) \in \sigma, \tag{58}$$

for $i = 1, \ldots, N$. The unknown coefficients $\{\alpha_{k,i}\}$ will also be written collectively as $\{\alpha_i\}$, so that $\rho_h(v_i) = \alpha_i$, $i = 1, \ldots, N_v$. To find the coefficients $\{\alpha_{k,i}\}$, solve the collocation system

$$2\pi\alpha_j - \sum_{i=1}^{N} \sum_{k=1}^{n_d} \alpha_{k,i} \int_{\Delta_i} l_{k,i}(Q) \frac{\partial}{\partial \mathbf{n}_Q}\left[\frac{1}{|v_j - Q|}\right] dS_Q = \hat{f}(v_j), \tag{59}$$

for $j = 1, \ldots, N_v$. The function $\hat{f}(P) \equiv 4\pi f(P)$, with $f(P)$ the Dirichlet data in (14). The basis function $l_{k,i}(Q)$ is defined by

$$l_{k,i}(m_i(s, t)) = l_k(s, t), \qquad (s, t) \in \sigma. \tag{60}$$

We refer abstractly to the integral equation as

$$(2\pi - \mathcal{K})\rho = \hat{f}. \tag{61}$$

The interpolation (57) defines a linear interpolation operator P_h from $C(S)$ onto the approximating subspace $\mathcal{X}_h$ of all piecewise polynomial functions of degree $\leq d$ on the triangulation $\{\Delta_i\}$. The collocation method can then be written as

$$(2\pi - P_h \mathcal{K})\rho_h = P_h \hat{f}. \tag{62}$$

[The functions in the approximating space $\mathcal{X}_h$ are assumed to be continuous. This is not necessary, and generalizations to piecewise continuous functions can be considered in the context of $L^\infty(S)$. For some discussion of this, see Ref. 48.] The following result is proven in Ref. 47.

Theorem 4.3. Assume $f \in C^{d+1}(S)$; and for simplicity, assume S is C^∞ at all points of its surface. Then $2\pi - P_h \mathcal{K}$ is invertible for all sufficiently small h, say $0 < h \leq h_0$, and these inverses are uniformly bounded as operators on $C(S)$ to $C(S)$. In addition, the approximate solution ρ_h satisfies

$$\|\rho - \rho_h\| \leq c [\|(I - P_h)\rho\|_\infty + \|(I + P_h)f\|_\infty] \tag{63}$$

$$\leq c \cdot h^{d+1}, \quad 0 < h \leq h_0, \tag{64}$$

with c a generic constant.

The proof of this theorem is based on showing $\{P_h \mathcal{K} \mid 0 < h \leq 1\}$ is a collectively compact and pointwise convergent sequence of operators on $C(S)$ to $C(S)$; and then the general approximation theory of Ref. 49 can be invoked to complete the proof. The bound (64) follows from (63) and standard results on multivariable interpolation of smooth functions by piecewise polynomial functions.

This theorem assumes that the exact surface S is used in all integration and interpolation in the problem. If instead we use the interpolating surface S_h defined in (50)–(52), then the analysis is more complicated. Assume that S_h is defined using piecewise polynomial interpolation of degree k. Then Wendland in Ref. 47, Theorem 4.5, states that the error in ρ_h satisfies

$$\|\rho - \rho_h\| \leq c \cdot h^{\min\{d+1, k+1\}}. \tag{65}$$

Results are also given on the needed accuracy in the numerical integrations of the quantities in system (59), showing that if the integrations are sufficiently accurate, then the order of convergence in (65) is not affected.

In discussing (65), Wendland refers to Refs. 26 and 46, and he states that the arguments presented in those papers concerning the effects of using $S \simeq S_h$ carry across to the more general integral-equation framework that he is considering. Also see Add. Ref. 2.

In the engineering literature, it has been common to use $d = 0$ (giving piecewise-constant approximating functions) and $k = 1$ (piecewise planar interpolation of the surface S_h), yielding what is called the *panel method*. As Wendland notes, the convergence rate will be improved to $O(h^2)$ by using $d = 1$, giving piecewise linear approximates ρ_h, to go along with the piecewise planar approximation S_h of S.

Crucial to the above discussion is that the surface S be smooth, contrary to many engineering problems. As discussed earlier in Section 2.4, much less is known about the behavior of the integral operator $\mathcal{K}$ and the solution ρ when S possesses edges or vertices. For the numerical treatment of such BIEs, there have been few rigorous mathematical analyses. One of the earliest is in Ref. 19. This paper discusses the collocation method with piecewise-constant approximants ($d = 0$), with no approximation of the surface. Convergence is shown under very general assumptions about the surface. The major limitation concerns the nonsmooth points P of the surface. Given such a point P, form the interior and exterior cones at P, using the tangent lines to the surface S at P. Then the paper assumes that either the interior or the exterior cone is convex, an assumption that eliminates a few, but not most, surfaces of interest. Less restrictive is the assumption that the solid angle satisfy

$$0 < \mathcal{I}(P) < 4\pi.$$

Another analysis of the collocation method is given in Refs. 48 and 50, with piecewise quadratic elements and piecewise quadratic isoparametric interpolation of the surface. The convergence is restricted to solid angles satisfying

$$\tfrac{4}{5}\pi < \mathcal{I}(P) < \tfrac{16}{5}\pi. \tag{66}$$

This is due to the method used to prove convergence, which is very similar to that used in Ref. 19. Also see Add. Ref. 2.

5. Numerical Integration

All BIE methods involve numerical integrations, usually of several kinds. These quadratures are often the most expensive part of a BIE method for solving Laplace's equation in three dimensions. If the numerical integrations are carried out to a higher accuracy than that present in the remainder of the calculation, then the BIE method is likely to be significantly less efficient than is possible.

The necessary numerical integrations are primarily affected by two factors. First, what is the form of the BIE being solved? If a direct method is being used, as in Section 2.1, then the right-hand side of the integral equation involves integrals that usually are evaluated numerically; see Eqs. (10) and (12). In contrast, indirect methods lead to right-hand sides that are simply the given data, as in Eqs. (16) and (19). The second factor affecting the needed numerical integrations is the type of numerical method being used. Is the method global or local, and is it of collocation or Galerkin type?

After solving the BIE for the unknown density function ρ on S, we usually want to evaluate the harmonic function $u(P)$ at points P inside the domain D of the Laplace equation. This means evaluating single- and/or double-layer potentials, as in Eqs. (9), (15), and (18). This is an area on which there has been very little research, especially for three-dimensional problems. The numerical integration methods used are usually the same as those used to evaluate the boundary integrals in the integral equation; these are described below. Nonetheless, if $u(P)$ is to be evaluated at many points $P \in D$, then the cumulative cost of the quadratures can be quite expensive. More research needs to be done on less expensive ways of evaluating single- and double-layer potentials when they are wanted at a large number of related points P.

For the global method described in Section 3.1, we refer the reader to the discussion in Ref. 36. The ideas in it can be used with the numerical integrations needed in other global BIE methods.

5.1. Boundary Element Integrations.

For local or boundary element methods, consider the approximating linear system. With Galerkin's method (29)–(30), the matrix elements involve double surface integrals, with each integration region a triangular element. The general form is

$$\int_{\Delta_i} \int_{\Delta_j} K(P, Q)\varphi(P)\psi(Q)\, dS(Q)\, dS(P), \tag{67}$$

with φ and ψ low-degree isoparametric polynomials and $K(P, Q)$ the kernel of the integral operator. With both single- and double-layer potentials, the kernel satisfies

$$K(P, Q) = O(|P - Q|^{-1}), \quad \text{as } Q \to P. \tag{68}$$

This will yield a singular integrand if $P = Q$ is possible, for example, if

$\Delta_j = \Delta_j$. For the right-hand sides in the Galerkin linear system (30), the entries are surface integrals of smooth functions over triangular elements.

For collocation methods [see (32)], the matrix elements are surface integrals over triangular regions, as illustrated in system (59). If the collocation point v_j is contained in the integration region Δ_i, then the surface integral has a singular integrand, as in (68).

To develop integration formulas over a triangular region Δ_i, we first use the mapping $m_i: \sigma \to \Delta_i$ to rewrite the integral as

$$I = \int_\sigma G(s, t)\, ds\, dt. \tag{69}$$

We want to consider three possible cases for such integrals.

Case 1. $G(s, t)$ is a well-behaved integrand. Then use any of a number of formulas for such integrals. We have generally used those from Refs. 51 and 52. Two such formulas are

$$\int_\sigma G(s, t)\, ds\, dt \doteq \tfrac{1}{6}[G(0, \tfrac{1}{2}) + G(\tfrac{1}{2}, 0) + G(\tfrac{1}{2}, \tfrac{1}{2})], \tag{70}$$

which has degree of precision 2, and

$$\int_\sigma G(s, t)\, ds\, dt \simeq \tfrac{9}{40} G(\tfrac{1}{3}, \tfrac{1}{3}) + \tfrac{1}{40}[G(0, 0) + G(0, 1) + G(1, 0)]$$

$$+ \tfrac{1}{15}[G(0, \tfrac{1}{2}) + G(\tfrac{1}{2}, 0) + G(\tfrac{1}{2}, \tfrac{1}{2})], \tag{71}$$

which has degree of precision 3. An integration formula with degree of precision p gives an integration accuracy of $O(h^{p+3})$ on a single integral over a triangular element Δ_i.

Case 2. $G(s, t)$ is nearly singular. This occurs with integrals

$$\int_{\Delta_i} K(P, Q)\psi(Q)\, dS(Q), \tag{72}$$

with P not in Δ_i, although close to it. The kernel function $K(P, Q)$ will generally be analytic in P and Q, but increasingly peaked as P approaches Δ_i. In order that the numerical integration will be sufficiently accurate, one must use an increasing number of quadrature node points as P approaches Δ_i. The manner of doing this has generally been ad hoc; and in many programs, this case has been ignored by including it with case (1), a poor strategy. For a theoretical discussion of the needed number of integration nodes, see Ref. 53.

In Ref. 50, we evaluated integral (72) by using a program with automatic error control, to obtain a sufficiently small error in a relative error sense. The program used the 7-point method T2:5-1, of degree of precision 5, from Ref. 51. Uniform subdivisions of σ were carried out until sufficient accuracy was obtained. The division into cases (1) and (2) was based on the distance from P to Δ_i; call it $d(P, \Delta_i)$. If $d(P, \Delta_i) \geq \varepsilon$, then case (1) was used; and if $0 < d(P, \Delta_i) < \varepsilon$, then case (2) was used. The method was adequate, but more work is needed to understand how best to handle cases (1) and (2).

Case 3. $G(s, t)$ is singular at a single point. This occurs in integral (72) when P is inside Δ_i. To treat this case, first assume the singularity occurs in σ at only $(s, t) = (0, 0)$. Other locations can be reduced to this case through a combination of splitting σ into smaller triangles and mapping these new triangles onto σ with $(0, 0)$ the singular point.

To handle the singularity is simpler than might first be thought possible. For the types of kernel $K(P, Q)$ in which we are interested, the variables of integration can be changed to obtain a new integral with a smooth integrand. The simplest such change of variables is

$$s = (1 - y)x, \qquad t = yx, \qquad 0 \leq x, y \leq 1. \tag{73}$$

Then the integral I in (69) becomes

$$I = \int_0^1 \int_0^1 xG((1 - y)x, yx)\, dy\, dx. \tag{74}$$

Assume in integral (72) that Δ_i is contained in a C^∞ section of S, and assume $\psi(Q)$ is an isoparametric polynomial over Δ_i. Let $G(s, t)$ arise from the parameterization of (72) over σ, and let $K(P, Q)$ be a single- or double-layer integral. Then it can be shown that the integrand in (74) will be C^∞ over the square $0 \leq x, y \leq 1$. The proof is relatively straightforward, especially for a single-layer kernel function; and the assumption of a C^∞ triangular element can also be reduced.

Using (73), we can use a simple product Gaussian quadrature formula to calculate (73). For increasing accuracy, simply increase the order of the Gaussian quadrature formula (product formulas based on one-variable Gaussian quadrature with only 2, 3, or 4 nodes are usually sufficient).

The change of variables (73) and other results on evaluating I are given in Ref. 54. This paper gives other alternatives to the use of (74) and product Gaussian quadrature; and it gives some numerical examples to illustrate the improved convergence [note: the work in the cited paper is for the triangle $0 \leq s \leq t \leq 1$, giving an alternative but equivalent formula to (73)]. In particular, they give the change of variables

$$s = x\left[1 - \tan\left(\frac{\pi}{4}y\right)\right], \; t = x\tan\left(\frac{\pi}{4}y\right), \tag{75}$$

for $0 \le x, y \le 1$. In their numerical examples, it gives better results than the use of (73). Their comparisons are strictly empirical, and it is not clear why (75) should perform better than (73). Also, both (73) and (75) give good results for what would otherwise be a difficult singular integral.

Other discussions of evaluating the boundary element integrals of (72) are given in Refs. 13, 55, and 15. A much more general framework for evaluating such integrals, for all BIEs (for more than just Laplace's equation) is given in Refs. 42 and 56. This work shows how to develop special quadrature formulas that take into account the particular type of singularity present in $K(P, Q)$. This includes difficult cases such as the integral operators arising from the normal derivative of a double-layer potential, needed in solving the Neumann problem for the Helmholtz equation $\Delta u + \lambda u = 0$. Results are given for both collocation and Galerkin methods.

6. Iterative Methods of Solution

The numerical methods for solving BIEs reduce to the solution of a finite system of linear equations. With Laplace's equation in the plane, these linear systems can usually be solved directly by Gaussian elimination. But with three-dimensional problems, the size of the linear system can easily be very large, prohibitively so for solution by direct methods.

There is not a well-developed theoretical framework for the iterative solution of linear systems arising from solving BIEs, although a wide variety of ad hoc techniques have been used quite successfully in the engineering literature. Most theoretical work on iterative methods has been for integral equations of the second kind. One of the earliest analyses was given in Ref. 19, in which convergence was shown for the classical successive approximations iteration.

A general framework for multigrid methods for solving second-kind BIEs is given in the book of W. Hackbusch (Ref. 57, Chapter 16); related methods are discussed in Refs. 58 and 59. A very extensive discussion of two-grid and multigrid iteration methods is given in Ref. 60, giving a convergence for iteration methods when the boundary S is smooth. The paper also contains an extensive survey of the literature. In Ref. 61, the numerical method for solving a BIE of the second kind is modified by small perturbations. The resulting method is equally accurate, but can be solved much more efficiently, generally by a multigrid method. In the latter case, they prove an operations cost of $O(h^{-2-\varepsilon})$, where $\varepsilon > 0$ can be chosen arbitrarily small. [Note that the number of linear equations to be solved is $O(h^{-2})$.] Two-grid methods for solving BIEs of the second kind are described in Refs. 62 and 50; and one such iteration is defined and illustrated below. A quite general account of all multilevel iteration methods for solving integral equations of the second kind is given in Ref. 63.

BIEs of the first kind have received much less attention. An early method is that of Landweber in Ref. 64, and it has been further analyzed and modified in Ref. 65. It has been used by Landweber in solving many BIEs arising in fluid flow calculations. More recently, people have suggested using the conjugate gradient method (see Ref. 66, Section 10.2), although no analyses of its use with BIEs, of the first or second kind, are known to this author.

6.1. A Two-Grid Method.
We denote the linear system to be solved by

$$(2\pi - A_M)\tilde{\rho}_M = \tilde{f}_M, \tag{76}$$

and we assume it arises from solving a second-kind BIE by the collocation method of (58)–(59). Let M denote the number of triangular elements in the decomposition of S, and let M_v denote the number of interpolation nodes $\{v_i^{(M)}\}$ used in the collocation method. We have $\tilde{\rho}_M, \tilde{f}_M \in \mathbb{R}^{M_v}$, with $\tilde{f}_{M,i} = f(v_i^{(M)})$, $i = 1, \ldots, M_v$. Once the vector $\tilde{\rho}_M$ is known, then

$$\rho(P) = \sum_{k=1}^{n_d} \tilde{\rho}_M(v_{i,k}^{(M)}) l_{k,i}^{(M)}(P), \qquad P \in \Delta_i, \tag{77}$$

for $i = 1, \ldots, M$. For notation, recall the remarks preceding (57).

We will approximate the inverse $(2\pi - A_M)^{-1}$ by using $(2\pi - A_N)^{-1}$, with $N < M$. We look for a fixed N such that the iteration defined below will converge for all $M > N$. First, some additional concepts and notation. To move between the linear systems of orders M_v and N_v, introduce the *restriction operator*

$$R_{MN} : \mathbb{R}^{M_v} \to \mathbb{R}^{N_v}, \tag{78a}$$

$$R_{MN}\tilde{g}_M(v_i^{(N)}) = \tilde{g}_M(v_i^{(N)}), \qquad i = 1, \ldots, N_v, \tag{78b}$$

and the *prolongation operator*

$$P_{NM} : \mathbb{R}^{N_v} \to \mathbb{R}^{M_v}, \tag{79a}$$

$$P_{NM}\tilde{g}_N(v_i^{(M)}) = \sum_{k=1}^{d_n} \tilde{g}(v_{j,k}^{(N)}) l_{k,j}^{(N)}(v_i^{(M)}), \qquad v_i^{(M)} \in \Delta_j, \tag{79b}$$

A Survey of Boundary Integral Equation Methods

for $i = 1, \ldots, M_v$. The definition of the restriction operator assumes that nodes for the order-N case are included in those for the order-M case; and the restriction of $\tilde{g}_M$ is simply the subvector of components corresponding to the nodes for the order-N case. The prolongation operator uses interpolation of the values in $\tilde{g}_N$ for the order-N case to predict values for the nodes of the order-M case.

Given an initial guess $\tilde{\rho}_M^{(0)}$, generate the iterates $\tilde{\rho}_M^{(\nu)}$, $\nu \geq 0$, as follows:

$$s_M^{(\nu)} := \frac{1}{2\pi}[\tilde{f}_M + A_M \tilde{\rho}_M^{(\nu)}], \tag{80}$$

$$\tilde{r}_M^{(\nu)} := (2\pi - A_M)s_M^{(\nu)} - \tilde{f}_M = (2\pi - A_M)(\tilde{s}_M^{(\nu)} - \tilde{\rho}_M), \tag{81}$$

$$\tilde{\rho}_M = \tilde{s}_M^{(\nu)} - (2\pi - A_M)^{-1}\tilde{r}_M^{(\nu)}, \tag{82a}$$

$$\tilde{\rho}_M^{(\nu+1)} := \tilde{s}_M^{(\nu)} - P_{NM}(2\pi - A_N)^{-1}R_{MN}\tilde{r}_M^{(\nu)}, \tag{82b}$$

for $\nu = 0, 1, \ldots$. The initial guess $\tilde{\rho}_M^{(0)}$ is usually based on interpolating a result $\tilde{\rho}_{M_1}$ obtained in a previous calculation with $M_1 < M$. In the language of multigrid methods, (80) is called the smoothing step and (82) is called the correction step. In calculating, do not calculate $(2\pi - A_N)^{-1}$, but instead solve the system

$$(2\pi - A_N)\tilde{\delta}_M^{(\nu)} = R_{MN}\tilde{r}_M^{(\nu)},$$

$$\tilde{\rho}_M^{(\nu+1)} := \tilde{s}_M^{(\nu)} - P_{NM}\tilde{\delta}_M^{(\nu)}.$$

For the iteration error,

$$\tilde{\rho}_M - \tilde{\rho}_M^{(\nu+1)} = \frac{1}{2\pi}[I - P_{NM}(2\pi - A_N)^{-1}R_{MN}(2\pi - A_M)]A_M(\tilde{\rho}_M - \tilde{\rho}_M^{(\nu)})$$

$$\equiv C_{NM}(\tilde{\rho}_M - \tilde{\rho}_M^{(\nu)}). \tag{83}$$

It can be shown that if S is a smooth surface, then

$$\sup_{M > N} \|C_{NM}\| \to 0, \quad \text{as } N \to \infty. \tag{84}$$

The matrix norm is the maximum of the absolute row sums, induced by the vector norm $\|\cdot\|_\infty$. It follows from (84) that if N is sufficiently large, then $\|C_{NM}\| < 1$ and the iteration will converge for all $M > N$. Moreover, the rate of convergence will be bounded away from one, uniformly in M.

Example 6.1. Solve the interior Dirichlet problem

$$\Delta u(P) = 0, \qquad P \in D,$$
$$u(P) = e^z \cos(y) + e^x \sin(z), \qquad P = (x, y, z) \in S, \tag{85}$$

where $\bar{D}$ is the ellipsoidal solid

$$x^2 + (y/1.5)^2 + (z/2)^2 \leq 1.$$

The collocation method (58)–(59) is used with isoparametric piecewise quadratic polynomials to define both ρ_M and S_h [note: $M = O(h^{-2})$].

We give the computed approximation to the ratio R,

$$R := \lim_{\nu \to \infty} \frac{\|\tilde{\rho}_M^{(\nu+2)} - \tilde{\rho}_M^{(\nu+1)}\|_\infty}{\|\tilde{\rho}_M^{(\nu+1)} - \tilde{\rho}_M^{(\nu)}\|_\infty}, \tag{86}$$

which should approximate $\|C_{NM}\|$. Table 1 contains these results for various values of N and M. The notation N_v gives the number of nodes (and equations) associated with the decomposition of S into N triangular elements.

Example 6.2. Solve the same Dirichlet problem, but with $\bar{D}$ defined by

$$x^2 + (y/2)^2 \leq z \leq 1,$$

a paraboloid. Here the surface S is only piecewise smooth, having an edge

Table 1. Ellipsoid Iteration Results

N	N_v	M	M_v	R
8	18	32	66	0.23
8	18	128	258	0.18
20	42	80	162	0.15
20	42	320	642	0.14

Table 2. Paraboloid Iteration Results

N	N_v	M	M_v	R
8	18	32	66	0.75
8	18	128	258	0.76
32	66	128	258	0.65

at $z = 1$. The convergence result (84) is no longer valid, and it is probably not true. The results of computing (85) are shown in Table 2. The values of R still indicate convergence, but at a slower rate than for the ellipsoidal case.

With most examples in which S is only piecewise smooth, the ratios (86) do not appear to decrease to zero as N increases. For some surfaces S, the values of ratio R do not decrease below 1.0 as N increases, thereby not yielding a convergent iteration. Such results for piecewise smooth surfaces should not be too surprising, as this same behavior has been observed in the two-dimensional case; see Refs. 59 and 67. In both cases, variants are proposed to compensate for the presence of corners; and something similar is almost certainly necessary with the three-dimensional case. Also see Add. Refs. 2-3, 10-12.

References

1. BANERJEE, P., AND WATSON, J., Editors, *Developments in Boundary Element Methods—4*, Elsevier Applied Sciences Publishers, New York, New York, 1986.
2. BESKOS, D., Editor, *Boundary Element Methods in Mechanics*, Elsevier Publishers, Amsterdam, Holland, 1986.
3. BREBBIA, C., Editor, *Topics in Boundary Element Research, Vol. 1: Basic Principles and Applications*, Springer-Verlag, Berlin, Germany, 1984.
4. BREBBIA, C., Editor, *Topics in Boundary Element Research, Vol. 2: Time-Dependent and Vibration Problems*, Springer-Verlag, Berlin, Germany, 1985.
5. BREBBIA, C., Editor, *Topics in Boundary Element Research, Vol. 3: Computational Aspects*, Springer-Verlag, Berlin, Germany, 1987.
6. BREBBIA, C., TELLES, J., AND WROBEL, L., *Boundary Element Techniques: Theory and Applications in Engineering*, Springer-Verlag, Berlin, Germany, 1984.
7. BREBBIA, C., AND WALKER, S., *Boundary Element Techniques in Engineering*, Newnes–Butterworths, London, England, 1980.
8. CRUSE, T., PITKO, A., AND ARMEN, H., Editors, *Advanced Topics in Boundary Element Analysis*, American Society of Mechanical Engineers, New York, New York, 1985.

9. DU, Q. H., Editor, *Boundary Elements*, Pergamon Press, London, England, 1986.
10. HESS, J., AND SMITH, A., *Calculation of Potential Flows about Arbitrary Bodies*, Progress in Aeronautical Sciences, Vol. 8, Edited by D. Küchemann, Pergamon Press, London, 1967.
11. INGHAM, D., AND KELMANSON, M., *Boundary Integral Equation Analyses of Singular, Potential, and Biharmonic Problems*, Springer-Verlag, Berlin, Germany, 1984.
12. LACHAT, J., AND WATSON, J., *Effective Numerical Treatment of Boundary Integral Equations*, International Journal for Numerical Methods for Engineering, Vol. 10, pp. 991–1005, 1976.
13. LEAN, M., AND WEXLER, A., *Accurate Numerical Integrations of Singular Boundary Element Kernels over Boundaries with Curvature*, International Journal for Numerical Methods in Engineering, Vol. 21, pp. 211–228, 1985.
14. SHAW, R., PERIAUS, J., CHAUDOUET, A., WU, J., MARINO, C., AND BREBBIA, C., Editors, *Innovative Numerical Methods in Engineering*, Springer-Verlag, Berlin, Germany, 1986.
15. WATSON, J., *Advanced Implementation of the Boundary Element Method for Two- and Three-Dimensional Elastostatics*, Developments in Boundary Element Methods—1, Edited by P. Banerjee and R. Butterfield, Elsevier Applied Sciences Publishers, London, England, 1979.
16. WATSON, J., *Hermitian Cubic and Singular Elements for Plane Strain*, Developments in Boundary Elements Methods—4, Edited by P. Banerjee and J. Watson, Elsevier Applied Sciences Publishers, New York, New York, 1986.
17. GRISVARD, P., *Elliptic Problems in Nonsmooth Domains*, Pitman Publishers, Boston, Massachusetts, 1985.
18. KELLOGG, O., *Foundations of Potential Theory*, Dover Publications, New York, New York, 1929.
19. WENDLAND, W., *Die Behandlung von Randwertaufgaben im $\mathbb{R}_3$ mit Hilfe von Einfach und Doppelschichtpotentialen*, Numerische Mathematik, Vol. 11, pp. 380–404, 1968.
20. MIKHLIN, S., *Mathematical Physics: An Advanced Course*, North-Holland, Amsterdam, Holland, 1970.
21. JASWON, M., AND SYMM, G., *Integral Equation Methods in Potential Theory and Elastostatics*, Academic Press, London, England, 1977.
22. GÜNTER, N., *Potential Theory*, Ungar, New York, New York, 1967.
23. POGORZELSKI, W., *Integral Equations and Their Applications, Vol. 1*, Pergamon Press, London, England, 1966.
24. JOHNSON, C., AND SCOTT, L. R., *An Analysis of Quadrature Errors in Second-Kind Boundary Integral Equations*, SIAM Journal of Numerical Analysis, Vol. 26, pp. 1356–1382, 1989.
25. WENDLAND, W., *Boundary Element Methods and Their Asymptotic Convergence*, Theoretical Acoustics and Numerical Techniques, Edited by P. Filippi, Springer-Verlag, Berlin, Germany, 1982.
26. NEDELEC, J., *Curved Finite Element Methods for the Solution of Singular Integral Equations on Surfaces in $\mathbb{R}^3$*, Computer Methods in Applied Mechanics and Engineering, Vol. 8, pp. 61–80, 1976.

27. ATKINSON, K., AND DE HOOG, F., *The Numerical Solution of Laplace's Equation on a Wedge*, IMA Journal of Numerical Analysis, Vol. 4, pp. 19-41, 1984.
28. CHANDLER, G., *Galerkin's Method for Boundary Integral Equations on Polygonal Domains*, Journal of the Australian Mathematical Society, Series B, Vol. 26, pp. 1-13, 1984.
29. CHANDLER, G., AND GRAHAM, I., *Product Integration-Collocation Methods for Noncompact Integral Operators*, Mathematics of Computing, Vol. 50, pp. 125-138, 1988.
30. COSTABEL, M., AND STEPHAN, E., *On the Convergence of Collocation Methods for Boundary Integral Equations on Polygons*, Mathematics of Computing, Vol. 49, pp. 461-478, 1987.
31. GRAHAM, I., AND CHANDLER, G., *High-Order Linear Functionals of Solutions of Second-Kind Integral Equations*, SIAM Journal on Numerical Analysis, Vol. 25, 1988.
32. ATKINSON, K., *The Numerical Evaluation of Particular Solutions for Poisson's Equation*, IMA Journal of Numerical Analysis, Vol. 5, pp. 319-338, 1985.
33. ATKINSON, K., *A Survey of Numerical Methods for the Solution of Fredholm Integral Equations of the Second Kind*, SIAM, Philadelphia, Pennsylvania, 1976.
34. ATKINSON, K., *The Numerical Solution of Laplace's Equation in Three Dimensions*, SIAM Journal on Numerical Analysis, Vol. 19, pp. 263-274, 1982.
35. RAGOZIN, D., *Constructive Polynomial Approximation on Spheres and Projective Spaces*, Transactions of the American Mathematical Society, Vol. 162, pp. 157-170, 1971.
36. ATKINSON, K., *Algorithm 629: An Integral Equation Program for Laplace's Equation in Three Dimensions*, ACM Transactions for Mathematical Software, Vol. 11, pp. 85-96, 1985.
37. ATKINSON, K., *The Numerical Solution of Laplace's Equation in Three Dimensions—2*, Numerical Treatment of Integral Equations, Edited by J. Albrecht and L. Collatz, Birkhäuser, Basel, Switzerland, 1980.
38. LIN, T. C., *The Numerical Solution of the Helmholtz Equation Using Integral Equations*, PhD Thesis, University of Iowa, Iowa City, Iowa, 1982.
39. SAAVEDRA, J., *Boundary Integral Equations for Nonsimply Connected Regions*, PhD Thesis, University of Iowa, Iowa City, Iowa, 1988.
40. WENDLAND, W., *Asymptotic Accuracy and Convergence*, Progress in Boundary Element Methods, Vol. 1, Edited by C. Brebbia, Wiley, New York, New York, 1981.
41. WENDLAND, W., *On Galerkin Collocation Methods for Integral Equations of Elliptic Boundary-Value Problems*, Numerische Behandlung von Integralgleichungen, Edited by J. Albrecht and L. Collatz, Birkhäuser, Basel, Switzerland, 1980.
42. SCHWAB, C., AND WENDLAND, W., *3D BEM and Numerical Integration*, Proceedings of the 7th Conference on Boundary Element Methods in Engineering, Edited by C. Brebbia, Springer-Verlag, Berlin, Germany, 1985.
43. WENDLAND, W., *On Some Mathematical Aspects of Boundary Element Methods for Elliptic Problems*, The Mathematics of Finite Elements and Applications—5, Edited by J. Whiteman, Academic Press, London, England, 1985.

44. GIROIRE, J., *Integral Equation Methods for Exterior Problems for the Helmholtz Equation*, Report No. 40, Center for Applied Mathematics, Ecole Polytechnique, Palaiseau, France, 1978.
45. JENG, G., AND WEXLER, A., *Isoparametric, Finite-Element, Variational Solution of Integral Equations for Three-Dimensional Fields*, International Journal for Numerical Methods in Engineering, Vol. 11, pp. 1455-1471, 1977.
46. GIROIRE, J., AND NEDELEC, J., *Numerical Solution of an Exterior Neumann Problem Using a Double-Layer Potential*, Mathematics of Computing, Vol. 32, pp. 973-990, 1978.
47. WENDLAND, W., *Asymptotic Accuracy and Convergence for Point Collocation Methods*, Topics in Boundary Element Research, Vol. 2: Time-Dependent and Vibration Problems, Edited by C. Brebbia, Springer-Verlag, Berlin, Germany, 1985.
48. ATKINSON, K., *Piecewise Polynomial Collocation for Integral Equations on Surfaces in Three Dimensions*, Journal of Integral Equations (Supplementary Issue), Vol. 9, pp. 24-48, 1985.
49. ANSELONE, P., *Collectively Compact Operator Approximation Theory*, Prentice-Hall, Englewood Cliffs, New Jersey, 1971.
50. ATKINSON, K., *Solving Integral Equations on Surfaces in Space*, Constructive Methods for the Practical Treatment of Integral Equations, Edited by G. Hämmerlin and K. Hoffman, Birkhäuser, Basel, Switzerland, 1985.
51. STROUD, A., *Approximate Calculation of Multiple Integrals*, Prentice-Hall, Englewood Cliffs, New Jersey, 1971.
52. LYNESS, J., AND JESPERSEN, D., *Moderate-Degree Symmetric Quadrature Rules for the Triangle*, Journal of the Institute for Mathematics and Applications, Vol. 15, pp. 19-32, 1975.
53. JOHNSON, C., *Finite Element Methods with Applications*, Cambridge University Press, Cambridge, England, 1987.
54. FAIRWEATHER, G., RIZZO, F., AND SHIPPY, D., *Computation of Double Integrals in the Boundary Integral Equation Method*, Advances in Computer Methods for Partial Differential Equations—3, Edited by R. Vichnevetsky and R. Stepleman, IMACS, New Brunswick, New Jersey, 1979.
55. PINA, H., *Numerical Integration*, Topics in Boundary Element Research, Vol. 3: Computational Aspects, Edited by C. Brebbia, Springer-Verlag, Berlin, Germany, 1987.
56. SCHWAB, C., AND WENDLAND, W., *On Numerical Quadrature in Boundary Element Methods*, Numerical Methods in Partial Differential Equations (to appear).
57. HACKBUSCH, W., *Multigrid Methods and Applications*, Springer-Verlag, Berlin, Germany, 1985.
58. SCHIPPERS, H., *Multigrid Methods for Boundary Integral Equations*, Numerische Mathematik, Vol. 46, pp. 351-363, 1985.
59. SCHIPPERS, H., *Theoretical and Practical Aspects of Multigrid Methods in Boundary Element Calculations*, Topics in Boundary Element Research, Vol. 3: Computational Aspects, Edited by C. Brebbia, Springer-Verlag, Berlin, Germany, 1987.

60. NOWAK, Z., *Use of the Multigrid Method for Laplacian Problems in Three Dimensions*, Multigrid Methods, Edited by W. Hackbusch and U. Trottenberg, Springer-Verlag, Berlin, Germany, 1982.
61. NOWAK, Z., AND HACKBUSCH, W., *On the Complexity of the Panel Method*, Report No. 8608, Institut für Informatik und Praktische Mathematik, Christian Albrecht Universität, Kiel, Germany, 1986.
62. ATKINSON, K., *Iterative Variants of the Nyström Method for the Numerical Solution of Integral Equations*, Numerische Mathematik, Vol. 22, pp. 17-31, 1973.
63. MANDEL, J., *On Multilevel Iterative Methods for Integral Equations of the Second Kind and Related Problems*, Numerische Mathematik, Vol. 46, pp. 147-157, 1985.
64. LANDWEBER, L., *An Iteration Formula for Fredholm Integral Equations of the First Kind*, American Journal of Mathematics, Vol. 73, pp. 615-624, 1951.
65. STRAND, O., *Theory and Methods Related to Singular-Function Expansion and Landweber's Iteration for Integral Equations of the First Kind*, SIAM Journal on Numerical Analysis, Vol. 11, pp. 798-825, 1974.
66. GOLUB, G., AND VAN LOAN, C., *Matrix Computations*, Johns Hopkins University Press, Baltimore, Maryland, 1983.
67. ATKINSON, K., AND GRAHAM, I., *An Iterative Variant of the Nyström Method for Boundary Integral Equations on Nonsmooth Boundaries*, The Mathematics of Finite Elements and Applications, Edited by J. Whiteman, Academic Press, New York, New York, 1987.

Additional References

1. ANGELL, T., KLEINMAN, R., AND KRAL, J., *Layer Potentials on Boundaries with Corners and Edges*, Casopis Pro Pestovani Matematiky, Vol. 113, pp. 387-402, 1988.
2. ATKINSON, K., *An Empirical Study of Boundary Element Methods for Integral Equations on Surfaces in Three Dimensions*, Technical Report in Computational Mathematics 1, University of Iowa, Iowa City, Iowa, 1989.
3. ATKINSON, K., AND GRAHAM, I., *Iterative Solution of Linear Systems Arising from the Boundary Integral Method*, 1989 (submitted).
4. BURTON, A., AND MILLER, G., *The Application of Integral Equation Methods to the Numerical Solution of Some Exterior Boundary-Value Problems*, Proceedings of the Royal Society, Series A, Vol. 323, pp. 201-210, 1971.
5. COSTABEL, M., *Principles of Boundary Element Methods*, Computer Physics Reports, Vol. 6, pp. 243-274, 1987.
6. COSTABEL, M., *Boundary Integral Operators on Lipschitz Domains: Elementary Results*, SIAM Journal on Mathematical Analysis, Vol. 19, pp. 613-626, 1988.
7. COSTABEL, M., AND STEPHAN, E., *An Improved Boundary Element Galerkin Method for Three-Dimensional Crack Problems*, Journal of Integral Equations and Operator Theory, Vol. 10, pp. 467-504, 1987.
8. COSTABEL, M., AND WENDLAND, W., *Strong Ellipticity of Boundary Integral Operators*, Zeitschrift für Reine und Angewandte Mathematik, Vol. 372, pp. 34-63, 1986.

9. DAUGE, M., *Elliptic Boundary-Value Problems on Corner Domains*, Springer-Verlag, Berlin, Germany, 1988.
10. HEBEKER, F., *Efficient Boundary Element Methods for Three-Dimensional Exterior Viscous Flows*, Numerical Methods for Partial Differential Equations, Vol. 2, pp. 273-297, 1986.
11. HEBEKER, F., *Characteristics and Boundary Elements for Three-Dimensional Nonstationary Navier-Stokes Flows*, Panel Methods in Mechanics, Edited by J. Ballmann, R. Eppler, and W. Hackbusch, GAMM Seminar, Kiel, Germany, 1987.
12. HEBEKER, F., *On the Numerical Treatment of Viscous Flows Against Bodies with Corners and Edges by Boundary Element and Multigrid Methods*, Numerische Mathematik (to appear).
13. KLEINMAN, R., AND ROACH, G., *Boundary Integral Equations for the Three-Dimensional Helmholtz Equation*, SIAM Review, Vol. 16, pp. 214-236, 1974.
14. KRAL, J., AND WENDLAND, W., *Some Examples Concerning Applicability of the Fredholm-Radon Method in Potential Theory*, Aplikace Matematiky, Vol. 31, pp. 293-308, 1986.
15. LEAN, M., FRIEDMAN, M., AND WEXLER, A., *Applications of the Boundary Element Method in Electrical Engineering Problems*, Developments in Boundary Element Methods—1, Edited by P. Banerjee and R. Butterfield, Elsevier Applied Sciences Publishers, London, England, 1979.
16. MIRANDA, C., *Partial Differential Equations of Elliptic Type*, Springer-Verlag, Berlin, Germany, 1970.
17. PETERSDORFF, T. V., AND STEPHAN, E., *Decompositions in Edge and Corner Singularities for the Solution of the Dirichlet Problem of the Laplacian in a Polyhedron*, Mathematische Nachrichten, 1989 (submitted).
18. STEPHAN, E., *Boundary Integral Equations for Screen Problems in $\mathbb{R}^3$*, Integral Equations and Operations Theory, Vol. 9, 1986.
19. STEPHAN, E., *A Boundary Integral Equation Method for Three-Dimensional Crack Problems in Elasticity*, Mathematical Methods in Applied Sciences, Vol. 8, 1986.
20. STEPHAN, E., *Boundary Integral Equations for Magnetic Screens in $\mathbb{R}^3$*, Proceedings of the Royal Society of Edinburgh, Vol. 102A, pp. 189-210, 1986.
21. STEPHAN, E., *Boundary Integral Equations for Mixed Boundary Value Problems in $\mathbb{R}^3$*, Mathematische Nachrichten, Vol. 134, pp. 21-53, 1987.
22. SCHMITZ, H., *Über das Singuläre Verhalten der Lösungen von Integralgleichungen auf Flächen mit Ecken*, PhD Thesis, Universität Stuttgart, 1989.
23. VOLK, K., *Zur Berechnung von Singulärfunktionen Dreidimensionaler Elastischer Felder*, PhD Thesis, Universität Stuttgart, 1989.
24. WENDLAND, W., VOLK, K., AND SCHMITZ, H., *A Boundary Element Method for Three-Dimensional Singularities of Elastic Fields*, Preprint, Universität Stuttgart, 1989.
25. WENDLAND, W., AND ZHU, J., *The Boundary Element Method for Three Dimensional Stokes Flows Exterior to an Open Surface*, Computers and Mathematics with Applications, 1989 (submitted).

2

Superconvergence

I. H. SLOAN

Abstract. Superconvergence properties of approximation methods for integral equations of the second kind are reviewed.

1. Introduction

The word "superconvergence," an evocative word indeed, was introduced in the early 1970s to describe an interesting phenomenon arising in the solution of two-point boundary-value problems and related partial differential equations, namely, that for certain piecewise-polynomial approximation methods, the order of convergence at the knots is higher than one might have expected from the order of the piecewise polynomials employed, and also higher than at other points of the interval. The phenomenon was observed with both the Galerkin method (Refs. 1-4) and the collocation method (Refs. 5-7), provided in the latter case that the collocation points are very carefully chosen.

Subsequently the usage has been extended to cover analogous phenomena for integral equations of the second kind,

$$y(t) - \int_a^b k(t,s)y(s)\,ds = f(t), \qquad t \in [a,b],$$

or in operator form

$$y - Ky = f, \tag{1}$$

under appropriate assumptions on k and f. This is the subject of the present chapter. Earlier reviews of part of the material are Refs. 8-9.

I. H. SLOAN • Department of Mathematics, The University of New South Wales, Kensington, New South Wales, Australia 2033.

Throughout most of this chapter k will be assumed to be, if not smooth, at least well enough behaved to make the integral operator K compact in some appropriate Banach space.

What do we mean by superconvergence in relation to Eq. (1)? Suppose that the equation is solved approximately by the Galerkin or collocation method, with the approximate solution y_h being a piecewise polynomial of order r (i.e., degree at most $r - 1$), on a partition with maximum mesh size h. Then the best result we can hope to achieve, in an appropriate L_p norm, is that $\|y_h - y\|_{L_p}$ be of order $O(h^r)$. Sometimes, however, certain quantities derived from y_h show a faster rate of convergence than this. *Any quantity computed from y_h that has an order of convergence higher than $O(h^r)$ is said to be superconvergent.* The usage can also extend to polynomial basis functions: in this case a quantity computed from y_h may be said to be superconvergent if it converges faster than the best polynomial approximation to y, in some suitable norm. For simplicity, we shall generally consider only piecewise-polynomial spaces in this paper.

Two different kinds of quantities computed from y_h seem to be interesting in practice. The first are linear functionals, such as (in appropriate spaces) $y_h(t)$ at some selected point t, or $\int_a^b y_h(s)v(s)\,ds$ for some continuous function v. For example, Richter (Ref. 10) has shown that the Galerkin method for certain piecewise-polynomial spaces can converge with one power more than normal at certain special points. The second are new (global) approximations to y itself. If y_h^* is a new approximation to y derived by some operation on y_h, then y_h^* is said to be "globally superconvergent" if in some appropriate norm $\|y_h^* - y\|$ has a higher order of convergence than $O(h^r)$.

Because we are here dealing with an equation of the second kind, which can be rewritten as $y = f + Ky$, an interesting global approximation derivable from y_h is its first iterate,

$$y_h^1 = f + Ky_h. \qquad (2)$$

It turns out that y_h^1 can indeed be globally superconvergent, under a wide variety of circumstances. We take up this question in Section 3. Some references are Refs. 11-23.

It is well known (see, for example, Refs. 24-25) that the Galerkin and collocation methods can be formulated in a way that emphasizes their similarity. This "projection-method" formulation, indicated in the next section, can also be carried through to the iterated Galerkin and collocation methods. This is the point of view we take in Section 3. We shall see, however, that in a certain sense superconvergence phenomena for the Galerkin and collocation methods have different origins. For that reason, we shall later find it convenient to consider the two cases separately, the Galerkin method being considered in Section 4 and the collocation method in Section 5.

One difference between the Galerkin and collocation cases is that the collocation method does not exhibit any superconvergence phenomena unless the collocation points are carefully chosen. Thus for the collocation method there is a whole area of concern, namely, the question of the choice of collocation points, that has no parallel in the Galerkin method. In Section 5 we shall see, for example, that for the case of discontinuous piecewise polynomials the Gauss quadrature points of appropriate order on each subinterval have very special attractions as collocation points. This result derives from Refs. 19-20, but is here proved in a different way.

The chapter concludes with a brief review in Section 6 of analogous superconvergence results for nonlinear integral equations.

Closely related to Eq. (1) is the eigenvalue problem

$$y(t) = \lambda \int_a^b k(t, s) y(s)\, ds, \qquad a \leq t \leq b. \tag{3}$$

In this chapter we do not discuss superconvergence aspects of the eigenvalue problem, but it is well known that many superconvergence considerations for second-kind equations have direct parallels for Eq. (3). The relation is well articulated in Ref. 26. Some papers concerned with superconvergence for the eigenvalue problem are Refs. 22, 27-28.

Superconvergence phenomena can also arise in the solution of Volterra integral equations (see Refs. 29-31), but the considerations in that case have a quite different flavor, associated with the initial-value character of the Volterra equation.

It should be said that phenomena similar to those considered here also arise with boundary integral equations of quite general kinds. For example, it is well known that the boundary integral method for elliptic problems on smoothly bounded regions may lead to "superapproximation" of solutions away from the boundary. For a discussion, see for example Ref. 21. For regions with corners, the preservation of rates of convergence and superconvergence by suitable "grading of the mesh" is the subject of much current research activity. The link between mesh grading and superconvergence for boundary integral equations that are formally of the second kind (but with K not compact!) is discussed in Ref. 32.

Finally, we may note that analogous superconvergence phenomena also arise in Cauchy singular integral equations, for example, in the airfoil equation; see Refs. 33-34.

2. Galerkin and Collocation Methods

The Galerkin and collocation methods both start out the same way: one seeks an approximation to y, the solution of Eq. (1), of the form $y_h \in S_h$ where S_h is a suitable finite-dimensional linear space. The space S_h might be a space of piecewise polynomials, in which case h is the length of the largest subinterval in the partition. Or it might be a space of polynomials, or trigonometric polynomials, in which case we may take h to be the reciprocal of the dimension of the space S_h.

The two methods differ in the way in which the approximation y_h is selected from the space S_h. Intuitively, one tries to make $y_h - Ky_h - f$ as small as possible. In the Galerkin method one seeks to achieve this by requiring

$$(y_h - Ky_h - f, \chi_h) = 0, \qquad \forall \chi_h \in S_h, \tag{4}$$

where (u, v) is the L_2 inner-product,

$$(u, v) = \int_a^b u(s)v(s)\,ds. \tag{5}$$

In the collocation method one chooses (carefully!) a set of distinct points $t_1, \ldots, t_n \in [a, b]$, where $n = n_h$ is the dimension of S_h, and imposes the conditions

$$y_h(t_j) - Ky_h(t_j) - f(t_j) = 0, \qquad j = 1, \ldots, n. \tag{6}$$

In practice, of course, one chooses a basis $\{u_1, u_2, \ldots, u_n\}$ for S_h, so that y_h is a linear combination,

$$y_h = \sum_{i=1}^n a_i u_i. \tag{7}$$

Then the coefficients $a_1, a_2, \ldots, a_n$ are determined in the Galerkin method by the set of linear equations

$$\sum_{i=1}^n [(u_i, u_j) - (Ku_i, u_j)] a_i = (f, u_j), \qquad j = 1, 2, \ldots, n, \tag{8}$$

and in the collocation method by

$$\sum_{i=1}^n [u_i(t_j) - Ku_i(t_j)] a_i = f(t_j), \qquad j = 1, 2, \ldots, n. \tag{9}$$

Superconvergence

We shall assume that the collocation points $t_1, t_2, \ldots, t_n$ and the space S_h are such that the $n \times n$ matrix $\{u_i(t_j)\}$ is nonsingular.

It then follows, as in Ref. 25, that (4) and (6) can both be written as projection methods of the form

$$P_h(y_h - Ky_h - f) = 0, \tag{10}$$

where P_h is a projection onto S_h. In the case of the Galerkin method P_h is the orthogonal projection, that is, $P_h v \in S_h$,

$$(P_h v - v, \chi_h) = 0, \quad \forall \chi_h \in S_h, \tag{11}$$

while for the collocation method it is the interpolatory projection, defined by $P_h v \in S_h$,

$$P_h v(t_j) = v(t_j), \quad j = 1, \ldots, n. \tag{12}$$

Since $P_h y_h = y_h$, (10) may be written as the second-kind equation

$$y_h - P_h K y_h = P_h f. \tag{13}$$

We may consider this as the defining equation for the Galerkin or collocation methods, with the appropriate choice of P_h.

To proceed further, one needs a suitable function space setting. We shall assume that f belongs to a Banach space B, with associated norm $\|\cdot\|$, and that K is a compact operator in this space, which does not have 1 as a characteristic value. Then Eq. (1) has a unique solution $y = (I - K)^{-1} f$ in this space. Moreover, $(I - K)^{-1}$ is necessarily a bounded operator in B, thus problem (1) is well posed.

For the Galerkin method B may be taken to be, for example, the space $L_2(a, b)$ of square-integrable functions on (a, b) with norm $\|v\|_{L_2} = (v, v)^{1/2}$. Or it may be the space $C[a, b]$ of continuous functions on $[a, b]$ with norm $\|v\|_\infty = \max|v(t)|$. If S_h contains bounded but discontinuous functions, the latter space may be replaced by the space $L_\infty(a, b)$ of essentially bounded functions on (a, b) with norm $\|v\|_{L_\infty} = \text{ess sup}|v(t)|$.

For the collocation method the simplest choice for the space B is $C[a, b]$. But if S_h contains discontinuous functions then the choice is more difficult, since, as pointed out in Ref. 35, the often proposed choice $L_\infty(a, b)$ is not satisfactory without further qualification. This is because the interpolatory projection P_h is not, as it stands, a well-defined linear operator on L_∞. A simple way of making the space L_∞ usable, and some alternative settings for the collocation method, are discussed in Ref. 35.

The following is a standard existence and stability result for projection methods (see, for example, Refs. 24-25; throughout the chapter c is a generic constant, which may take different values at its different occurrences).

Theorem 2.1. Let K be a compact operator on the Banach space B, not having 1 as a characteristic value, and assume $f \in B$. If $S_h \subset B$ and if the projections P_h satisfy

$$\|P_h K - K\| \to 0, \quad \text{as } h \to 0, \tag{14}$$

then for h sufficiently small, Eq. (13) has a unique solution $y_h \in S_h$, which satisfies

$$\|y_h - y\| \leq c\|P_h y - y\|, \tag{15}$$

with c independent of f and h.

For a proof, see the references cited above. The main point in the proof is that, because of (14), $(I - P_h K)^{-1}$ exists as a (uniformly) bounded operator on B for h sufficiently small. The estimate (15) then follows from the identity

$$y_h - y = (I - P_h K)^{-1}(P_h y - y). \tag{16}$$

A sufficient condition for (14) in the theorem to hold is that

$$\|P_h v - v\| \to 0, \quad \text{as } h \to \infty, \quad \forall v \in B. \tag{17}$$

This is because of the general proposition that pointwise convergence of bounded linear operators is necessarily uniform on compact sets. However, condition (17) is not necessary, and is not always appropriate. For example, Ref. 36 discusses polynomial collocation in a setting in which (14) holds, but (17) does not.

Now we are ready to look at superconvergence phenomena arising in the iterated Galerkin and collocation methods.

3. Iterated Projection Method

As in the preceding sections, we let $y_h \in S_h$ denote either the Galerkin or collocation approximation to the exact solution of Eq. (1). Then the iterated Galerkin or collocation approximation is defined by

$$y_h^1 = f + K y_h. \tag{18}$$

Superconvergence

In this section we are concerned with (global) superconvergence properties of y_h^1.

In the preceding section, we have seen that y_h satisfies the second-kind equation

$$y_h - P_h K y_h = P_h f. \tag{19}$$

From this and Eq. (18) we have immediately

$$y_h = P_h y_h^1, \tag{20}$$

so that (18) yields an integral equation for y_h^1,

$$y_h^1 - K P_h y_h^1 = f. \tag{21}$$

It is easy to see that the solutions of Eqs. (19) and (21) are in one-to-one correspondence, with the correspondence given explicitly in one direction by (18), and in the other by (20). In the preceding section the analysis was based on the integral equation for y_h. But it is just as possible to base it on the integral equation for y_h^1. That is the path we follow in this section.

The analysis is simplest under the assumption that $\|KP_h - K\| \to 0$. Under this assumption we are able to prove, by arguments similar to those used to prove Theorem 2.1, a very general superconvergence result for y_h^1; see Theorem 3.1 below. However, the reader should be warned: the two propositions that follow show that although this assumption may be appropriate for the Galerkin method, it is definitely *not* so for the collocation method.

Proposition 3.1. Let K be a compact operator on a Hilbert space H, and let P_h be the orthogonal projection from H onto a finite-dimensional subspace S_h, where $\{S_h\}$ is such that

$$\|P_h v - v\| \to 0, \quad \text{as } h \to 0, \quad \forall v \in H. \tag{22}$$

Then

$$\|KP_h - K\| \to 0, \quad \text{as } h \to 0.$$

Proof. Because the Hilbert space norm of an operator A is the same as that of its adjoint A^*, and because the orthogonal projection P_h is self-adjoint, we have

$$\|KP_h - K\| = \|P_h K^* - K^*\|. \tag{23}$$

Moreover, the compactness of K implies that K^* is compact. Then the result follows from (22) by the proposition that pointwise convergence of bounded linear operators is necessarily uniform on compact sets. □

Proposition 3.2. Let K be a compact operator on the space $C[0, 1]$, and let P_h be an interpolatory projection, as defined in (12), from $C[0, 1]$ onto a finite-dimensional space $S_h \subset C[0, 1]$. Then

$$\|KP_h - K\| \geq \|K\|. \tag{24}$$

This well-known result is easily proved by considering the fixed operator $KP_h - K$ acting on a suitable sequence of continuous functions $v_{h,i}$, with $\|v_{h,i}\|_\infty = 1$, each of which the first term does not see at all—that is, each function $v_{h,i}$ is constructed so that it vanishes at every collocation point $t_1, \ldots, t_n$. From this it follows that $\|(KP_h - K)v_{h,i}\|_\infty = \|Kv_{h,i}\|_\infty$. The remainder of the proof is left to the reader.

The following result, demonstrating global superconvergence of y_h^1 under the assumption $\|KP_h - K\| \to 0$, is taken from Ref. 12; see also Ref. 13.

Theorem 3.1. Let K be a compact operator on the Banach space B, not having 1 as a characteristic value, and assume $f \in B$. If the projection $P_h : B \to S_h$ satisfies

$$\|KP_h - K\| \to 0, \quad \text{as } h \to 0, \tag{25}$$

then for h sufficiently small Eq. (21) has a unique solution $y_h^1 \in B$, which satisfies

$$\|y_h^1 - y\| \leq c\|KP_h - K\| \inf_{\chi_h \in S_h} \|\chi_h - y\| \tag{26}$$

$$\leq c\|KP_h - K\| \cdot \|y_h - y\|, \tag{27}$$

with c independent of f and h.

Proof. From the identity

$$(I - K)^{-1}(I - KP_h) = I + (I - K)^{-1}(K - KP_h),$$

it follows that, if h is small enough to ensure

$$\|(I - K)^{-1}\| \cdot \|K - KP_h\| \leq \rho < 1,$$

Superconvergence

then $(I - KP_h)^{-1}$ exists and is uniformly bounded as an operator on B. Under this condition Eq. (21) has the unique solution $y_h^1 = (I - KP_h)^{-1}f \in B$, and we have

$$y_h^1 - y = (I - KP_h)^{-1}f - y = (I - KP_h)^{-1}(f - y + KP_h y)$$
$$= (I - KP_h)^{-1}(KP_h - K)y. \tag{28}$$

Now we come to the point at which the "trick," if there may be said to be one, is done: because P_h is a projection onto S_h, for arbitrary $\chi_h \in S_h$ we may write

$$y_h^1 - y = (I - KP_h)^{-1}(KP_h - K)(y - \chi_h), \tag{29}$$

and hence

$$\|y_h^1 - y\| \le c\|KP_h - K\| \cdot \|y - \chi_h\|.$$

Because χ_h is arbitrary, the desired result (26) now follows immediately. □

The conclusion is that y_h^1 converges to y faster than any approximation to y from the space S_h (and therefore in particular faster than y_h), by a factor proportional to $\|KP_h - K\|$. Thus the iterated approximation y_h^1 is superconvergent; and moreover the superconvergence factor is uniform in f for $f \in B$.

For the case of piecewise-polynomial spaces S_h the bounds in the theorem can be used to calculate explicit superconvergence rates for the iterated Galerkin method, as first observed, for the case $B = C[a, b]$, by Chandler (Ref. 15). The following is a special case of a result of Sloan and Thomée quoted in Ref. 9. In it we meet for the first time the Sobolev norms,

$$\|v\|_{W_p^m} = \sum_{j=0}^{m} \|D^{(j)}v\|_{L_p}, \tag{30}$$

for m an arbitrary nonnegative integer, $1 \le p \le \infty$, and $D^{(j)}$ the jth derivative. The simplest results are obtained by setting $p = q = 2$.

Proposition 3.3. Let P_h be the L_2 orthogonal projection onto S_h, a space of piecewise polynomials of order r (i.e., degree $\le r - 1$). For fixed p satisfying $1 \le p \le \infty$, assume that S_h has the approximation property

$$\|P_h v - v\|_{L_q} \le ch^s \|v\|_{W_q^s}, \qquad s = 0, 1, \ldots, r, \tag{31}$$

where c is independent of h and v, and q is given by $1/p + 1/q = 1$. Moreover, assume that K^* satisfies

$$\|K^*v\|_{W_q^m} \le c\|v\|_{L_q}, \tag{32}$$

for some integer m satisfying $1 \le m \le r$. Then

$$\|KP_h - K\|_{L_p \to L_p} \le ch^m.$$

Proof. For simplicity we prove the result only for $p = q = 2$, leaving to the reader (or see Ref. 9) the easy generalization to arbitrary p. Since the L_2 norm of an operator is the same as that of its adjoint, we have

$$\|KP_h - K\|_{L_2 \to L_2} = \|P_h K^* - K^*\|_{L_2 \to L_2} = \sup_{v \in L_2} \frac{\|(P_h K^* - K^*)v\|_{L_2}}{\|v\|_{L_2}}$$

$$\le ch^m \sup_{v \in L_2} \frac{\|K^*v\|_{W_2^m}}{\|v\|_{L_2}} \le ch^m. \qquad \square$$

This result suggests that the origin of the superconvergence behavior is the smoothing produced by the adjoint operator K^*: specifically, inequality (32) assumes that K^*v, for arbitrary $v \in L_q$, has m derivatives in L_q. Since

$$K^*v(s) = \int_a^b k(t, s)v(t)\, dt, \tag{33}$$

the smoothing property (32) holds if, for instance, $k(t, s)$ has continuous partial derivatives with respect to s of all orders up to the mth.

In the most favorable case we have $m = r$ in Proposition 3.3, and, correspondingly, $\|KP_h - K\| = O(h^r)$. In this situation Theorem 3.1 tells us that if the Galerkin method is of the maximum order $O(h^r)$, then the iterated Galerkin method is of order $O(h^{2r})$. The potential of the iterated variant to double the rate of convergence in the most favorable case was first observed, it seems, by Chandler; see Refs. 8, 14–15.

The following result is the formal result of combining Theorem 3.1 and Proposition 3.3 (in the hypotheses we may now omit the assumption that K is compact, because the smoothing property (32) ensures that K^* is compact in L_q, from which it follows that K is compact in L_p).

Theorem 3.2. For fixed p satisfying $1 \leq p \leq \infty$ let $f \in L_p$, and let K be a bounded operator on L_p such that $I - K$ is one-to-one. With q defined by $1/p + 1/q = 1$, assume that K^* satisfies (32) for some integer m satisfying $1 \leq m \leq r$. Further, let P_h be the L_2 orthogonal projection onto a space S_h of piecewise polynomials of order r, for which the approximation property (31) holds. Then for h sufficiently small (21) has a unique solution $y_h^1 \in L_p(a, b)$, which satisfies

$$\|y_h^1 - y\|_{L_p} \leq ch^m \|y_h - y\|_{L_p}. \tag{34}$$

For a C^∞ kernel, such as

$$k(t, s) = 1/[1 + (t - s)^2], \tag{35}$$

we may choose $m = r$ and $p = \infty$ in the theorem, and so obtain a superconvergence factor of $O(h^r)$ in the uniform norm. On the other hand, for the weakly singular kernel

$$k(t, s) = \log|t - s|,$$

it is shown in Ref. 23 that in (32) we may choose $m = 1$ and any value of q in $1 < q < \infty$ (corresponding to p satisfying $1 < p < \infty$ in Theorem 3.2), but that the result (32) is *not* valid with $m = 1$ and $q = 1$ or ∞ (corresponding to $p = \infty$ or 1). For a discussion of analogous two-dimensional examples, see Refs. 9, 23.

We shall have more to say about the iterated Galerkin method in Section 4, where a different argument, not based on the projection P_h, is used to obtain similar results, as well as superconvergence results for *derivatives* of y_h^1 (such results can also be obtained by the present method—see Ref. 9).

3.1. Iterated Kantorovich Method.

Before leaving iterated projection methods, there is a variant, introduced in Ref. 38, that deserves mention. The idea behind "Kantorovich regularization" (Ref. 37) is that if the inhomogeneous term f in Eq. (1) is not a smooth function, one might do better to write

$$y = f + z \tag{36}$$

and seek a numerical approximation only for the term $z \equiv Ky$, while treating the badly behaved function f exactly. Intuitively, this is appealing if the kernel is smooth compared to f, since then z will be smoother than f, and so will presumably be easier to approximate than y itself.

The new unknown function $z = Ky$ clearly satisfies

$$z - Kz = Kf, \tag{37}$$

whose only difference from (1) lies in the new inhomogeneous term Kf. This is preumably smoother than f, though in practice harder to calculate.

The Kantorovich method is the approximation scheme in which Eq. (37) is solved by the Galerkin method, and then the exact f is added as in (36) to obtain an approximation to y. Thus if P_h denotes the L_2 orthogonal projection onto a finite-dimensional space S_h, then the Kantorovich method is defined by

$$z_h - P_h K z_h = P_h K f, \tag{38}$$

$$\hat{y}_h = f + z_h, \tag{39}$$

with the caret symbol used to distinguish the Kantorovich approximation from the Galerkin approximation y_h.

Following Ref. 38, we now introduce the iterated Kantorovich approximation $\hat{y}_h^1 = f + K\hat{y}_h$. Can we expect, on the analogy of the iterated Galerkin method, that this iteration will produce superconvergence? It is easy to see that the answer is yes: in fact we have immediately

$$\hat{y}_h^1 = f + z_h^1, \tag{40}$$

where

$$z_h^1 = K\hat{y}_h = Kf + Kz_h, \tag{41}$$

so that, as the notation suggests, z_h^1 is the first iterate of the Galerkin approximation to (37); and

$$\hat{y}_h^1 - y = z_h^1 - z, \tag{42}$$

from which we see that the error in $\hat{y}_h^1$ is just the error in the iterated Galerkin approximation z_h^1, to which our earlier analysis is applicable.

Thus, Theorem 3.1 yields immediately the following theorem.

Theorem 3.3. See Ref. 38. Let K be a compact operator on a Banach space B, not having 1 as a characteristic value, and assume $f \in B$. If the projection $P_h: B \to S_h$ satisfies (25), then for h sufficiently small (40) and (41) yield a unique approximation $\hat{y}_h^1 \in B$, satisfying

$$\|\hat{y}_h^1 - y\| \le c\|KP_h - K\| \inf_{\chi_h \in S_h} \|\chi_h - z\| \le c\|KP_h - K\| \cdot \|\hat{y}_h - y\|,$$

where $z = y - f = Ky$.

3.2. An Example.

The following example, taken from Ref. 38, provides an opportunity to compare the performance of four methods: the Galerkin, iterated Galerkin, Kantorovich, and iterated Kantorovich methods.

The equation is

$$y(t) - \lambda \int_0^1 \frac{1}{1 + (t-s)^2} y(s) \, ds = t^{1/2}, \qquad t \in [0, 1], \qquad (43)$$

with λ not a characteristic value. As S_h we choose the space of continuous piecewise-linear functions on the uniform mesh $0, 1/n, \ldots, 1 - 1/n, 1$, with $h = 1/n$. For simplicity, the error analysis is carried out in the space $B = L_2$.

From Theorem 2.1, the Galerkin approximation y_h is of optimal order. But because the exact solution y is of the form

$$y(t) = t^{1/2} + z(t),$$

where z is a C^∞ function, the optimal rate of convergence is just that of the best L_2 approximation to the nonsmooth function $t^{1/2}$. In this way it is shown in Ref. 38 that

$$\|y_h - y\|_{L_2} \le c(y)h,$$

and also that this estimate is sharp with respect to order.

For the iterated Galerkin method, because the kernel is smooth, we may take $m = r = 2$ in (32), and so deduce from Theorem 3.2

$$\|y_h^1 - y\|_{L_2} \le ch^2 \|y_h - y\|_{L_2} \le c(y)h^3.$$

The Kantorovich approximation $\hat{y}_h$, because it seeks to approximate only the smooth function z, is of optimal order, that is,

$$\|\hat{y}_h - y\|_{L_2} = \|z_h - z\|_{L_2} \le c(z)h^2;$$

and it is shown in Ref. 38 that this estimate too is sharp with respect to order.

Finally, as expressed by Theorem 3.3, the iterated Kantorovich method has the same factor of the improvement over the Kantorovich method as the iterated Galerkin method has over the Galerkin method: setting $m = 2$ in (32), we have from Proposition 3.3 and Theorem 3.3

$$\|\hat{y}_h^1 - y\|_{L_2} \leq ch^2 \|\hat{y}_h - y\|_{L_2} \leq c(z)h^4.$$

Thus for this particular example we have, for the Galerkin and Kantorovich methods,

$$\|y_h - y\|_{L_2} = O(h)$$

and

$$\|\hat{y}_h - y\|_{L_2} = O(h^2),$$

with the order being sharp in both cases; and for the iterated Galerkin and iterated Kantorovich methods,

$$\|y_h^1 - y\|_{L_2} = O(h^3)$$

and

$$\|\hat{y}_h^1 - y\|_{L_2} = O(h^4).$$

It should be said, however, that for a less smooth kernel the benefits produced by the iteration may be less impressive. As Proposition 3.3 makes clear, everything depends on the smoothing properties of K^*, as expressed by (32).

4. Linear Functionals of the Galerkin Approximation

In practice, one is often interested not so much in the values $y(t)$ of the solution itself, as in certain quantities that depend (linearly) on y. Often such quantities can be expressed as integrals of the form

$$\int_a^b y(t)g(t)\, dt \equiv (y, g), \tag{44}$$

where g is more or less smooth. We shall see that the convergence of (y_h, g) to (y, g), where y_h is the Galerkin approximation, can be faster than the pointwise convergence of y_h to y—that is, it can be superconvergent. Later, in subsection 4.3, we shall describe a different approximation to (y, g), proposed recently in Ref. 39, which converges even faster than (y_h, g).

4.1. Approximation of (y, g) by (y_h, g).
In essence the argument is simple. If the quantity in which we are interested is (y, g), let w be the solution (assuming for the moment that such a solution exists) of the adjoint equation

$$w(s) - \int_a^b k(t, s) w(t)\, dt = g(s), \qquad s \in [a, b], \tag{45}$$

or in operator form

$$(I - K^*) w = g. \tag{46}$$

Then, the error in the Galerkin approximation (y_h, g) is

$$(y_h, g) - (y, g) = (y_h - y, (I - K^*) w) = ((I - K)(y_h - y), w)$$
$$= ((I - K)(y_h - y), w - \chi_h), \tag{47}$$

for arbitrary $\chi_h \in S_h$, where in the last step we have used the very definition of the Galerkin method, together with $(I - K) y = f$. Thus if $\|\cdot\|$ denotes the L_2 norm, we have

$$|(y_h, g) - (y, g)| \leq (1 + \|K\|) \|y_h - y\| \inf_{\chi_h \in S_h} \|w - \chi_h\|, \tag{48}$$

in which the last factor provides the superconvergence.

The above argument, borrowed from the finite-element literature, embodies Nitsche's trick—see Ref. 21. It was first used in this context by Chandler (Refs. 8, 14-15).

So far we have not considered the question of the existence of the function w in Eq. (45), nor have we estimated the rate of superconvergence in (48). The following result estimates the rate of superconvergence through an expression analogous to (48), but with the setting generalized to L_p spaces with a general value of p. The theorem is adapted from Ref. 23 or Ref. 40. Again we see, as in the preceding section, that superconvergence derives from the smoothing produced by the adjoint of K.

Theorem 4.1. For fixed p in $1 \leq p \leq \infty$, assume $f \in L_p$, and let K be a bounded operator on L_p such that $I - K$ is one-to-one. With q defined by $1/p + 1/q = 1$, assume that K^* satisfies

$$\|K^* v\|_{W_q^m} \leq c \|v\|_{L_q}, \tag{49}$$

for some fixed positive integer $m \leq r$ and some c independent of v, and assume also $g \in W_q^m$. Further, let S_h be a space of piecewise polynomials of order r, with the approximation property

$$\inf_{\chi_h \in S_h} \|v - \chi_h\|_{L_q} \leq ch^s \|v\|_{W_q^s}, \qquad s = 0, 1, \ldots, r, \qquad (50)$$

and assume that the Galerkin approximation $y_h \in S_h$ exists uniquely. Then

$$|(y_h, g) - (y, g)| \leq ch^m \|y_h - y\|_{L_p} \cdot \|g\|_{W_q^m}. \qquad (51)$$

Proof. It follows from (49) that K^* maps L_q into W_q^m, and hence (since W_q^m is compactly imbedded in L_q) that K^* is a compact operator on L_q. Thus K is a compact operator on L_p, which by assumption does not have one as a characteristic value. Since $g \in W_q^m \subset L_q$, it now follows from the Fredholm alternative that (45) has a unique solution $w \in L_q$. But since $w = g + K^* w$, in which both terms on the right belong to W_q^m, it follows that $w \in W_q^m$. By the foregoing argument $I - K^*$ maps W_q^m onto W_q^m, from which it follows that $(I - K^*)^{-1}$ is a bounded operator on W_q^m, and hence

$$\|w\|_{W_q^m} \leq c\|g\|_{W_q^m}.$$

Now the result follows easily from (47): we have

$$|(y_h, g) - (y, g)| \leq \|(I - K)(y_h - y)\|_{L_p} \cdot \inf_{\chi_h \in S_h} \|w - \chi_h\|_{L_q}$$

$$\leq (1 + \|K\|_{L_p}) \|y_h - y\|_{L_p} \cdot ch^m \|w\|_{W_q^m}$$

$$\leq ch^m \|y_h - y\|_{L_p} \cdot \|g\|_{W_q^m},$$

and the result is proved. □

Sometimes it is convenient to restate the conclusion of the theorem in the language of negative norms. As in Ref. 23, for $m \geq 0$ we may define the negative norm $\|\cdot\|_{W_p^{-m}}$ by

$$\|v\|_{W_p^{-m}} = \sup_{z \in W_q^m} \frac{|(v, z)|}{\|z\|_{W_q^m}}. \qquad (52)$$

Then (51) may be stated in the equivalent form

$$\|y_h - y\|_{W_p^{-m}} \leq ch^m \|y_h - y\|_{L_p}. \qquad (53)$$

The content is the same as (51), but sometimes the negative-norm formulation is the more convenient.

4.2. The Iterated Galerkin Method Revisited. The foregoing results may now be applied to the iterated Galerkin method, thus providing a different method of attack from that used in Section 3. From the definition

$$y_h^1 = f + Ky_h,$$

we have

$$y_h^1 - y = K(y_h - y), \qquad (54)$$

and hence

$$y_h^1(t) - y(t) = (y_h - y, k_t), \qquad (55)$$

where

$$k_t(s) = k(t, s). \qquad (56)$$

Since the right-hand side of Eq. (55) is of the form $(y_h - y, g)$, the results earlier in the section may be used to provide pointwise convergence results for y_h^1, provided k_t is suitably smooth (roughly speaking, requiring k_t to be smooth is equivalent to a smoothness assumption on the image of the adjoint operator K^*). This is the path followed by Chandler in Refs. 8, 14.

Rather than follow that path directly, we quote a result from Refs. 23 and 40. This result, which builds directly on Theorem 4.1, emphasizes again the primary role played by the smoothing property of the adjoint operator K^*. Note that one may obtain a *uniform*-norm (super)convergence result for the iterated Galerkin method by specializing to the case $p = \infty$, or $q = 1$.

Theorem 4.2. For fixed p in $1 \le p \le \infty$, assume $f \in L_p$, and let K be a bounded operator on L_p such that $I - K$ is one-to-one. With q defined by $1/p + 1/q = 1$, assume that K^* satisfies (49) for some fixed positive integer $m \le r$. Further, let S_h be a space of piecewise polynomials of order r, with the approximation property (50), and assume that the Galerkin approximation $y_h \in S_h$ exists uniquely. Then the iterated Galerkin approximation satisfies

$$\|y_h^1 - y\|_{L_p} \le ch^m \|y_h - y\|_{L_p}. \qquad (57)$$

Proof. From the duality property of the L_p and L_q norms, we have

$$\|y_h^1 - y\|_{L_p} = \sup_{v \in L_q} \frac{|(y_h^1 - y, v)|}{\|v\|_{L_q}} = \sup_{v \in L_q} \frac{|(K(y_h - y), v)|}{\|v\|_{L_q}}$$

$$= \sup_{v \in L_q} \frac{|(y_h - y, K^*v)|}{\|v\|_{L_q}}.$$

Theorem 4.1 now gives immediately

$$\|y_h^1 - y\|_{L_p} \le ch^m \|y_h - y\|_{L_p} \sup_{v \in L_q} \frac{\|K^*v\|_{W_q^m}}{\|v\|_{L_q}} \le ch^m \|y_h - y\|_{L_p},$$

where in the last step we have used (49). □

We now state a superconvergence result, taken from Ref. 23, for the *derivatives* of y_h^1. The condition (59) in the theorem, contrary perhaps to first appearances, is stronger than our previously stated smoothing condition (49), if p and q are related by $1/p + 1/q = 1$. This is because the condition in the theorem clearly implies

$$\|Kz\|_{L_p} \le c\|z\|_{W_p^{-m}}, \tag{58}$$

which is shown in Ref. 23 to be precisely equivalent to (49), with the same value of c. It follows, then, that Theorem 4.2 is recovered by setting $l = 0$. It is shown in Ref. 23 that the condition (59) in the theorem is satisfied if and only if the operators $D^\beta (D^\alpha K)^*$ are bounded in L_q for all $\alpha \le l$, $\beta \le m$; and that this is the case if, for instance, the integral operators with kernels $D_t^\alpha D_s^\beta k(t, s)$ are bounded in L_p.

Theorem 4.3. For fixed p in $1 \le p \le \infty$, assume $f \in L_p$, and let K be a bounded operator on L_p such that $I - K$ is one-to-one. Assume that K satisfies

$$\|Kz\|_{W_p^l} \le c\|z\|_{W_p^{-m}}, \tag{59}$$

for some integers m and l satisfying $0 < m \le r$, $0 \le l \le r$ and some c independent of z. Further, let S_h be a space of piecewise polynomials of order r, with the approximation property (50), where $1/p + 1/q = 1$, and assume that the Galerkin approximation $y_h \in S_h$ exists uniquely. Then the iterated Galerkin approximation satisfies

$$\|y_h^1 - y\|_{W_p^l} \le ch^m \|y_h - y\|_{L_p}. \tag{60}$$

Superconvergence

Proof. We have, using (59) and Theorem 4.1 in the form (53),

$$\|y_h^1 - y\|_{W_p^l} = \|K(y_h - y)\|_{W_p^l} \le c\|y_h - y\|_{W_p^{-m}}$$
$$\le ch^m \|y_h - y\|_{L_p}. \qquad \square$$

Note that the form (59) for the smoothness assumption on K makes the proof particularly easy!

Reference 23 also generalizes the results of Theorem 4.2 in three other ways. First, it establishes results not only for one-dimensional integral equations, but also for certain integral equations over the boundaries of two- and three-dimensional regions with smooth boundaries. Second, for some equations of this kind *uniform* error estimates are obtained, even in situations where the estimates (49) or (59), though holding for $1 < p < \infty$, fail for $p = \infty$, or equivalently $q = 1$. And third, results analogous to Theorems 4.2 and 4.3 are obtained also for higher iterates of the Galerkin method, defined by

$$y_h^i = f + Ky_h^{i-1}, \qquad i \ge 1, \tag{61}$$

with $y_h^0 = y_h$.

4.3. HMP Method for Approximating (y, g).

Very recently Hebeker, Mika, and Pack (Ref. 39) proposed a different way of approximating the functional

$$Q = (y, g) = \int_a^b y(s)g(s)\,ds. \tag{62}$$

This employs not only y_h, the Galerkin approximation to y, but also w_h, the Galerkin approximation to w, where w is the solution of the adjoint equation

$$w - K^*w = g. \tag{63}$$

In terms of the iterated Galerkin approximations $y_h^1 = f + Ky_h$ and $w_h^1 = g + K^*w_h$, the approximation is

$$Q_h = (y_h, g) + (y_h^1 - y_h, w_h^1),$$

or

$$Q_h = (f + Ky_h, g) + (f + Ky_h - y_h, K^*w_h). \tag{64}$$

The merit of this approximation derives from the identity

$$Q - Q_h = ((I - K)K(y - y_h), w - w_h), \tag{65}$$

which follows easily from (62) and (64), with the aid of nothing more than the defining equations of K^*, y, and w. From this we obtain the following theorem.

Theorem 4.4. For fixed p in $1 \le p \le \infty$, assume $f \in L_p$ and $g \in L_q$, where $1/p + 1/q = 1$, and let K be a bounded operator on L_p such that $I - K$ is one-to-one. Assume that K satisfies (59) for some integers m and l satisfying $0 < m \le r$, $0 \le l \le r$ and some c independent of z. Further, let S_h be a space of piecewise polynomials of order r, with the approximation property (50), together with the same property with q replaced by p, and assume that the Galerkin approximations $y_h \in S_h$ and $w_h \in S_h$ to $y = (I - K)^{-1}f$ and $w = (I - K^*)^{-1}g$, respectively, exist uniquely. Then, with Q and Q_h given by (62) and (64),

$$|Q - Q_h| \le ch^{m+l}\|y - y_h\|_{L_p}\|w - w_h\|_{L_q}, \tag{66}$$

with c independent of f, g, and h.

Proof. From (65) and (52) we have

$$|Q - Q_h| \le \|(I - K)K(y - y_h)\|_{W_p^l}\|w - w_h\|_{W_q^{-l}}$$

$$\le c\|K(y - y_h)\|_{W_p^l}\|w - w_h\|_{W_q^{-l}}$$

$$\le c\|y - y_h\|_{W_p^{-m}}\|w - w_h\|_{W_q^{-l}} \le ch^{m+l}\|y - y_h\|_{L_p}\|w - w_h\|_{L_q},$$

where we have used (59) and Theorem 4.1 in the form (53). □

The most favorable result occurs, of course, if l and m in (59) both have the maximum possible value r. If for simplicity we set $p = q = 2$, and assume f, $g \in W_2^r$, then the theorem yields

$$|Q - Q_h| \le ch^{2r}\|y - y_h\|_{L_2}\|w - w_h\|_{L_2} \le ch^{4r}\|y\|_{w_2^r}\|w\|_{w_2^r}. \tag{67}$$

Thus the best order of convergence of Q_h is $O(h^{4r})$, compared to a best possible result of $O(h^{2r})$ for the direct approximation (y_h, g), a striking result indeed.

This result may also be used, as pointed out in Ref. 39, to obtain an $O(h^{4r})$ approximation to $y(t)$ at an arbitrary point t, since

$$y(t) = f(t) + \int_a^b k(t, s)y(s)\,ds = f(t) + (y, k_t), \tag{68}$$

in which the second term is just Q for the special case $g = k_t$. The approximation of this term by Q_h then leads, under suitable assumptions, to an $O(h^{4r})$ approximation to $y(t)$. For further details see Ref. 39.

4.4. Discrete Galerkin and Iterated Galerkin Methods. In practice it is often necessary to evaluate some or all of the integrals that occur in the Galerkin and iterated Galerkin methods numerically. This leads to "discrete" versions of the corresponding approximations.

In more detail, let S_h have dimension $n_h = n$ and a basis $\{u_1, u_2, \ldots, u_n\}$. Then the Galerkin and iterated Galerkin approximations become

$$y_h = \sum_{i=1}^{n} a_i u_i, \tag{69}$$

$$y_h^1 = f + \sum_{i=1}^{n} a_i K u_i, \tag{70}$$

where $a_1, \ldots, a_n$ satisfy the linear equations

$$\sum_{i=1}^{n} [(u_i, u_j) - (K u_i, u_j)] a_i = (f, u_j), \quad j = 1, \ldots, n.$$

Thus there are several different kinds of integral to be performed: the various inner products (Ku_i, u_j), (u_i, u_j), and (f, u_j); the integrals Ku_i within the first of these inner products; and finally the integrals Ku_i within the iterated approximation (70). In practice some or all of these might be done numerically.

Several authors (Refs. 15, 22, 41-43) have pursued the question of how accurately, in various circumstances, each of these integrals needs to be evaluated in order to preserve the exact rates of convergence. Note that the question is somewhat complicated, because the required accuracy depends on what one wants to calculate in the end: for example, if one wants eventually to calculate pointwise values of y_h^1, then it is clear that an appropriately high order should be used even in the intermediate calculation of the Galerkin approximation.

Spence and Thomas (Ref. 22) have carried out an analysis of this kind, using the formalism of prolongation and restriction operators. Some improvements to these results have been obtained by Joe (Refs. 41-42). The latter paper summarizes the results that can be achieved by this approach, and the relation to the earlier results of Chandler (Ref. 15).

The discrete Galerkin and iterated Galerkin methods have also been studied recently by Atkinson and Bogomolny (Ref. 43) using an approach squarely based on the matrix approximations themselves. The reader is referred to the original papers and Chapter 3 for further details.

5. Iterated Collocation Method

In this section we return to the iterated collocation method, defined by

$$y_h^1 = f + Ky_h, \tag{71}$$

where $y_h \in S_h$ is the collocation approximation, satisfying

$$y_h(t_j) - Ky_h(t_j) = f(t_j), \qquad j = 1, \ldots, n. \tag{72}$$

We may recall from Section 3 that y_h^1 satisfies the second-kind equation

$$y_h^1 - KP_h y_h^1 = f, \tag{73}$$

where P_h is the interpolatory projection onto S_h defined by (12).

A useful if elementary remark is that the iterated collocation approximation coincides with the collocation approximation at the collocation points: for we have, from (71) and (72),

$$y_h^1(t_j) = f(t_j) + Ky_h(t_j) = y_h(t_j), \qquad j = 1, \ldots, n.$$

It follows that superconvergence of the iterated collocation method in the uniform norm, if it occurs, must result in superconvergence of the collocation method itself at the collocation points. For this and other reasons we shall always consider y_h^1 in the space $C[a, b]$.

Another useful observation is the following: Eq. (73), which can be written explicitly as

$$y_h^1(t) - \int_a^b k(t, s)(P_h y_h^1)(s)\, ds = f(t), \qquad t \in [a, b],$$

is a special case of the product-integration method (Refs. 24, 44–45). This point of view is interesting, because of the possibility that the exact integration in this equation may lead to a higher order of convergence than is achieved by the underlying interpolation process, through the cancellation of oscillatory interpolation errors. In other words, there is the possibility of superconvergence.

On the other hand, the iterated collocation method is sometimes *not* superconvergent. An important example of this is piecewise-linear collocation for the case of collocation at the breakpoints. In this case it is easy to see that there is in general no substantial cancellation of the errors in the associated piecewise-linear interpolation. Hence in this case the order of convergence of the iterated collocation method is in general no better than the $O(h^2)$ order associated with piecewise-linear interpolation itself.

We now begin the error analysis, based on Eq. (73). As remarked in Section 3, Theorem 3.1 is of no help for the iterated collocation method, because in this case $\|KP_h - K\| \not\to 0$. But all is not lost, because of the following theorem (which, though not new, may not have appeared before in quite this form):

Theorem 5.1. Let K be a compact operator on a Banach space B, not having one as a characteristic value, and assume $f \in B$. Further, assume that P_h is a bounded linear operator on B satisfying

$$\|P_h v - v\| \to 0, \qquad \forall v \in B. \tag{74}$$

Then, for h sufficiently small, Eq. (73) has a unique solution $y_h^1 \in B$ which satisfies

$$\|y_h^1 - y\| \leq c \|KP_h y - Ky\|. \tag{75}$$

Proof. From (74) and the uniform boundedness theorem it follows that

$$\sup_h \|P_h\| = m < \infty.$$

Since K is compact and the set $\{P_h\}$ is bounded, the set $\{KP_h\}$ is collectively compact (see Ref. 46). Since also

$$\|KP_h v - Kv\| \leq \|K\| \cdot \|P_h v - v\| \to 0, \qquad \text{as } h \to \infty, \qquad \forall v \in B, \tag{76}$$

the conditions of Anselone's collectively compact theory (Ref. 46) are all satisfied. Thus for h sufficiently small the approximate inverses $(I - KP_h)^{-1}$ exist and are uniformly bounded on B. The estimate (75) then follows from the identity (28). □

Actually, the conclusions of Theorem 5.1 hold more widely, since the assumption (74) can be replaced by any conditions which ensure both the collective compactness of $\{KP_h\}$ and $\|KP_h v - Kv\| \to 0$, $\forall v \in B$. More general conditions of this kind are used in Ref. 47 to analyze the iterated collocation method for the case of a space S_h of *global* polynomials, with collocation at the zeros of the first-kind Chebyshev polynomials $T_n(x)$ shifted to $[a, b]$.

5.1. Iterated Collocation for Piecewise-Polynomial Spaces S_h.

The simplest such case is that in which S_h is a space of piecewise-constant functions, on a uniform mesh (at the breakpoints we may assign values to the elements of S_h by requiring the functions to be, say, continuous from the right).

Piecewise-constant interpolation, for example in the uniform norm, is an $O(h)$ approximation. However, as shown in Ref. 48, the iterated collocation approximation can in this case give (in the uniform norm) $O(h^2)$ convergence, *provided* that the collocation points are taken at the midpoints of each subinterval. From our earlier discussion, it then follows that the piecewise-constant collocation method will itself be $O(h^2)$ superconvergent at the collocation points. For precise conditions, see Ref. 48, Section 6.

More generally, Chatelin and Lebbar (Refs. 19-20) have studied collocation with respect to discontinuous piecewise polynomials of arbitrary order r (i.e., degree $r - 1$), with the r collocation points on each interval being chosen to be the zeros of the rth degree Legendre polynomial (or equivalently the abscissas for r-point Gauss quadrature) shifted to that interval. In general terms, it was shown in Refs. 19-20 that $\|y_h^1 - y\|_\infty$ can in this case be of order $O(h^{2r})$, compared to at best $O(h^r)$ for the uniform error of the collocation method itself, provided that k_t for fixed t has r continuous derivatives, and that the exact solution y has $2r$ continuous derivatives.

Subsequently a modest improvement of these results for the case of collocation at the Gauss points has appeared; see Ref. 49 (in essence, the conditions on k_t and y need now hold only in the L_1 sense). The latter results are stated precisely, and then proved, in the next subsection, with some effort being made to simplify the arguments of Ref. 49.

For the case of collocation with respect to *continuous* piecewise polynomials, Joe (Ref. 50) has carried out a similar analysis to that in Ref. 49, but with the collocation points now taken to be the Lobatto quadrature points (which by definition include the end points) on each subinterval. In this case the maximum order of convergence that can be achieved is $O(h^{2r-2})$. In particular, in the continuous piecewise-linear case (i.e., $r = 2$)

the iterated collocation method is not superconvergent at all. The reader is referred to Ref. 50 for further details.

5.2. Collocation at the Gauss Points. For each positive integer N, we choose a partition of $[a, b]$ into N subintervals,

$$a = s_0 < s_1 < s_2 < \cdots < s_N = b,$$

and define, for $i = 0, 1, \ldots, N - 1$,

$$I_i = (s_i, s_{i+1}),$$

$$h_i = |I_i| = s_{i+1} - s_i,$$

$$h = \max_i h_i,$$

with h assumed to converge to 0 as $N \to \infty$.

Then, for fixed $r \geq 1$, the space S_h is taken to be the space of (discontinuous) functions on $[a, b]$, whose restriction to the subinterval I_i is a polynomial of degree $r - 1$ or less. At the interior breakpoints the elements of S_h are assumed, for definiteness, to be continuous from the right.

The collocation points on the subinterval I_i are taken to be the zeros of $p_{r,i}$, the rth degree Legendre polynomial shifted to I_i.

The following result is taken from Ref. 49, but the proof is arguably simpler than that in Ref. 49.

In the theorem, it is assumed that $k_t(s) = k(t, s)$ satisfies

$$k_t \in L_1, \quad \lim_{\tau \to t} \|k_\tau - k_t\|_{L_1} = 0, \quad \forall t \in [a, b]. \tag{77}$$

It is well known that these conditions ensure that K is a compact linear operator on the space $C = C[a, b]$.

Theorem 5.2. See Ref. 49. Let $f \in C$. Assume that the kernel satisfies (77), and that K, which is compact on C, does not have 1 as a characteristic value. Moreover, for fixed integers l, m satisfying $0 < l \leq 2r$ and $0 < m \leq r$, assume that $y \in W_1^l$ and

$$k_t \in W_1^m, \quad \forall t \in [a, b], \quad \|k_t\|_{W_1^m} \leq c < \infty, \tag{78}$$

with c independent of t. Finally, let S_h be the space of (discontinuous) piecewise polynomials described above, and let the collocation points on the subinterval I_i be the zeros of $p_{r,i}$, the rth-degree Legendre polynomial transferred to that interval. Then for h sufficiently small (73) has a unique solution y_h^1, which satisfies

$$\|y_h^1 - y\|_{L_\infty} \leq c h^{\min(l, m+r)} \|y\|_{W_1^l}. \tag{79}$$

The proof that follows is based on the Bramble–Hilbert lemma (Ref. 51). For the convenience of the reader, we state the lemma in a simplified form, which is immediately applicable to our purposes.

Lemma 5.1. See Ref. 51. Let a be a positive integer, and I be an interval of length h. Further, let F be a linear functional on $W_1^a(I)$ satisfying both

$$F(u) = 0, \quad \text{for } u \in P_{a-1}, \tag{80}$$

and

$$|F(u)| \leq c \sum_{j=0}^{a} h^{j-1} \|u^{(j)}\|_{L_1(I)}, \tag{81}$$

with c independent of both u and h. Then,

$$|F(u)| \leq c h^{a-1} \|u^{(a)}\|_{L_1(I)}. \tag{82}$$

Proof of Theorem 5.2. Since Theorem 5.1 is applicable with $B = L_\infty$ we have immediately that y_h^1 exists, and satisfies

$$\|y_h^1 - y\|_{L_\infty} \leq c \|KP_h y - Ky\|_{L_\infty}, \tag{83}$$

if h is sufficiently small. It only remains to estimate the right-hand side.

For any fixed $t \in [a, b]$, we may write

$$(KP_h y - Ky)(t) = (k_t, P_h y - y)$$

$$= \sum_{i=0}^{N-1} \int_{s_i}^{s_{i+1}} k_t(s)(P_h y - y)(s) \, ds. \tag{84}$$

Superconvergence 61

On the interval I_i we represent k_t by its Taylor formula, with remainder, of the appropriate order:

$$k_t(s) = \sum_{j=0}^{m-1} \frac{(s-s_i)^j}{j!} k_t^{(j)}(s_i) + \frac{1}{(m-1)!} \int_{s_i}^{s} (s-\sigma)^{m-1} k_t^{(m)}(\sigma) \, d\sigma.$$

In this way we find immediately, from (84),

$$(KP_h y - Ky)(t) = A_t(y) + R_t(y), \tag{85}$$

where

$$A_t(y) = \sum_{i=0}^{N-1} A_{t,i}(y), \qquad R_t(y) = \sum_{i=0}^{N-1} R_{t,i}(y), \tag{86}$$

$$A_{t,i}(y) = \sum_{j=0}^{m-1} \frac{k_t^{(j)}(s_i)}{j!} F_i^j(y), \tag{87}$$

with

$$F_i^j(u) = \int_{s_i}^{s_{i+1}} (s-s_i)^j (P_h u - u)(s) \, ds, \qquad j = 0, 1, \ldots, m-1, \tag{88}$$

and

$$R_{t,i}(y) = \frac{1}{(m-1)!} \int_{s_i}^{s_{i+1}} \left(\int_{s_i}^{s} (s-\sigma)^{m-1} k_t^{(m)}(\sigma) \, d\sigma \right) (P_h y - y)(s) \, ds. \tag{89}$$

To estimate $A_t(y)$ we make use of the Bramble-Hilbert lemma, applied to the linear functional F_i^j. We first observe that $F_i^j(u)$ certainly vanishes if $u \in P_{r-1}$, since in that situation the interpolating polynomial on the interval I_i is exact, i.e., $P_h u = u$. Now suppose that u is a polynomial of degree d, with $r \le d \le 2r - j - 1$. Then, because the polynomial $P_h u - u$ vanishes at the zeros of the rth degree polynomial $p_{r,i}$, we may write it as

$$(P_h u - u)(s) = p_{r,i}(s) v_i(s), \qquad s \in I_i,$$

where v_i is a polynomial of degree $\le r - j - 1$. Thus for u a polynomial of degree between r and $2r - j - 1$ we obtain

$$F_i^j(u) = \int_{s_i}^{s_{i+1}} (s-s_i)^j v_i(s) p_{r,i}(s) \, ds,$$

which vanishes because $(s - s_i)^j v_i(s)$ is a polynomial of degree at most $r - 1$, and hence orthogonal on I_i to $p_{r,i}$. We conclude that

$$F_i^j(u) = 0, \quad \text{if } u \in P_{2r-j-1}, \quad j = 0, \ldots, m. \tag{90}$$

Moreover, from (88) we have the obvious bound

$$|F_i^j(u)| \leq h_i^j \|P_h u - u\|_{L_1(I_i)}$$
$$\leq c h_i^{j+s} \|u^{(s)}\|_{L_1(I_i)}, \quad j = 0, \ldots, m, \quad s = 0, \ldots, r, \tag{91}$$

where in the last step we have used a standard property of polynomial interpolation. We deduce, using the Bramble–Hilbert lemma with $a = l - j$, if $j \leq l$, that

$$|F_i^j(u)| \leq \begin{cases} c h_i^j \|u\|_{L_1(I_i)}, & \text{if } j > l, \\ c h_i^l \|u^{(l-j)}\|_{L_1(I_i)}, & \text{if } j \leq l, \end{cases}$$
$$\leq c h_i^l \|u\|_{W_1^l(I_i)}.$$

As a result (87) yields

$$|A_{t,i}(y)| \leq \sum_{j=0}^{m-1} \frac{\|k_t^{(j)}\|_{L_\infty}}{j!} c h_i^l \|y\|_{W_1^l(I_i)} \leq c h_i^l \|k_t\|_{W_\infty^{m-1}} \|y\|_{W_1^l(I_i)}.$$

Finally, by summing over i, and using the imbedding of W_1^m in W_∞^{m-1}, or

$$\|v\|_{W_\infty^{m-1}} \leq c \|v\|_{W_1^m}, \tag{92}$$

we have

$$|A_t(y)| \leq c h^l \|k_t\|_{W_1^m} \|y\|_{W_1^l}. \tag{93}$$

For the second term of (85) we have, from (89),

$$|R_{t,i}(y)| \leq c h_i^m \|k_t^{(m)}\|_{L_1(I_i)} \|P_h y - y\|_{L_\infty(I_i)} \leq c h_i^m \|k_t\|_{W_1^m(I_i)} \|P_h y - y\|_{L_\infty},$$

and hence

$$|R_t(y)| \leq c h^m \|k_t\|_{W_1^m} \|P_h y - y\|_{L_\infty}. \tag{94}$$

The last factor in (94), which is the interpolation error in the uniform approximation of y, is of order $O(h^r)$ if $l > r$, but may be of lower order if $l \leq r$. More precisely, on defining $\delta = \min(l - 1, r)$, we have

$$\|P_h y - y\|_{L_\infty} \leq c h^\delta \|y^{(\delta)}\|_{L_\infty} \leq c h^\delta \|y\|_{W_\infty^{l-1}} \leq c h^\delta \|y\|_{W_1^l},$$

so that (94) gives

$$|R_t(y)| \le ch^{m+\delta}\|k_t\|_{W_1^m}\|y\|_{W_1^l}. \tag{95}$$

On combining (93) and (95), we obtain from (85)

$$|(KP_h y - Ky)(t)| \le ch^{\min(l,m+r)}\|k_t\|_{W_1^m}\|y\|_{W_1^l},$$

and the desired result now follows from (83) and the assumed uniform bound on $\|k_t\|_{W_1^m}$. □

5.3. Iterated Collocation Method versus Iterated Galerkin Method. According to Theorems 3.2 and 5.2, the iterated collocation method and iterated Galerkin method can both yield an error of order $O(h^{2r})$ if S_h is a piecewise-polynomial space of order r, and if the circumstances are suitable. Thus in favorable cases the iterated collocation method will perform as well as the iterated Galerkin method, while being much easier and cheaper to implement. However, there are two important qualifications. The first is that the iterated collocation method can achieve this high order only for very special choices of both the space S_h (discontinuous piecewise polynomials) and the collocation points (Gauss points). The second, which we address here, is that Theorem 5.2 for the iterated collocation method imposes very strong regularity conditions on the exact solution, if the maximum order of convergence is to be achieved.

Thus, while Theorem 3.2 requires y to have just r continuous derivatives for y_h^1 to achieve maximum order superconvergence, Theorem 5.2 requires it to have, in round terms, $2r - 1$ continuous derivatives. This suggests that in practical situations, where the regularity of y is often limited, the iterated collocation method may in fact *not* perform as well as the iterated Galerkin method.

But is the severe regularity requirement that Theorem 5.2 imposes on y really necessary? Or is it perhaps just an artifact of the proof? This question was investigated in Ref. 49, where it was found that in fact the regularity requirement in Theorem 5.2 cannot be significantly weakened without falsifying the theorem. This was done by constructing an example for which the result in the theorem is almost sharp. The example is

$$y(t) - \lambda \int_0^1 y(s)\,ds = f(t), \qquad t \in [0, 1],$$

with $\lambda \neq 1$, and with f chosen so that $y(t) = t^{r-1+\varepsilon}$, with ε a small positive number. The partition is assumed uniform. For this example it is shown in Ref. 49 that the iterated collocation method has a uniform error of exact order $O(h^{r+\varepsilon})$, arbitrarily close to the order $O(h^r)$ predicted by Theorem 5.2 if in it we take $l = r$. On the other hand, it is shown in Ref. 49 that the iterated Galerkin method gives for this example a uniform error of order $O(h^{2r})$ [in fact this follows from Eq. (55), together with Theorem 4.1 with $p = 1$, $q = \infty$, and $m = r$, and Theorem 2.1 with $B = L_1$]. Thus the iterated collocation method does indeed place greater demands on the regularity of y.

5.4. Discrete Collocation and Iterated Collocation Methods.

The discrete versions of the methods arise if the values of $Ku_i(t_j)$ required for the collocation matrix, or the values of $Ku_i(t)$ required for constructing $y_h^1(t)$, are evaluated numerically. The discrete collocation and iterated collocation approximations have been considered in Refs. 15, 30, 41, 52, the general considerations being broadly similar to those described above for the Galerkin and iterated Galerkin methods. The reader is referred to the original literature and Chapter 3 for further details.

6. Nonlinear Integral Equations

The literature on superconvergence for nonlinear integral equations is small, and until recently possibly nonexistent.

Atkinson and Potra (Ref. 53) recently considered the iterated Galerkin and iterated collocation methods for the general (Urysohn) integral equation

$$y(t) = \int_a^b F(t, s, y(s)) \, ds, \qquad t \in [a, b], \tag{96}$$

and have shown that properties analogous to those obtained earlier in this paper hold also for the nonlinear case, if F is well enough behaved.

It should be said, however, that there is a well-recognized practical difficulty in the way of using the usual Galerkin or collocation methods or their iterative variants for nonlinear problems. Consider, for example, the collocation method, in which we seek an approximate solution of the form

$$y_h = \sum_{i=1}^n a_i u_i,$$

such that

$$y_h(t_j) = \int_a^b F(t_j, s, y_h(s))\, ds, \qquad j = 1, \ldots, n,$$

or

$$\sum_{i=1}^n a_i u_i(t_j) = \int_a^b F\left(t_j, s, \sum_{i=1}^n a_i u_i(s)\right) ds, \qquad j = 1, \ldots, n. \tag{97}$$

The problem is that this is not, as it stands, a closed set of algebraic nonlinear equations for $a_1, \ldots, a_n$, because of the occurrence of integrals in (97). In practice these nonlinear equations would need to be solved iteratively, with the integrals evaluated *numerically* at each iteration. Because the integrals must be evaluated numerically, a proper analysis should include the discretization errors—and indeed such an extension of the analysis of Ref. 53, at least for the Galerkin case, has very recently appeared (Ref. 54).

A different generalization of the collocation and iterated collocation methods, arguably closer in spirit to the linear case, has recently been proposed for the particular case of the (Hammerstein) equation

$$y(t) - \int_a^b k(t, s) g(s, y(s))\, ds = f(t), \qquad t \in [a, b], \tag{98}$$

by Kumar and Sloan (Ref. 55). In this method the collocation approximation is made not to $y(t)$, but to the related function defined by

$$z(t) = g(t, y(t)), \qquad t \in [a, b]. \tag{99}$$

Since y is given in terms of z by

$$y(t) = f(t) + \int_a^b k(t, s) z(s)\, ds, \tag{100}$$

it is clear that z satisfies the equation

$$z(t) = g\left(t, f(t) + \int_a^b k(t, s) z(s)\, ds\right). \tag{101}$$

The solutions of the latter equation are in one-to-one correspondence with those of Eq. (98), with the correspondence given in one direction by (99) and in the other by (100)—see Ref. 55.

The first (collocation) stage of approximation in the method of Ref. 55 is to seek $z_h \in S_h$ such that

$$z_h(t_j) = g\left(t_j, f(t_j) + \int_a^b k(t_j, s) z_h(s)\, ds\right), \qquad j = 1, \ldots, n. \qquad (102)$$

If we write

$$z_h = \sum_{i=1}^n b_i u_i,$$

then these collocation equations *do* give a closed set of nonlinear algebraic equations for the unknowns $b_1, \ldots, b_n$, namely,

$$\sum_{i=1}^n b_i u_i(t_j) = g\left(t_j, f(t_j) + \sum_{i=1}^n v_i(t_j) b_i\right), \qquad j = 1, \ldots, n, \qquad (103)$$

where

$$v_i(t) = \int_a^b k(t, s) u_i(s)\, ds, \qquad t \in [a, b]. \qquad (104)$$

Thus the only integrals that have to be evaluated are those that occur in the collocation method for the *linear* integral operator with kernel $k(t, s)$. These integrals can be evaluated once and for all, even perhaps analytically in suitable cases, just as in the linear case.

The approximation to y is then obtained by substituting z_h into the right-hand side of (100). That is, the approximation to y is

$$y_h^1(t) = f(t) + \int_a^b k(t, s) z_h(s)\, ds, \qquad t \in [a, b]. \qquad (105)$$

Our notation here is based on the fact that in the linear case $g(s, y(s))$ becomes $y(s)$, so that z becomes y, z_h becomes just the collocation approximation y_h, and y_h^1 becomes just the iterated collocation approximation.

The interesting question, then, is whether the superconvergence results for the linear case, given for discontinuous piecewise polynomials in Theorem 5.2, carry over to this nonlinear extension of the iterated collocation method. Kumar (Refs. 56–57) has shown that indeed they do, under reasonable conditions on the function g. The interested reader is referred to the original literature for details.

Finally, a discrete version of the method of Ref. 55 has very recently been analyzed, in Refs. 57–58.

7. Concluding Remarks

Many aspects of superconvergence for one-dimensional integral equations of the second kind are by now well understood. On the other hand, this knowledge seems to have had little effect as yet on practical computations. With the rapid development in our understanding of the discrete versions of the various methods, real opportunities now seem to exist for the practical exploitation of superconvergence.

Acknowledgments

The author has benefited from discussing questions relating to superconvergence with many persons, including Drs. C. Chandler, F. Chatelin, I. Graham, S. Joe, S. Kumar, Q. Lin, P. Rabinowitz, V. Thomée, W. Wendland, R. F. Xie, and Y. Yan. Part of this review was written while the author was supported as Gastprofessor at the University of Stuttgart by the Deutsches Forschungsgemeinschaft. The author is indebted to all of the above, and to the Australian Research Grants Scheme for its consistent support.

References

1. THOMÉE, V., *Spline Approximation for Difference Schemes for the Heat Equation*, The Mathematical Foundations of the Finite Element Method, with Applications to Partial Differential Equations, Edited by A. K. Aziz, Academic Press, New York, New York, 1972.
2. THOMÉE, V., AND WENDROFF, B., *Convergence Estimates for Galerkin Methods for Variable Coefficient Initial-Value Problems*, SIAM Journal on Numerical Analysis, Vol. 11, pp. 1059-1068, 1974.
3. DOUGLAS, J., JR., AND DUPONT, T., *Galerkin Approximations for the Two-Point Boundary-Value Problem Using Continuous Piecewise Polynomial Spaces*, Numerische Mathematik, Vol. 22, pp. 99-109, 1974.
4. DUPONT, T., *A Unified Theory of Superconvergence for Galerkin Methods for Two-Point Boundary-Value Problems*, SIAM Journal on Numerical Analysis, Vol. 13, pp. 362-369, 1976.
5. DE BOOR, C., AND SWARTZ, B., *Collocation at Gaussian Points*, SIAM Journal on Numerical Analysis, Vol. 10, pp. 582-606, 1973.
6. DE BOOR, C., AND SWARTZ, B., *Local Piecewise Polynomial Projection Methods for an ODE Which Give High-Order Convergence at Knots*, Mathematics of Computation, Vol. 36, pp. 21-33, 1981.
7. DOUGLAS, J., JR., AND DUPONT, T., *Collocation Methods for Parabolic Equations in a Single Space Variable*, Springer-Verlag, Berlin, Germany, 1974.

8. CHANDLER, G. A., *Superconvergence for Second-Kind Integral Equations*, The Application and Numerical Solution of Integral Equations, Edited by R. S. Anderssen, F. R. De Hoog, and M. A. Lukas, Sijthoff and Noordhoff, Alphen aan den Rijn, Holland, 1980.
9. SLOAN, I. H., *Superconvergence and the Galerkin Method for Integral Equations of the Second Kind*, Treatment of Integral Equations by Numerical Methods, Edited by C. T. H. Baker and G. F. Miller, Academic Press, London, England, 1982.
10. RICHTER, G. R., *Superconvergence of Piecewise Polynomial Galerkin Approximations for Fredholm Integral Equations of the Second Kind*, Numerische Mathematik, Vol. 31, pp. 63–70, 1978.
11. SLOAN, I. H., BURN, B. J., AND DATYNER, N., *A New Approach to the Numerical Solution of Integral Equations*, Journal of Computational Physics, Vol. 18, pp. 92–105, 1975.
12. SLOAN, I. H., *Error Analysis for a Class of Degenerate-Kernel Methods*, Numerische Mathematik, Vol. 25, pp. 231–238, 1976.
13. SLOAN, I. H., *Improvement by Iteration for Compact Operator Equations*, Mathematics of Computation, Vol. 30, pp. 758–764, 1976.
14. CHANDLER, G. A., *Global Superconvergence of Iterated Galerkin Solutions for Second Kind Integral Equations*, Australian National University, Technical Report, 1978.
15. CHANDLER, G. A., *Superconvergence of Numerical Solutions to Second-Kind Integral Equations*, Australian National University, PhD Thesis, 1979.
16. LIN, Q., *Some Problems about the Approximate Solution for Operator Equations*, Acta Mathematica Sinica, Vol. 22, pp. 219–230, 1979 (in Chinese).
17. GRAHAM, I. G., *The Numerical Solution of Fredholm Integral Equations of the Second Kind*, University of New South Wales, PhD Thesis, 1980.
18. GRAHAM, I. G., *Galerkin Methods for Second Kind Integral Equations with Singularities*, Mathematics of Computation, Vol. 39, pp. 519–533, 1982.
19. CHATELIN, F., AND LEBBAR, R., *The Iterated Projection Solution for the Fredholm Integral Equation of Second Kind*, Journal of the Australian Mathematical Society, Series B, Vol. 22, pp. 439–451, 1981.
20. CHATELIN, F., AND LEBBAR, R., *Superconvergence Results for the Iterated Projection Method Applied to a Fredholm Integral Equation of the Second Kind and the Corresponding Eigenvalue Problem*, Journal of Integral Equations, Vol. 6, pp. 71–91, 1984.
21. HSIAO, G. C., AND WENDLAND, W. L., *The Aubin–Nitsche Lemma for Integral Equations*, Journal of Integral Equations, Vol. 3, pp. 299–315, 1981.
22. SPENCE, A., AND THOMAS, K. S., *On Superconvergence Properties of Galerkin's Method for Compact Operator Equations*, IMA Journal of Numerical Analysis, Vol. 3, pp. 253–271, 1983.
23. SLOAN, I. H., AND THOMÉE, V., *Superconvergence of the Galerkin Iterates for Integral Equations of the Second Kind*, Journal of Integral Equations, Vol. 9, pp. 1–23, 1985.
24. ATKINSON, K. E., *A Survey of Numerical Methods for the Solution of Fredholm Integral Equations of the Second Kind*, SIAM, Philadelphia, Pennsylvania, 1976.

25. IKEBE, Y., *The Galerkin Method for the Numerical Solution of Fredholm Integral Equations of the Second Kind*, SIAM Review, Vol. 14, pp. 465-491, 1972.
26. CHATELIN, F., *Spectral Approximation of Linear Operators*, Academic Press, New York, New York, 1983.
27. SLOAN, I. H., *Iterated Galerkin Method for Eigenvalue Problems*, SIAM Journal on Numerical Analysis, Vol. 13, pp. 753-760, 1976.
28. CHATELIN, F., *Sur les Bornes d'Erreur a Posteriori pour les Éléments Propres d'Opérateurs Linéaires*, Numerische Mathematik, Vol. 32, pp. 233-246, 1979.
29. BRUNNER, H., *The Application of the Variation of Constants Formulas in the Numerical Analysis of Integral and Integrodifferential Equations*, Utilitas Mathematica, Vol. 19, pp. 255-290, 1981.
30. BRUNNER, H., *Iterated Collocation Methods and Their Discretizations for Volterra Integral Equations*, SIAM Journal on Numerical Analysis, Vol. 21, pp. 1132-1145, 1984.
31. BRUNNER, H., AND VAN DER HOUWEN, P. J., *The Numerical Solution of Volterra Equations*, North-Holland, Amsterdam, Holland, 1986.
32. CHANDLER, G. A., *Superconvergent Approximations to the Solution of a Boundary Integral Equation on Polygonal Domains*, SIAM Journal on Numerical Analysis, Vol. 23, pp. 1214-1229, 1986.
33. GOLBERG, M. A., LEA, M., AND MIEL, G., *A Superconvergence Result for the Generalized Airfoil Equation with Application to the Flap Problem*, Journal of Integral Equations, Vol. 5, pp. 175-186, 1983.
34. MIEL, G., *Rates of Convergence and Superconvergence of Galerkin's Method for the Generalized Airfoil Equation*, Numerical Solution of Singular Integral Equations, Edited by A. Gerasoulis and R. Vichnevetsky, IMACS, 1984.
35. ATKINSON, K. E., GRAHAM, I. G., AND SLOAN, I. H., *Piecewise Continuous Collocation for Integral Equations*, SIAM Journal on Numerical Analysis, Vol. 20, pp. 172-186, 1983.
36. PHILLIPS, J. L., *The Use of Collocation as a Projection Method for Solving Linear Integral Equations*, SIAM Journal on Numerical Analysis, Vol. 9, pp. 14-28, 1972.
37. KANTOROVICH, L. V., *Functional Analysis and Applied Mathematics*, Uspekhi Matematicheskikh Nauk, Vol. 3, pp. 89-185, 1948. English translation: N.B.S. Report No. 1509 (1952).
38. SLOAN, I. H., *Four Variants of the Galerkin Method for Integral Equations of the Second Kind*, IMA Journal of Numerical Analysis, Vol. 4, pp. 9-17, 1984.
39. HEBEKER, F. K., MIKA, J., AND PACK, D. C., *Application of the Superconvergence Properties of the Galerkin Approximation to the Calculation of Upper and Lower Bounds for Linear Functionals of Solutions of Integral Equations*, IMA Journal of Applied Mathematics, Vol. 38, pp. 61-70, 1987.
40. SLOAN, I. H., *The Iterated Galerkin Method for Integral Equations of the Second Kind*, Proceedings of the Centre for Mathematical Analysis, Vol. 5, Edited by B. Jefferies and A. McIntosh, Australian National University, Canberra, Australia, 1984.
41. JOE, S., *The Numerical Solution of Second Kind Fredholm Integral Equations*, University of New South Wales, Australia, PhD Thesis, 1985.

42. JOE, S., *Discrete Galerkin Methods for Fredholm Integral Equations of the Second Kind*, IMA Journal of Numerical Analysis, Vol. 7, pp. 149-164, 1987.
43. ATKINSON, K. E., AND BOGOMOLNY, A., *The Discrete Galerkin Method for Integral Equations*, Mathematics of Computation, Vol. 48, pp. 595-616, 1987.
44. DE HOOG, F. R., AND WEISS, R., *Asymptotic Expansions for Product Integration*, Mathematics of Computation, Vol. 27, pp. 295-306, 1973.
45. SCHNEIDER, C., *Produktintegration zur Losung Schwachsingularer Integralgleichungen*, ZAMM, Vol. 61, pp. T317-T319, 1981.
46. ANSELONE, P. M., *Collectively Compact Operator Approximation Theory and Applications to Integral Equations*, Prentice Hall, Englewood Cliffs, New Jersey, 1971.
47. SLOAN, I. H., AND BURN, B. J., *Collocation with Polynomials for Integral Equations of the Second Kind: a New Approach to the Theory*, Journal of Integral Equations, Vol. 1, pp. 77-94, 1979.
48. SLOAN, I. H., NOUSSAIR, E., AND BURN, B. J., *Projection Methods for Equations of the Second Kind*, Journal of Mathematical Analysis and Applications, Vol. 69, pp. 84-103, 1979.
49. GRAHAM, I. G., JOE, S., AND SLOAN, I. H., *Iterated Galerkin versus Iterated Collocation for Integral Equations of the Second Kind*, IMA Journal of Numerical Analysis, Vol. 5, pp. 355-369, 1985.
50. JOE, S., *Collocation Methods using Piecewise Polynomials for Second Kind Integral Equations*, Journal of Computational and Applied Mathematics, Vols. 12 & 13, pp. 391-400, 1985.
51. BRAMBLE, J. H., AND HILBERT, S. R., *Bounds for a Class of Linear Functionals with Applications to Hermite Interpolation*, Numerische Mathematik, Vol. 16, pp. 362-369, 1971.
52. JOE, S., *Discrete Collocation Methods for Second Kind Fredholm Integral Equations*, SIAM Journal on Numerical Analysis, Vol. 22, pp. 1167-1177, 1985.
53. ATKINSON, K. E., AND POTRA, F., *Projection and Iterated Projection Methods for Nonlinear Integral Equations*, SIAM Journal on Numerical Analysis, Vol. 24, pp. 1352-1373, 1987.
54. ATKINSON, K. E., AND POTRA, F., *The Discrete Galerkin Method for Nonlinear Integral Equations*, Journal of Integral Equations and Applications, Vol. 1, pp. 17-54, 1988.
55. KUMAR, S., AND SLOAN, I. H., *A New Collocation-Type Method for Hammerstein Integral Equations*, Mathematics of Computation, Vol. 48, pp. 585-593, 1987.
56. KUMAR, S., *Superconvergence of a Collocation-Type Method for Hammerstein Equations*, IMA Journal of Numerical Analysis, Vol. 7, pp. 313-325, 1987.
57. KUMAR, S., *A New Collocation-Type Method for the Numerical Solution of Hammerstein Equations*, University of New South Wales, Australia, PhD Thesis, 1987.
58. KUMAR, S., *A Discrete Collocation-Type Method for Hammerstein Equations*, SIAM Journal on Numerical Analysis, Vol. 25, pp. 328-341, 1988.

3

Perturbed Projection Methods for Various Classes of Operator and Integral Equations

M. A. GOLBERG

Abstract. We develop a convergence theory for various perturbed projection methods for a class of operator equations which includes Fredholm, Volterra, and many Cauchy singular equations. Using this theory, we rederive a number of recent convergence results of Brunner, Spence, and Thomas, and Joe, for discrete Galerkin and collocation methods and their iterates, where the perturbations are induced by numerical integration errors. These results are compared to those obtained by Atkinson and Bogomolny by nonperturbation methods. Although the perturbation approach is more general, the nonperturbation method appears to give sharper error estimates for specific classes of problems. A conjecture is made as to the origin of this discrepancy.

1. Introduction

As we have seen elsewhere in this volume, projection methods, such as Galerkin and collocation methods, are used extensively for the solution of all types of integral equations, with the basic theory for many of these techniques being essentially complete. However, in much of the analysis that has appeared in the past two decades it has been assumed that all of the operations, such as integration and function evaluation, are performed exactly, which is certainly not possible in virtually all realistic situations. Practically, it then becomes important to determine the effects of numerical integration, function approximations, etc., on the convergence properties of these algorithms. Until recently, however, this problem seems to have

---

M. A. GOLBERG • Department of Mathematical Sciences, University of Nevada, Las Vegas, Nevada 89154.

been somewhat neglected (Refs. 1-4). For instance, in Baker's treatise on the numerical solution of integral equations (Ref. 1), no mention is made concerning the effect of integration errors on Galerkin and collocation methods in a book of about 1000 pages. On the other hand, in the past decade there has been renewed interest in this subject, to some extent motivated by the problem of determining integration rules which preserve the rate of convergence of spline-based Galerkin and collocation methods for Fredholm and Volterra integral equations of the second kind (Refs. 3-8). In addition, as will be seen in Chapter 5, this problem has recently been taken up in the convergence analysis of algorithms for Cauchy singular integral equations (Refs. 9-10) and related work has been done for very general classes of elliptic pseudodifferential equations (which contain Fredholm integral equations, but not all Cauchy singular ones) by Wendland and his co-workers (Refs. 11-14).

Needless to say, with so many different problems in this area, researchers have used different methods of analysis, not all of which have led to the same results. For instance, in analyzing the effect of quadrature errors in Galerkin methods for Fredholm equations, Chandler has used classical perturbation theory (Ref. 15), Spence and Thomas in Ref. 3 and Joe in Ref. 8 have used the technique of restriction and prolongation operators, Atkinson and Bogomolny developed a nonperturbation approach using the notion of discrete projections (Ref. 4), while Brunner in Ref. 6 uses an approach relating the discretization error to the residual via resolvent and general variation of constants formulas. In this chapter we survey some of these methods, and apply them primarily to the analysis of a variety of projection methods for solving Fredholm and Volterra equations of the second kind. In Chapter 5 some of these ideas will be applied to Galerkin methods for Cauchy singular equations. Our emphasis will be on the development of perturbation methods, as they appear to be the most general, and perhaps the most convenient way for dealing with numerical integration and other errors. However, as will be seen in Sections 4 and 6, the "brute force" application of this technique does not necessarily yield the best results, and subtle modifications need to be made in order to obtain them. Since the same equation is being analyzed in two different ways, a more careful investigation of the application of perturbation theory should ultimately yield equivalent results. This, at least to our knowledge, has not been done. For instance, to determine the precision of the integration rules which preserve the convergence rate of piecewise polynomial Galerkin methods for solving one-dimensional Fredholm integral equations, Chandler used perturbation theory to show that rules of precision $2l - 2$ are sufficient if polynomials of degree $\leq l - 1$, $l \geq 1$, are used to approximate the solution [this same result was obtained by Joe using an essentially

Perturbed Projection Methods

equivalent approach via restriction and prolongation operators (Ref. 8)]. On the other hand, Atkinson and Bogomolny show that it suffices to use rules of precision $l - 1$ as long as the weights are positive (Ref. 4).

In Section 2 we present a brief discussion of a basic projection method and a number of variants for solving operator equations of the form $(H - K)u = f$, where H is a bounded invertible operator from a Banach space X to a Banach space Y and K is compact. When $X = Y$ and $H = I$, the identity operator, the basic theory reduces to that discussed by Sloan in Chapter 2.

As will be seen in Chapter 5, this generalization has important applications in the analysis of numerical methods for Cauchy singular equations, and has been used in other contexts as well (Ref. 16). In Section 3 perturbed versions of these methods will be analyzed under fairly general conditions on the defining projections. Our desire there, among other things, is to keep the level of mathematics fairly elementary, essentially requiring only Banach's lemma, because this eventually leads to relatively simple proofs of a number of important algorithms (such as the Nyström method using Gaussian quadrature) without the use of collectively compact operator theory (similar ideas will be used in Chapter 5 for Cauchy singular equations).

In Section 4 we use some of the results of Section 3 to rederive those of Joe (Ref. 8), Chandler (Ref. 15), and Spence and Thomas (Ref. 3), concerning the convergence of discrete Galerkin methods for Fredholm equations when integration errors are present. In Section 5 we outline the method of Atkinson and Bogomolny which is used to sharpen some of the results obtained in Section 4. In Section 6 we consider collocation methods and discuss Brunner's approach for obtaining superconvergence results. This can be compared to the technique used by Sloan in Section 5 of Chapter 2 and that of Chatelin and Lebbar in Ref. 17. In Section 7 we briefly consider the case where H is positive definite which occurs in the analysis of boundary integral equations.

2. Projection Methods and Their Variants

In this section we consider a number of numerical methods for solving operator equations of the form

$$Hu = Ku + f, \tag{1}$$

where H is a bounded invertible operator with a bounded inverse from a Banach space X to a Banach space Y and K is compact. When H is the identity operator I, then as is well known, Eq. (1) can be used to represent Fredholm and Volterra integral equations, while $H \neq I$ corresponds to differential equations (Ref. 16) and to a wide variety of singular integral equations (Ref. 14). Since many aspects of solving singular equations will be taken up in Chapters 5 and 6, our emphasis here will be on the case $H = I$ and applications to Fredholm and Volterra equations of the second kind. Such equations are of the form

$$u(x) = \int_{D_x} k(x, y) u(y) \, d\sigma(y) + f(x), \qquad (2)$$

where $x \in D_x$ and D_x is either a subset of m-dimensional Euclidean space R^m with a nonempty interior, or an $(m-1)$-dimensional "surface." Quantity D_x may depend on x, and $d\sigma(y)$ is the element of volume, area, or length on D_x. If $D_x \equiv D$ does not depend on x, then (1) is a Fredholm equation, otherwise we call it a Volterra equation.

2.1. Notation. If X is a Banach space, $\|x\|_X$ will denote the norm on X, and if $L(X, Y)$ is the space of bounded linear operators from X to a Banach space Y, then $\|A\|$ will denote the induced operator norm of $A \in L(X, Y)$. When $X = Y$, $L(X, Y)$ will be denoted by $L(X)$. On occasion, when standard function spaces are employed, other notation for norms will be used. For example, for spaces of bounded functions on a set D, then

$$\|f\|_\infty = \sup_{x \in D} |f(x)| \qquad (3)$$

will denote the sup or maximum norm, while

$$\|f\|_p = \left[\int_D |f(y)|^p \, d\sigma(y) \right]^{1/p}, \qquad p \geq 1, \qquad (4)$$

will denote the usual p-norms of f.

When X is an inner-product space, then $\langle x, y \rangle$ will denote the inner product of x and y and the induced norm, $\|x\|_X = (\langle x, x \rangle)^{1/2}$.

Dim(W) will denote the dimension of a finite-dimensional vector space W.

Throughout the chapter c will generally be used to indicate a constant, which may vary in meaning from place to place.

If X is an inner-product space and $K \in L(X)$, then its adjoint, if it exists, will be denoted by K^*.

2.2. Projection Methods. To solve Eq. (1) numerically, let X_n, $n \geq 1$, be a sequence of finite-dimensional subspaces of X with $\dim(X_n) = N(n) \equiv N \geq 1$, $N(n_2) > N(n_1)$ if $n_2 > n_1$, and look for an approximation $u_n \in X_n$ satisfying

$$P_n H u_n = P_n K u_n + P_n f, \tag{5}$$

where $P_n g$ is the projection of $g \in Y$ onto $Y_n \subseteq Y$, i.e., $P_n^2 = P_n$, and $\dim(Y_n) = \dim(X_n) = N$. If $H u_n \in Y_n$, then $P_n H u_n = H u_n$ (because P_n is a projection) and (5) becomes

$$H u_n = P_n K u_n + P_n f. \tag{6}$$

In particular, if $X = Y$, and $H = I$ the identity operator, then $I u_n = u_n$, and u_n satisfies

$$u_n = P_n K u_n + P_n f. \tag{7}$$

In this case the theory of such methods is well known (Refs. 1-2, 4, 8), and here we present a generalization of these ideas to cover methods corresponding to (6). The restriction $H X_n = Y_n$ simplifies matters considerably in comparison to the general case (5). In fact, in that generality not a great deal seems to be known, because there only appear to be a few known sufficient conditions on H and the projections $P_n: Y \to Y_n$ and $Q_n: X \to X_n$ so that $P_n H Q_n$ converges to H. One such situation corresponds to taking $X = Y$ as Hilbert spaces, H is positive definite, $X_n = Y_n$ and $P_n = Q_n$, the orthogonal projection of X onto X_n. Then (6) represents Galerkin's method applied to a wide variety of differential and integral equations (Ref. 14). We shall return to this case in Section 7. For now we consider only methods defined by (6). Although the condition that $HX_n = Y_n$ may seem to be quite restrictive, there are many equations and methods for which this is the case (Refs. 18-20).

A sequence of approximations $\{u_n\}$ for solving (1) satisfying (6) will be called a projection method.

To utilize (6) numerically, let $\{\phi_1, \ldots, \phi_N\}$ be a basis for X_n and let $\{l_1, \ldots, l_N\}$ be a set of linearly independent functionals on Y defining P_n (Ref. 2). Then

$$u_n = \sum_{j=1}^{N} a_{j,n} \phi_j, \tag{8}$$

where $\{a_{j,n}\}_{j=1}^{N}$ are determined by solving the linear equations

$$\sum_{j=1}^{N} a_{j,n} l_k(H\phi_j) - \sum_{j=1}^{N} l_k(K\phi_j) a_{j,n} = l_k(f), \quad k = 1, 2, \ldots, N. \tag{9}$$

To simplify notation the subscript n on $a_{j,n}$ will be omitted from now on.

As has already been shown in Chapters 1 and 2 there are a large number of methods which fit into this framework, and here we give a series of examples for the case $H = I$. Some of these have been discussed in previous chapters but we repeat them here for future reference. Examples for $H \neq I$ may be found in Chapters 5 and 6.

Example 2.1. Let $D = [a, b] \subseteq R$ be an interval, where $-\infty < a < b < \infty$, and let $X = C[a, b] \equiv C$ be the Banach space of real-valued continuous functions on X with the sup norm. Let $k(x, y)$ be defined on $[a, b] \times [a, b]$ such that

(i) $$\sup_{x \in D} \int_a^b |k(x, y)| \, dy < \infty, \tag{10}$$

and

(ii) $$\lim_{h \to 0} \int_a^b |k(x + h, y) - k(x, y)| \, dy = 0. \tag{11}$$

Then

$$Ku(x) = \int_a^b k(x, y) u(y) \, dy \tag{12}$$

defines a compact operator on X (Ref. 1). In this case (2) becomes the Fredholm equation

$$u(x) = \int_a^b k(x, y) u(y) \, dy + f(x), \tag{13}$$

where $f(x) \in X$ and we assume that (13) has a unique solution for each f.

To solve (13) numerically, let X_n be the set of polynomials of degree $\leq n - 1$, and let $\{\phi_k\}_{k=1}^n$ be a basis for X_n (here $N = n$). A family of projection methods for (13) can be defined as follows.

Let $w(x)$ be a nonnegative integrable weight function on $[a, b]$ such that

$$\langle f, g \rangle = \int_a^b w(x) f(x) g(x) \, dx \tag{14}$$

defines an inner product on C. Then, as is well known,

$$l_k(g) = \langle g, \phi_k \rangle, \qquad k = 1, 2, \ldots, n, \tag{15}$$

are linearly independent functionals on C, and

$$u_n = \sum_{j=1}^{n} a_j \phi_j \qquad (16)$$

can be obtained by solving

$$\sum_{j=1}^{n} \langle \phi_j, \phi_k \rangle a_j - \sum_{j=1}^{n} \langle K\phi_j, \phi_k \rangle a_j = \langle f, \phi_k \rangle, \qquad k = 1, 2, \ldots, n. \qquad (17)$$

Since $\Phi = [\langle \phi_k, \phi_j \rangle]_{(j,k)=1}^{n}$ is positive definite, Φ^{-1} exists and

$$P_n u = \sum_{j=1}^{n} \sum_{k=1}^{n} \Phi_{jk}^{-1} \langle u, \phi_k \rangle \phi_j \qquad (18)$$

(where Φ_{jk}^{-1} is the jkth element of Φ^{-1}) is well defined. A little algebra shows that P_n is the orthogonal projection of X onto X_n and that (16)-(17) are equivalent to

$$u_n = P_n K u_n + P_n f. \qquad (19)$$

In this case the projection method is called Galerkin's method.

Particularly important special cases of this class of methods correspond to taking $a = -1$, $b = 1$, $w(x) = 1$, and $\phi_k(x)$ the Legendre polynomial of degree $k - 1$, or $w(x) = 1/(1 - x^2)^{1/2}$ and $\phi_k(x) = T_{k-1}(x)$ the Chebyshev polynomial of degree $k - 1$. This latter method has been studied extensively in the past ten years by Delves and his colleagues (Refs. 21-22), because suitable discretizations of the inner products $\langle K\phi_j, \phi_k \rangle$ and $\langle f, \phi_k \rangle$ allow one to use fast Fourier transform techniques, which generally makes the arithmetic comparable to that of the Nyström method, and is capable of handling many types of discontinuous kernels as well (Ref. 22).

Because C is not complete with respect to the inner product $\langle \cdot, \cdot \rangle$, the analysis of Galerkin's method in this case is usually carried out in the space $L_w^2[a, b]$, the space of Lebesgue square-integrable functions on $[a, b]$ (Refs. 1-2). In this situation the conditions (10)-(11) on $k(x, y)$ are usually replaced by

$$\int_a^b \int_a^b [w(x)k^2(x, y)/w(y)]\, dx\, dy < \infty, \qquad (20)$$

which implies that K is Hilbert-Schmidt and thus compact (Ref. 1).

In this framework P_n, $n \geq 1$, are bounded, and $\|P_n\| = 1$, $n \geq 1$ [if we work directly in C, then $\{P_n\}$ are generally unbounded (Ref. 2), which is the reason that the L_2 framework has traditionally been preferred].

Example 2.2. Here we consider Galerkin's method for solving (13) using piecewise polynomial rather than polynomial approximations. Specifically, assume that X is the space of Riemann integrable functions on $[a, b]$, $-\infty < a < b < \infty$, with the norm

$$\|f\|_\infty = \max_{x \in [a,b]} |f(x)|, \tag{21}$$

where $k(x, y)$ satisfies (10)-(11), so that K is compact.

For $n \geq 1$ and $h = (b - a)/n$ define a partition $\mathscr{P}_n$ of $[a, b]$ by letting $t_k = a + kh$, $k = 0, 1, 2, \ldots, n$, so that $[a, b] = \{\bigcup_{k=0}^{n-1} [t_k, t_{k+1})\} \cup Z_n$, where $Z_n = \{t_k\}_{k=1}^n$. Let X_n be the space of possibly discontinuous piecewise polynomial functions on $[a, b]$ of degree $\leq l - 1$, $l \geq 1$, relative to P_n. If an inner product is defined on X via (21) with $w(x) = 1$, then $\mathscr{P}_n$, the orthogonal projection of X onto X_n, again defines a Galerkin method. In this case $\{P_n\}$ are uniformly bounded not only on $L^2[a, b]$, but on X as well (Ref. 8). Thus it is possible to use standard projection theory with bounded projections (Ref. 2) to analyze the convergence of $\{u_n\}$.

Example 2.3. The methods in Examples 2.1 and 2.2 are particular cases of Galerkin's method for solving Eq. (2) in the Fredholm case. Here, we assume that the integral defines a bounded but not necessarily compact operator on X, a suitable space of functions on D (examples where K is contracting, i.e., $\|K\| < 1$, may be found in Refs. 5-6). If $\{X_n\}_{n=1}^\infty$ is a sequence of subspaces of X, and $\{\phi_k\}_{k=1}^N$ is a basis for X_n, then $u_n = \sum_{j=1}^N a_j \phi_j$ can be obtained by solving equations analogous to (17) where $\langle \cdot, \cdot \rangle$ is an inner product on X. As before u_n is given by a projection method, with P_n, the orthogonal projection of X onto X_n, being given by (18).

Example 2.4. Let $X = C[a, b]$, $-\infty < a < b < \infty$, with the sup norm, and again assume that $k(x, y)$ satisfies (10)-(11). Let X_n be the set of polynomials of degree $\leq n - 1$ and assume that $\{x_k\}_{k=1}^n$ are n distinct points in $[a, b]$. Define u_n by

$$u_n = \sum_{j=1}^n a_j \phi_j, \tag{22}$$

where $\{a_j\}_{j=1}^n$ are determined by solving

$$\sum_{j=1}^n \phi_j(x_k) a_j - \sum_{j=1}^n a_j K\phi_j(x_k) = f(x_k), \quad k = 1, 2, \ldots, n. \tag{23}$$

Perturbed Projection Methods

Because the matrix $\psi = [\phi_j(x_k)]_{(j,k)=1}^n$ is invertible, (22)-(23) define a projection method where P_n is the operator which maps $u \in X$ onto its polynomial interpolant on $\{x_k\}_{k=1}^n$. In particular,

$$P_n u = \sum_{j=1}^n \sum_{k=1}^n \psi_{jk}^{-1} u(x_k)\phi_j(x), \qquad (24)$$

where ψ_{jk}^{-1} is the jkth element of the inverse of ψ. Here the defining linear functionals are $l_k(u) = u(x_k)$, $k = 1, 2, \ldots, n$, and (22)-(23) is a collocation method.

In this case it is well known (and nontrivial) that $\{P_n\}$ are generally not uniformly bounded, because $P_n u$ does not necessarily converge to u for arbitrary choices of collocation points. This makes the analysis of polynomial collocation methods somewhat more delicate than Galerkin methods, even in the one-dimensional case. For multidimensional problems, as pointed out by Atkinson in Chapter 1, almost nothing is known about global collocation methods for Eq. (2), although collocation methods have generally been the method of choice for solving such problems.

If $\{x_k\}_{k=1}^n$ are the Gaussian quadrature points for a weight function $w(x)$, then (2)-(23) can be analyzed in $L_w^2[a, b]$ because the Erdös-Turan theorem guarantees mean-square convergence of $P_n u$ to u, if $u \in C[a, b]$ (Ref. 23). Choosing $w(x) = 1/(1 - x^2)^{1/2}$ and $\{x_k\}_{k=1}^n$ as the zeros of $T_n(x)$ gives a particularly important special case of this class of methods (Ref. 21).

Example 2.5. For multidimensional problems collocation methods are generally preferred to Galerkin ones, because only m-dimensional (usually $m = 2$ or 3) integrals rather than $2m$-dimensional ones need to be evaluated. In such problems the kernel is often discontinuous and, as shown by Graham in Refs. 24-25 using piecewise constant approximations [as is frequently done in practice (Ref. 26)], can often be close to optimal. Because of its importance, we briefly describe the method for Eq. (2) where D is the closure of an open bounded set in R^m with a smooth boundary.

Following Ref. 25 let $D = \bigcup_{j=1}^N \bar{D}_j$, where D_j are open, pairwise disjoint, simply connected subsets of D, such that each D_j contains its centroid ξ_j in its interior. Let $X = R(D)$ be the space of Riemann integrable functions on D and let X_n be the subspace of piecewise constant functions relative to the partition defined by $\{D_j\}_{j=1}^N$. Then, modulo sets of measure zero, we can take the characteristic functions

$$\phi_k = \begin{cases} 1, & x \in D_k, \\ 0, & \text{otherwise,} \end{cases}$$

as a basis for X_n, and u satisfying Eq. (2) can be approximated by

$$u_n = \sum_{j=1}^{N} a_j \phi_j, \qquad (25)$$

where $\{a_j\}_{j=1}^{N}$ are determined by solving

$$a_k - \sum_{j=1}^{N} a_j \int_{\bar{D}_j} k(\xi_k, y)\, dy = f(\xi_k), \qquad k = 1, 2, \ldots, N, \qquad (26)$$

where $dy = dy_1 \ldots dy_n$ and (26) holds because $\phi_j(\xi_k) = \delta_{jk}$.
When $k(x, y)$ satisfies

(i) $$\sup_{x \in D} \int_D |k(x, y)|\, dy < \infty, \qquad (27)$$

(ii) $$\lim_{h \to 0} \int_D |k(x+h, y) - k(x, y)|\, dy = 0, \qquad (28)$$

and $f(x)$ is continuous, it was shown in Ref. 25 that u_n converges uniformly to u. Also if $u \in C^1(D)$, then $\|u - u_n\|_\infty = O(h)$, where h is the maximum of the diameters of $\{D_j\}$, and it is assumed that $h \to 0$ as $N \to \infty$. Moreover, under slightly stronger conditions on k, the Sloan iterate $u_n^s = f + Ku_n$ superconverges, i.e., $\|u - u_n^s\|_\infty = O(h^2)$ as long as the integrations in (26) are performed exactly [it is for this latter result that choosing the collocation points as the centroids is crucial, and it generalizes a previous one-dimensional result of Sloan, Noussair, and Burn (Ref. 27)].

In the specific numerical example analyzed by Graham in Ref. 25, the numerical integrations in (26) were performed analytically and the superconvergence observed numerically. It was also argued in Ref. 25 that a related Galerkin method is generally preferable to the collocation method described here, because less smoothness of k is required in order for the superconvergence of u_n^s to occur. However, if the integrals need to be computed numerically, then the same smoothness requirements on the Galerkin and collocation methods may be necessary for superconvergence to hold. In this case collocation will again be preferable to Galerkin's method. We return to this point in Sections 4 and 5.

Last, we observe that the projections $\{P_n\}$ defining (25)-(26) have $\|P_n\| = 1$ and so the convergence of $\{u_n\}$ may be analyzed by standard methods.

2.3. Convergence Analysis of Projection Methods.

Here we present some convergence theorems for the projection methods given by (6). For the most part, these are straightforward generalizations of well-known results for the case $H = I$ and so many of the details will be omitted. Further discussion of some of the proofs will be found in Chapter 5.

Throughout the remainder of the chapter we assume that Eq. (1) has a unique solution, a sufficient condition being that the homogeneous equation $Hu = Ku$ has only zero as a solution (Ref. 18).

Theorem 2.1. Let u_n be given by (6). If $\|f - P_n f\|_Y \to 0$ and $\|K - P_n K\| \to 0$, $n \to \infty$, then for all $n \geq n_0$, (6) has a unique solution given by

$$u_n = (H - P_n K)^{-1} P_n f, \tag{29}$$

and $u_n \to u$, $n \to \infty$. Moreover, we have the error estimates

$$\|u - u_n\|_X \leq c \|Hu - P_n Hu\|_Y \tag{30}$$

and

$$\|u - u_n\|_X \leq c(\|f - P_n f\|_Y + \|K - P_n K\|). \tag{31}$$

Proof. The existence of u_n follows from that of $(H - P_n K)^{-1}$, and the existence of the latter follows by using Banach's lemma (Ref. 2). The convergence of u_n then follows from (31), which in turn follows from (30). To prove (30) we have

$$u - u_n = (H - K)^{-1} f - (H - P_n K)^{-1} P_n f$$

$$= (H - P_n K)^{-1}(H - P_n H)(H - K)^{-1} f$$

$$= (H - P_n K)^{-1}(H - P_n H)u, \tag{32}$$

and taking norms in (32) using the uniform boundedness of $\{(H - P_n K)^{-1}\}$, $n \geq n_0$, gives (30). □

Note that Theorem 2.1 gives sufficient conditions for the convergence of u_n that apparently need to be verified for every pair (K, f). Clearly it would be useful to have conditions which do not require this to be done. The simplest of these appears to be that

$$\|g - P_n g\|_Y \to 0, \quad \forall g \in Y. \tag{33}$$

It then follows from the compactness of K that $\|K - P_n K\| \to 0$ (Ref. 2) and then Theorem 2.1 holds. Moreover, it follows from the principle of uniform boundedness that $\{P_n\}$ are uniformly bounded (Ref. 2), and then (30) can be used to give an error estimate based on the approximation properties of $\{Y_n\}$. It is these latter properties which usually can be verified in practice and lead to practical rates of convergence. We summarize these observations in Theorem 2.2.

Theorem 2.2. If $\{P_n\}$ satisfy (33), then u_n defined by (6) exists for all n sufficiently large and

$$\|u - u_n\|_X \leq c \inf_{s_n \in Y_n} \|Hu - s_n\|_Y. \tag{34}$$

Proof. From Theorem 2.1, u_n exists for $n \geq n_0$ and (30) holds. Now let $s_n \in Y_n$; then

$$\|Hu - P_n Hu\|_Y = \|Hu - s_n - P_n(Hu - s_n)\|_Y, \tag{35}$$

where (35) follows because $P_n = s_n$. Taking norms in (35) gives

$$\|Hu - P_n Hu\|_Y \leq (1 + \|P_n\|)\|Hu - s_n\|_Y \leq c\|Hu - s_n\|_Y,$$

and since s_n is arbitrary and $\{\|P_n\|\}$ are uniformly bounded, (34) is true. □

In some circumstances it is preferable to obtain convergence rates in terms of the approximating properties of X_n. In that case we have the estimates

$$\inf_{s_n \in X_n} \|u - s_n\|_X \leq \|u - u_n\|_X \leq c \inf_{s_n \in X_n} \|u - s_n\|_X. \tag{36}$$

When $H = I$ estimate (36) is well known, and as pointed out in Chapter 2 we then say that the convergence of u_n is quasi-optimal.

Since $u_n \in X_n$, the left-hand inequality in (36) is obvious. To obtain the right-hand side, use (30) to get

$$\|u - u_n\|_X \leq c\|Hu - P_n Hu\|_Y \leq c\|H\|\|u - H^{-1} P_n Hu\|_X$$

$$\leq c\|u - Q_n u\|_X, \tag{37}$$

where $Q_n = H^{-1} P_n H$ is a projection onto X_n. Since $\{P_n\}$ are uniformly bounded, so are $\{Q_n\}$, and letting $s_n \in X_n$ in (37) gives

$$\|u - u_n\|_X \leq c\|u - s_n - Q_n(u - s_n)\|_X$$

$$\leq c(1 + \|Q_n\|)\|u - s_n\|_X \leq c\|u - s_n\|_X, \tag{38}$$

so that

$$\|u - u_n\|_X \le c \inf_{s_n \in X_n} \|u - s_n\|_X.$$

Since verifying (33) directly may be tedious, it is useful to have other conditions which guarantee its validity.

Theorem 2.3. If $\{P_n\}$ are uniformly bounded, and $\bigcup_{n \ge 1} Y_n$ is dense in Y, then $\|g - P_n g\|_Y \to 0, n \to \infty$.

Proof. See Ref. 2. □

If K is compact and either (33) or Theorem 2.3 is true, then $\|K - P_n K\| \to 0$ and the projection approximations $\{u_n\}$ converge. Although these conditions do not cover all the cases that arise in practice (for instance, polynomial collocation), they are sufficient for the purposes of the chapter, and from now on we will assume that at least one set of these conditions holds.

Using Theorems 2.1-2.3 we can easily establish the convergence of a number of the algorithms discussed in Examples 2.1-2.5.

Example 2.6. In Example 2.1 we considered Galerkin's method for solving (13) using a polynomial basis when $k(x, y)$ is a Hilbert-Schmidt kernel. In this case P_n is the orthogonal projection onto X_n and as we have observed previously $\|P_n\| = 1, n \ge 1$. Since polynomials are dense in $L^2_w[a, b]$ (Ref. 23), it follows from Theorems 2.1-2.3 that u_n converges to u in $L^2_w[a, b]$ and the error estimate (36) is true. Under additional smoothness conditions on k and f one can prove uniform convergence of u_n (Ref. 18). For the arguments along this line we refer the reader to Section 19 of Chapter 5.

Example 2.7. In Example 2.2 we considered Galerkin's method for (13) using a piecewise polynomial basis. In this case, using X as the space of Riemann integrable functions on $[a, b]$, X is complete with respect to the norm given by (21). In addition, the orthogonal projections $\{P_n\}$ are uniformly bounded on X and piecewise polynomials are dense (Ref. 28). If $k(x, y)$ satisfies (10)-(11) and f is Riemann integrable, then all the conditions of Theorem 2.3 are satisfied and so u_n converges uniformly to u. Moreover, if u is r times continuously differentiable, $r \ge 1$, then it follows from (36) that $\|u - u_n\|_\infty \le c h^{\min(r,l)}$ provided that the integrals $\{\langle K\phi_j, \phi_k \rangle\}$ and $\{\langle f, \phi_k \rangle\}$ are evaluated exactly (Ref. 4).

Example 2.8. In Ref. 25, Graham developed a Galerkin method for solving Eq. (2) in the Fredholm case using piecewise constant approximations as in Example 2.5. Assuming D has a Lipshitz continuous boundary, then $X = R(D)$, the space of Riemann integrable functions on D is well defined. If we use the sup norm on X, then X is complete, and $k(x, y)$ satisfying (27)-(28) is compact. Again it is straightforward to verify that P_n, the orthogonal projection onto X_n, is bounded with $\|P_n\| = 1$ (Ref. 25). As piecewise constant functions are dense in X, the conditions of Theorem 2.3 are satisfied and the Galerkin approximations $\{u_n\}$ converge uniformly to u. If u is continuously differentiable, it follows from (36) using Taylor series approximations that $\|u - u_n\|_\infty \leq ch$.

The convergence of the collocation method in Example 2.5 follows along similar lines (Refs. 24-25) and we shall take this up in more detail in Section 6.

2.4. Some Variants of the Projection Method. As can be seen from Examples 2.7 and 2.8, a projection method may converge rather slowly if either the solution to Eq. (2) is not smooth and/or the elements of X_n do not provide a high degree of approximation. To improve the rates of convergence when either of these circumstances prevails, a number of devices have been proposed. In particular, when $H = I$, the Sloan iterate $u_n^s = f + Ku_n$ has received considerable attention in the past decade, as discussed in detail in Chapter 2. We now indicate how this idea and a number of related ones can be extended to solve Eq. (1) when $H \neq I$.

2.4.1. The Sloan Iterate. If u_n satisfies (6), then the Sloan iterate u_n^s is defined by solving

$$Hu_n^s = f + Ku_n. \tag{39}$$

Since H is invertible, u_n is well defined, although its practical calculation may be difficult (Refs. 18, 29). When $H = I$, then $u_n^s = f + Ku_n$, as defined in Chapter 2. From (39) it follows that u_n^s converges to u if u_n converges, and the convergence is quasi-optimal. However, as shown in Chapter 2, when $H = I$, under appropriate conditions u_n^s exhibits superconvergence, and here we extend some of these results to the case $H \neq I$. For this we need a lemma analogous to Theorem 5.1 in Chapter 2.

Lemma 2.1. Let $Q_n = H^{-1}P_nH$, $n \geq 1$, where $\{P_n\}$ satisfy the condition in Theorem 2.3. Then $\{Q_n\}$ is a sequence of uniformly bounded, pointwise convergent projections on X. In addition, if K is compact then for all n sufficiently large, $n \geq n_0$, $(H - KQ_n)^{-1}$ exists and the sequence $\{(H - KQ_n)^{-1}\}$, $n \geq n_0$, is uniformly bounded.

Proof. The uniform boundedness of $\{Q_n\}$ follows immediately from the boundedness of H, H^{-1}, and the uniform boundedness of $\{P_n\}$.

If $g \in X$, then $g - Q_n g = g - H^{-1} P_n H g = H^{-1}(Hg - P_n Hg)$, and $Q_n g \to g$, because $P_n H g \to H g$.

To prove that $(H - KQ_n)^{-1}$, $n \geq n_0$, exists, write $H - KQ_n = H(I - H^{-1}KQ_n) = H(I - MQ_n)$ where $M = H^{-1}K$ is compact. Thus to complete the proof of the theorem it suffices to show that $(I - MQ_n)^{-1}$ exists for all $n \geq n_0$. Using collectively compact operator theory the result now follows from Theorem 5.1 in Chapter 2 since $\{MQ_n\}$ is a sequence of collectively compact operators.

For those readers who may not be familiar with that theory, one can give an independent proof using the argument in Theorem 1.10 of Ref. 30. The key point is that it is no longer true in general that $\{MQ_n\}$ converges *uniformly*, even though it converges pointwise, so that Banach's lemma cannot be immediately used on $H - MQ_n$. However, as we now show, $(M - MQ_n)MQ_n$ converges uniformly to zero, and this suffices to allow the use of Banach's lemma.

Observe that $(M - MQ_n)MQ_n = M(M - Q_n M)Q_n$, so that $\|(M - MQ_n)MQ_n\| \leq c\|M - Q_n M\|$ because $\{Q_n\}$ are uniformly bounded. Since $\{Q_n\}$ are pointwise convergent and M is compact, $\|M - Q_n M\| \to 0$ and so $\|(M - MQ_n)MQ_n\| \to 0$.

Now let $L_n = MQ_n$ and $B_n = I + (I - M)^{-1}L_n$. Then

$$B_n(I - L_n) = I - (I - M)^{-1}(L_n - M)L_n = I - A_n. \tag{40}$$

Since $\|(M - L_n)L_n\| \to 0$, there exists an n_0 such that for all $n \geq n_0$, $\|A_n\| < 1$. Thus, for $n \geq n_0$, $(I - A_n)^{-1}$ exists by Banach's lemma, and $I - L_n = I - MQ_n$ has a left inverse $(I - A_n)^{-1}B_n$. Thus $I - L_n$ is one-to-one. Since L_n is compact, by the Fredholm alternative it is onto, so that $(I - L_n)^{-1} = (I - A_n)^{-1}B_n$ exists for $n \geq n_0$. Taking norms, using the uniform boundedness of $(I - A_n)^{-1}$ and $\{L_n\}$, $n \geq n_0$, shows that $\{(I - L_n)^{-1}\}$, $n \geq n_0$, are uniformly bounded. □

Using Lemma 2.1 we now obtain error estimates for u_n^s which lead to superconvergence under appropriate circumstances.

Theorem 2.4. For all n sufficiently large

$$u - u_n^s = (H - KQ_n)^{-1}(K - KQ_n)u, \tag{41}$$

where $Q_n = H^{-1}P_n H$.

Proof. Operating with P_n on (39) gives

$$P_n H u_n^s = P_n f + P_n K u_n = H u_n, \tag{42}$$

so that
$$u_n = H^{-1}P_n H u_n^s = Q_n u_n^s, \tag{43}$$

and substituting (43) into (39) yields
$$H u_n^s = f + K Q_n u_n^s. \tag{44}$$

From Lemma 2.1, $(H - KQ_n)^{-1}$ exists for $n \geq n_0$ so that
$$u_n^s = (H - KQ_n)^{-1} f. \tag{45}$$

Thus, for $n \geq n_0$,
$$\begin{aligned} u - u_n^s &= (H - K)^{-1} f - (H - KQ_n)^{-1} f \\ &= (H - KQ_n)^{-1}(K - KQ_n) u. \end{aligned}$$ □

Corollary 2.1. If $\{Q_n\}$ are such that $\|K - KQ_n\| \to 0$, then
$$\|u - u_n^s\|_X \leq c \|K - KQ_n\| \|u - u_n\|_X, \tag{46}$$

so that u_n^s superconverges to u.

Proof. Using the fact that Q_n is a projection onto X_n, $(K - KQ_n)u = (K - KQ_n)(u - u_n)$, and using this in (41) gives
$$u - u_n^s = (H - KQ_n)^{-1}(K - KQ_n)(u - u_n). \tag{47}$$

Taking norms in (47) using the uniform boundedness of $\{(H - KQ_n)^{-1}\}$ yields (46).

Now if $\|K - KQ_n\| \to 0$, then (46) shows that the convergence of u_n^s is better than quasi-optimal by the factor $\|K - KQ_n\|$. On using the definition of superconvergence in Chapter 2, u_n^s superconverges to u. □

When $H = I$, Corollary 2.1 gives Theorem 3.1 of Chapter 2. In Chapter 5 we give an example of Theorem 2.4 when $H \neq I$, for a Galerkin method for solving Cauchy singular equations. We refer the reader to Section 19 of that chapter for further details.

As shown in Chapter 2 for $H = I$, the best superconvergence results essentially occur when K and u are in the same smoothness class. Usually, but not always, this means that K and f are in the same smoothness class. When this is not the case, then other strategies can be employed to improve the convergence rate of a particular projection method. In Ref. 31 and Chapter 2, Sloan examines the use of Kantorovich regularization, and we extend this idea to the case $H \neq I$ next. An application will be found in Chapter 5.

2.4.2. Kantorovich Regularization.
When the kernel of K is smoother than f, we consider solving Eq. (1) in the following way. Let u_0 be the unique solution to

$$Hu_0 = f \tag{48}$$

and let

$$z = u - u_0 \tag{49}$$

(when $H = I$, $u_0 = f$ and the method we describe reduces to the classical Kantorovich regularization for equations of the second kind). Then

$$(H - K)z = (H - K)u - (H - K)u_0 = f - Hu_0 + Ku_0 = Ku_0,$$

so that z satisfies

$$Hz = Kz + Ku_0. \tag{50}$$

If (50) is solved by the projection method (X_n, P_n), then an approximation u_n^k to u is defined by

$$u_n^k = z_n^k + u_0, \tag{51}$$

where z_n^k solves

$$Hz_n^k = P_n K z_n^k + P_n K u_0. \tag{52}$$

Theorem 2.5. If $\{P_n\}$ satisfy the conditions of Theorem 2.3, then for all $n \geq n_0$, z_n^k is well defined and converges to u. Moreover,

$$\|u - u_n^k\|_X \leq c\|Hz - P_n Hz\|_Y. \tag{53}$$

Proof. From Theorem 2.1, z_n^k exists for $n \geq n_0$, and using (49) and (51)

$$u_n^k - u = z - z_n^k, \tag{54}$$

so the convergence of u_n^k follows from that of z_n^k. However, $z_n^k \to z$ so that $u_n^k \to u$.

From (52) and (54), $z_n^k = (H - P_n K)^{-1} P_n K u_0$ and $z = (H - K)^{-1} K u_0$, so that

$$u - u_n^k = z - z_n^k = (H - P_n K)^{-1}(H - P_n H)(H - K)^{-1} K u_0$$

$$= (H - P_n K)^{-1}(H - P_n H)z, \tag{55}$$

and taking norms in (55) gives (53). □

As for the case $H = I$, $Hz = Kz + Ku_0$ will generally be in the same smoothness class as K, and if K is smoother than f, then Hz will be smoother than u, and u_n^k can be expected to converge faster than u_n. By iterating u_n^k this convergence can be expected to improve again.

2.4.3. Iterated Kantorovich Regularization.
If u_n^k is the regularized approximation to u, then

$$H\hat{u}_n^k = f + Ku_n^k \tag{56}$$

defines the iterated Kantorovich approximation $\hat{u}_n^k$. When $H = I$, $\hat{u}_n^k$ agrees with the definition given by Sloan in Ref. 31 and in Chapter 2. Since $u_n^k = z_n^k + u_0$,

$$H\hat{u}_n^k = f + Kz_n^k + Ku_0 = f + H\hat{z}_n^k, \tag{57}$$

where $\hat{z}_n^k$ is the Sloan iterate of z_n^k. Thus

$$Hu - H\hat{u}_n^k = Ku + f - f - H\hat{z}_n^k = Ku - H\hat{z}_n^k.$$

But $z = H^{-1}Ku$, as can be verified from (48)–(49), so that

$$Hu - H\hat{u}_n^k = Hz - H\hat{z}_n^k \tag{58}$$

and

$$u - \hat{u}_n^k = z - \hat{z}_n^k. \tag{59}$$

Since $\hat{z}_n^k$ is the Sloan iterate of z_n^k, it follows from (41), with u replaced by z, that for all n sufficiently large,

$$z - \hat{z}_n^k = (H - KQ_n)^{-1}(K - KQ_n)z,$$

and

$$u - \hat{u}_n^k = (H - KQ_n)^{-1}(K - KQ_n)(z - Q_n z).$$

From (59) it follows that $\hat{u}_n^k \to u$ and

$$\|u - \hat{u}_n^k\|_X \le c\|K - KQ_n\|\|z - Q_n z\|_X. \tag{60}$$

Perturbed Projection Methods

Thus, if K is smoother than f, and $\|K - KQ_n\| \to 0$, we expect that $\hat{u}_n^k$ will converge faster than u_n^s. As shown for Eq. (43) of Chapter 2, where regularization is appropriate, u_n converges the slowest, followed by u_n^k, u_n^s, and $\hat{u}_n^k$ in that order. More generally, it is shown in Ref. 31 that for the piecewise polynomial Galerkin method described in Example 2.2, $\|u - u_n\|_\infty \le ch^{\rho^*}$, $\|u - u_n^k\|_\infty \le ch^{r^*}$, $\|u - u_n^s\|_\infty \le ch^{\rho^*+r^*}$, and $\|u - \hat{u}_n^k\|_\infty \le ch^{2r^*}$, if $k \in C^r([a, b] \times [a, b])$, $f \in C^\rho[a, b]$, $\rho < r$, $\rho^* = \min(\rho, l)$, and $r^* = \min(r, l)$. Similar results can be shown to hold for polynomial Galerkin methods, for both Fredholm and a variety of Cauchy singular equations.

2.4.4. Accelerated Projection Methods. As we have seen, iterating a projection approxmation can lead to substantial improvement in the convergence rate of a basic projection method, sometimes with essentially no increase in cost (Ref. 32). This suggests that further iteration will yield increasingly accurate approximations. Unfortunately, that is not always the case (Ref. 32). However, in Ref. 33 Dellwo and Friedman introduced a sequence of approximations which builds on a basic projection method to produce approximations $u_{n,N}$ which have the property that under very general conditions on K, $\|u - u_{n,N}\|_X = O(\|u - u_n\|_X^N)$, $n \to \infty$. Since examining the full theory would take us too far afield, we consider only the first element in their sequence. As we shall see, it already has the convergence rate of the iterated Kantorovich approximation.

Assume that $H = I$ and let $\{P_n\}$ satisfy the conditions in Theorem 2.3. Then define u_n^d as the solution to

$$u_n^d = f + (K - KP_n)f + KP_n u_n^d. \tag{61}$$

To see how u_n^d is related to u_n, operate on both sides of (61) with P_n giving

$$P_n u_n^d = P_n f + P_n K f - P_n K P_n f + P_n K P_n u_n^d.$$

Letting $w_n^d = P_n u_n^d$, w_n^d satisfies

$$w_n^d = P_n f + P_n(Kf - KP_n)f + P_n K w_n^d = P_n g_n + P_n K w_n^d, \tag{62}$$

so that

$$u_n^d = g_n + K w_n^d. \tag{63}$$

Thus u_n^d may be regarded as the first Sloan iterate of w_n^d, where w_n^d is a projection approximation for Eq. (1) with $H = I$ and f replaced by $g_n = f + (K - KP_n)f$.

Theorem 2.6. If $\{P_n\}$ satisfy the conditions in Theorem 2.3, then u_n^d converges to u, and

$$\|u - u_n^d\|_X \leq c \|K - KP_n\| \|Ku - P_n Ku\|_X. \tag{64}$$

Proof. From Lemma 2.1, $(I - KP_n)^{-1}$ exists for sufficiently large n so that (61) has the unique solution

$$u_n^d = (I - KP_n)^{-1} g. \tag{65}$$

Thus

$$\begin{aligned}
u - u_n^d &= (I - K)^{-1} f - (I - KP_n)^{-1} P_n g_n \\
&= (I - K)^{-1} f - (I - KP_n)^{-1} f - (I - KP_n)^{-1}(K - KP_n) f \\
&= (I - KP_n)^{-1}[(K - KP_n)u - (K - KP_n)f] \\
&= (I - KP_n)^{-1}[Ku - KP_n u - (K - KP_n)(u + Ku)] \\
&= (I - KP_n)^{-1}[(K - KP_n)Ku] \\
&= (I - KP_n)^{-1}(K - KP_n)(I - P_n)Ku, \tag{66}
\end{aligned}$$

and taking norms in (66) gives (64). Since $(I - P_n)Ku \to 0$ and

$$\|K - KP_n\| \leq \|K\|(1 + \|P_n\|) \leq c, \tag{67}$$

(66) shows that u_n^d converges to u with an error estimate identical to that of the Kantorovich iterate. □

As the results of this section and chapter show, simple iterative modifications of a basic projection method can lead to substantial improvements in convergence, and this theoretical property has been borne out in a number of calculations (Refs. 3, 6, 25, 33). However, as we mentioned in the Introduction, these results assume that all the necessary analytic and algebraic operations are performed without error. As this can rarely be done, it is important to determine the effect of various approximations on the convergence properties of projection methods and their variants. In the following section we develop a general theory for doing this, and then various applications and modifications will be discussed in Sections 4–7. Further results can be found in Chapter 5.

3. Perturbed Projection Methods

Let $\{X_n\}$ and $\{P_n\}$ be chosen as in Section 2 and let $Y_n = HX_n$ as before. In addition, let $\{R_n\}$ be a sequence of linear operators from $X_n \to Y_n$ and let $r_n \in Y_n$. If $v_n \in X_n$ satisfies

$$Hv_n = P_n K v_n + R_n v_n + P_n f + r_n, \tag{68}$$

then $\{v_n\}$ defines a perturbed projection method for solving Eq. (1). Quantities R_n and r_n are referred to as the perturbations, and typically represent errors in evaluating $P_n K u_n$ and $P_n f$, respectively. When $H = I$, the convergence theory for $\{v_n\}$ is well known, and for $H \neq I$ it has been discussed by Prössdorf in Ref. 34, Junghanns and Silbermann in Ref. 20, and Miel in Refs. 19, 35.

Definition 3.1. Let $R_n : X_n \to Y_n$ be a linear operator. Then $\|R_n\|_n = \text{lub}_{\{\|x_n\|_X = 1\}}\{\|R_n x_n\|_Y, x_n \in X_n\}$ is called the induced operator norm of R_n.

Theorem 3.1. Let $\{P_n\}, \{X_n\}, \{Y_n = HX_n\}$ be defined as before, and assume that $\|R_n\|_n \to 0$ and $\|r_n\|_X \to 0$, $n \to \infty$. Then v_n converges to u and

$$\|u - v_n\|_X \leq c(\|u - u_n\|_X + \|R_n\|_n + \|r_n\|_Y). \tag{69}$$

Proof. According to Theorem 1 of Ref. 35, for all $n \geq n_0$, $A_n = (H - P_n K - R_n)|_{X_n}$ has an inverse $A_n^{-1} : Y_n \to X_n$ and the norms $\|A_n^{-1}\|_n$ are uniformly bounded [this requires essentially two applications of Banach's lemma (Ref. 35)]. Thus, for $n \geq n_0$, v_n exists and

$$v_n = A_n^{-1} P_n f + A_n^{-1} r_n. \tag{70}$$

Hence

$$u - v_n = u - u_n + u_n - v_n, \tag{71}$$

so that

$$\|u - v_n\|_X \leq \|u - u_n\|_X + \|u_n - v_n\|_X. \tag{72}$$

Now

$$u = (H - P_n K)^{-1} P_n f,$$

so that

$$u_n - v_n = (H - P_n K)^{-1} P_n f - (H - P_n K - R_n)^{-1}(P_n f + r_n)$$

$$= -A_n^{-1} R_n u_n - A_n^{-1} r_n.$$

Since $u_n \in X_n$ converges, $\{u_n\}$ are uniformly bounded and

$$\|u_n - v_n\|_X \le \|A_n^{-1}\|_n \|R_n\|_n \|u_n\|_X + \|A_n^{-1}\|_n \|r_n\|_Y$$

$$\le c(\|R_n\|_n + \|r_n\|_X). \tag{73}$$

Thus $\|u_n - v_n\|_X \to 0$, and since $u_n \to u$ it follows from (72) that $v_n \to u$. Moreover, using (72) and (73)

$$\|u - v_n\|_X \le \|u - u_n\|_X + c(\|R_n\|_n + \|r_n\|_Y)$$

$$\le c(\|u - u_n\|_X + \|R_n\|_n + \|r_n\|_Y). \quad \square$$

Definition 3.2. Let $\{u_n\}$ converge to u in X. If $\{v_n\} \subseteq X$, we say that $\{v_n\}$ is consistent with $\{u_n\}$ iff for all n sufficiently large, $\|u - v_n\|_X \le c\|u - u_n\|_X$ where c does not depend on n. In particular v_n converges to u (essentially v_n converges at the same rate as u_n).

From Theorem 3.1 we see that a sufficient condition for $\{v_n\}$ and $\{u_n\}$ to be consistent is that $\|R_n\|_n = O(\|u - u_n\|_X)$ and $\|r_n\|_Y = O(\|u - u_n\|_X)$. However, examining the proof one sees that the possibly weaker condition $\|R_n u_n\|_Y = O(\|u - u_n\|_X)$ is sufficient as well. This observation is consistent (no pun intended) with one made by Joe in Ref. 8 in his analysis of the effect of quadrature errors on piecewise polynomial Galerkin methods for solving Fredholm equations. We shall return to this point in Section 4. We now consider the effect of perturbations on u_n^s, u_n^k, $\hat{u}_n^k$, and u_n^d.

3.1. The Perturbed Sloan Iterate.

Let u_n satisfy (68), then the perturbed Sloan iterate is defined by

$$H v_n^s = f + K v_n + \delta_n, \tag{74}$$

where δ_n is a perturbation which typically represents errors in evaluating $K v_n$. Now it follows from (1) and (74) that

$$H(u - v_n^s) = K(u - v_n) - \delta_n, \tag{75}$$

so that

$$u - v_n^s = H^{-1} K(u - v_n) - H^{-1} \delta_n, \tag{76}$$

and

$$\|u - v_n^s\|_X \le c(\|u - v_n\|_X + \|\delta_n\|_Y). \tag{77}$$

Perturbed Projection Methods 93

Thus $v_n^s \to u$ if $\{v_n\}$ is consistent with $\{u_n\}$ and $\|\delta_n\|_Y \to 0$. In particular (77) shows that v_n^s appears to converge no faster than $\|u - v_n\|_X$. However, in considering u_n^s we know that in certain cases it superconverges, and one can expect that v_n^s will do so as well. To prove this, where applicable, we need to obtain a different error estimate for $\|u - v_n^s\|_X$.

Theorem 3.2. If v_n^s is given by (74) and $\{P_n\}$ satisfy the conditions of Theorem 2.3, then

$$\|u - v_n^s\|_X \le \|u - u_n^s\|_X + \|H^{-1}K(u_n - v_n) - H^{-1}\delta_n\|_X. \tag{78}$$

Proof.

$$u - v_n^s = u - u_n^s + u_n^s - v_n^s \tag{79}$$

and

$$Hu_n^s - Hv_n^s = f + Ku_n - f - Kv_n - \delta_n = K(u_n - v_n) - \delta_n, \tag{80}$$

so that

$$u_n^s - v_n^s = H^{-1}K(u_n - v_n) - H^{-1}\delta_n. \tag{81}$$

Substituting (81) into (79) gives (78). □

As we shall see in the next section, careful analysis of $\alpha_n = \|H^{-1}K(u_n - v_n) - H^{-1}\delta_n\|_X$ often gives $\alpha_n = O(\|u - u_n^s\|_X)$, establishing the superconvergence of v_n^s to u.

3.2. The Perturbed Kantorovich Method. The perturbed Kantorovich approximation v_n^k is defined by

$$v_n^k = w_n^k + u_0 + \delta_n, \tag{82}$$

where $u_0 = H^{-1}f$ is given in Section 2, and w_n^k satisfies

$$Hw_n^k = P_n K w_n^k + R_n w_n^k + P_n K u_0 + r_n, \tag{83}$$

the perturbed version of (52). When $H = I$, then $u_0 = f$ and $\delta_n = 0$, if δ_n represents the error in evaluating $H^{-1}f$, which generally depends on n, $\delta_n \to 0$ (otherwise the method won't converge).

Theorem 3.3. If $\|R_n\|_n \to 0$, $\|r_n\|_Y \to 0$, and $\|\delta_n\|_X \to 0$, $n \to \infty$, then $v_n^k \to u$ and

$$\|u - v_n^k\|_X \le \|u - u_n^k\|_X + c(\|R_n\|_n + \|r_n\|_Y + \|\delta_n\|_X). \tag{84}$$

Proof. Using the same argument as in Theorem 3.1, it follows from (83), and the facts that $\|R_n\|_n \to 0$ and $\|r_n\|_Y \to 0$, that $w_n^k \to z$. Thus $u_n^k \to z + u_0 = u$, because $\delta_n \to 0$.

To get (84) observe that

$$u - v_n^k = u - u_n^k + u_n^k - v_n^k$$
$$= u_n - u_n^k + z_n^k + u_0 - w_n^k - u_0 - \delta_n = u_n - u_n^k + z_n^k - w_n^k - \delta_n.$$

Now arguing as in Theorem 3.1,

$$z_n^k - w_n^k = -A_n^{-1} R_n z_n^k - A_n^{-1} r_n,$$

so that

$$u - v_n^k = u_n - u_n^k - A_n^{-1}(R_n z_n^k + r_n) - \delta_n, \qquad (85)$$

and taking norms in (85) gives (84). □

3.3. The Perturbed Kantorovich Iterate. The perturbed Kantorovich iterate is defined by

$$H\hat{v}_n^k = f + Kv_n^k + \beta_n, \qquad (86)$$

where v_n^k is given by (82).

Theorem 3.4. If the conditions of Theorem 3.3 hold, and $\|\beta_n\|_Y \to 0$, $n \to \infty$, then $\hat{v}_n^k \to u$ and

$$\|u - \hat{v}_n^k\|_X \leq c(\|u - u_n^k\|_X + \|K(z_n^k - w_n^k) - \delta_n - \beta_n\|_Y). \qquad (87)$$

Proof. Since $v_n^k \to u$ and $\beta_n \to 0$, it follows immediately from (86) that $\hat{v}_n^k \to u$. To get (87), write

$$u - \hat{v}_n^k = u - \hat{u}_n^k + \hat{u}_n^k - \hat{v}_n^k$$
$$= u - \hat{u}_n^k + H^{-1}f + H^{-1}Ku_n^k - H^{-1}f - H^{-1}Kv_n^k - H^{-1}\beta_n$$
$$= u - \hat{u}_n^k + H^{-1}K(u_n^k - v_n^k) - H^{-1}\beta_n. \qquad (88)$$

But $u_n^k = z_n^k + u_0$ and $v_n^k = w_n^k + u_0 + \delta_n$, so that $u_n^k - v_n^k = z_n^k - w_n^k - \delta_n$, and

$$u - \hat{v}_n^k = u - \hat{u}_n^k + H^{-1}K(z_n^k - w_n^k) - H^{-1}K\delta_n - H^{-1}\beta_n. \qquad (89)$$

Taking norms in (89) gives (87). □

3.4. The Perturbed Dellwo–Friedman Method.

From Section 2 recall that the Dellwo-Friedman method is defined by

$$u_n^d = g_n + Kw_n^d, \qquad (90)$$

where w_n^d satisfies

$$w_n^d = P_n g_n + P_n K w_n^d,$$

and $g_n = f + (K - KP_n)f$. The perturbed version of this method is given by solving

$$\hat{w}_n^d = P_n K \hat{w}_n^d + R_n \hat{w}_n^d + P_n g_n + r_n, \qquad (91)$$

and then

$$\hat{u}_n^d = g_n + K\hat{w}_n^d + \delta_n. \qquad (92)$$

Theorem 3.5. If $\|R_n\|_n \to 0$, $\|r_n\|_X \to 0$, $\|\delta_n\|_X \to 0$, and the conditions of Theorem 2.3 are satisfied, then $\hat{u}_n^d \to u$ and

$$\|u - \hat{u}_n^d\|_X \leq \|u - u_n^d\|_X + \|K(w_n^d - \hat{w}_n^d) - \delta_n\|_X. \qquad (93)$$

Proof. Again, using the same arguments as in Theorem 3.1, we find that $\hat{w}_n^d \to u$, and from (92) and the assumptions of the theorem $\hat{u}_n^d \to f + Ku = u$. Also

$$u - \hat{u}_n^d = u - u_n^d + u_n^d - \hat{u}_n^d = u - u_n^d + g_n + Kw_n^d - g_n - K\hat{w}_n^d - \delta_n$$

$$= u - u_n^d + K(w_n^d - \hat{w}_n^d) - \delta_n, \qquad (94)$$

and taking norms in (94) gives (93). □

4. Galerkin's Method with Quadrature Errors

The results in Theorems 3.1–3.4 give a general scheme for analyzing the effects of perturbations on the convergence of the projection method (6) and a number of important variants. In the remainder of the chapter we specialize to the situation where the errors arise from numerical integration in a number of Galerkin and collocation methods. For Fredholm and

Volterra equations, the problem which currently appears to be of most interest is the determination of conditions on the quadrature methods which guarantee the consistency of the resulting approximations with those when the integrations are performed exactly. In particular, finding conditions which guarantee the superconvergence of the resulting Sloan iterate seems to have motivated much of the resurgence of interest in this problem (Refs. 3-4, 6).

In this section we show how the results of Theorems 3.1 and 3.2 can be used to reproduce the results of Joe in Ref. 8 without the formalism of restriction and prolongation operators. Our approach extends that of Chandler in Ref. 15 and shows that the perturbation approach, carefully applied, is capable of yielding important results. In our opinion, this method is more natural than that of Spence-Thomas (Ref. 3) and Joe (Ref. 8), and should enable one to study the connection between perturbation theory and the technique of Atkinson and Bogomolny (Ref. 4) (to be discussed in Section 5) in a unified fashion.

Although the theory presented in Section 3 is capable of dealing with integral equations with discontinuous kernels where product integration techniques are used to evaluate the necessary integrals (Ref. 2), for now we restrict ourselves to the case where $k(x, y)$ and $f(x)$ in Eq. (2) are continuous.

Assume now that (2) in the Fredholm case is solved by a Galerkin method using piecewise polynomial approximations $u_n \in X_n$ relative to a partition $D = \bigcup_{j=1}^{L(n)} D_j$, $n = 1, 2, \ldots$, of D. If $\{\phi_k\}_{k=1}^{N}$ is a basis of X_n, then as in Examples 2.2 and 2.3 u_n is given by

$$u_n = \sum_{j=1}^{N} a_j \phi_j, \tag{95}$$

where $\{a_j\}_{j=1}^{N}$ are determined by solving

$$\sum_{j=1}^{N} a_j \langle \phi_j, \phi_k \rangle - \sum_{j=1}^{N} a_j \langle K\phi_j, \phi_k \rangle = \langle f, \phi_k \rangle, \qquad k = 1, 2, \ldots, N, \tag{96}$$

and

$$\langle f, g \rangle = \int_D f(y) g(y) \, d\sigma(y) \tag{97}$$

is an inner product on $X = R(D)$, the space of Riemann integrable functions on D with respect to the sup norm (from now on assume that D is the closure of a bounded open set in R^m and K defines a compact operator on X)

$$\|f\|_\infty = \operatorname*{lub}_{x \in D} |f(x)|. \tag{98}$$

It then follows that u_n satisfies (7) where P_n is the orthogonal projection of X onto X_n. Under very general conditions on the partitions $\{D_j\}$ and the approximating subspaces X_n, $\{P_n\}$ are uniformly bounded on X and the subspaces X_n are dense in X (Refs. 4, 8, 28). If this is true, then the conditions of Theorems 2.2 and 2.3 are satisfied and the sequence of approximations $\{u_n\}$ given by (95)-(96) is well defined for $n \geq n_0$ and u_n converges uniformly to u. In Examples 4.1-5.2 we consider specific cases of this setup—for now we just assume that the conditions of Theorems 2.2 and 2.3 concerning $\{P_n\}$ and $\{X_n\}$ are met.

In practice, numerical implementation of (95)-(96) requires the evaluation of the integrals $\{\langle \phi_j, \phi_k \rangle\}$, $\{\langle K\phi_j, \phi_k \rangle\}$, and $\{\langle f, \phi_k \rangle\}$, which usually must be done numerically. Here we assume that $\{\langle \phi_j, \phi_k \rangle\}_{(j,k)=1}^N$ are evaluated exactly, as is the case in Refs. 3, 8, 15—although the perturbation formalism is capable of handling approximations to these as well. Integrals of the form $\int_D f(x)\, dx$ are approximated by a quadrature rule Q_M,

$$\int_D f(x)\, d\sigma(x) \simeq \sum_{p=1}^{M(n)} w_p f(x_p) = Q_M(f), \tag{99}$$

where $\{w_p\}$ and $\{x_p\}$ are the weights and nodes of Q_M, and double integrals $\int_{D \times D} f(x, y)\, d\sigma(x)\, d\sigma(y)$ are approximated by the product rule

$$Q_M \times Q_M(f) \simeq \sum_{m=1}^{M(n)} \sum_{p=1}^{M(n)} w_p w_m f(x_p, y_m) \tag{100}$$

(this condition is not necessary for our approach, but appears to be essential for the results in Ref. 4).

Using (99) and (100) in (96) u_n is approximated by

$$v_n = \sum_{j=1}^N b_j \phi_j, \tag{101}$$

where $\{b_j\}_{j=1}^N$ are obtained by solving

$$\sum_{j=1}^N \langle \phi_j, \phi_k \rangle b_j - \sum_{j=1}^N b_j \left[\sum_{m=1}^{M(n)} \sum_{p=1}^{M(n)} w_p w_m k(x_p, y_m) \phi_k(x_p) \phi_j(y_m) \right]$$

$$= \sum_{p=1}^{M(n)} w_p f(x_p) \phi_k(x_p), \quad k = 1, 2, \ldots, N. \tag{102}$$

We shall refer to (101)-(102) as a discrete Galerkin method.

Defining $K_n : X \to X$ by

$$(K_n u)(x) = \sum_{m=1}^{M(n)} w_m k(x, y_m) u(y_m) \qquad (103)$$

and

$$(\hat{P}_n u)(x) = \sum_{j=1}^{N} \sum_{k=1}^{N} \sum_{p=1}^{M(n)} \Phi_{jk}^{-1} w_p \phi_k(x_p) u(x_p) \phi_j(x), \qquad (104)$$

some tedious algebra shows that v_n satisfies the operator equation

$$v_n = \hat{P}_n K_n v_n + \hat{P}_n f \qquad (105)$$

(the expression for $\hat{P}_n$ arises by discretizing the inner products $\langle u, \phi_k \rangle$ in the expression for P_n. It may be regarded as an "approximation" to P_n—see Ref. 4 for further discussion of this point—but in general is *not* a projection itself). Letting $R_n = \hat{P}_n K_n - P_n K$ and $r_n = \hat{P}_n f - P_n f$, v_n defines a perturbed projection method

$$v_n = P_n K v_n + R_n v_n + P_n f + r_n. \qquad (106)$$

If the quadrature rules $\{Q_M\}$ are such that $\|R_n\|_n \to 0$ and $\|r_n\|_\infty \to 0$, $n \to \infty$, then v_n converges uniformly to u and from (69)

$$\|u - v_n\|_\infty \le \|u - u_n\|_\infty + c(\|R_n u_n\|_\infty + \|r_n\|_\infty). \qquad (107)$$

In particular, $\{v_n\}$ is consistent with $\{u_n\}$ if $\|R_n u_n\|_\infty = O(\|u - u_n\|_\infty)$ and $\|r_n\|_\infty = O(\|u - u_n\|_\infty)$. Sufficient conditions for this to be true in the case where D is the closure of a bounded open set in R^m, $m \ge 1$, are given in Ref. 8 and in Theorem 4.1 we rederive these results from (107)—but first some additional observations and notation.

Using the expressions (17), (102)–(103) for P_n, $\hat{P}_n$, and K_n, we find that

$$P_n K v_n = \sum_{j=1}^{N} \sum_{k=1}^{N} \Phi_{jk}^{-1} \langle K v_n, \phi_k \rangle \phi_j \qquad (108)$$

and

$$\hat{P}_n K_n v_n = \sum_{j=1}^{N} \sum_{k=1}^{N} \sum_{m=1}^{M(n)} \sum_{p=1}^{M(n)} \Phi_{jk}^{-1} w_p w_m k(x_p, y_m) v_n(y_m) \phi_k(x_p) \phi_j(x), \qquad (109)$$

so that

$$R_n v_n = -\sum_{j=1}^{N} \sum_{k=1}^{N} \Phi_{jk}^{-1} E_k(k v_n \phi_k) \phi_j, \tag{110}$$

where $E_k(k v_n \phi_k) \equiv E_k$ is the numerical integration error in evaluating $\langle K v_n, \phi_k \rangle$ by $Q_M \times Q_M(k v_n \phi_k)$. We now use (110) to derive bounds on $\|R_n w_n\|_\infty$, $w_n \in X_n$, and $\|R_n\|_n$ which are crucial for the error analysis in Theorem 4.1.

Letting

$$\beta_j = \sum_{k=1}^{N} \Phi_{jk}^{-1} E_k, \tag{111}$$

and

$$w_n = \sum_{j=1}^{N} b_j \phi_j,$$

$$R_n w_n = -\sum_{j=1}^{N} \beta_j \phi_j, \tag{112}$$

so that

$$|R_n w_n| \le \max_j |\beta_j| \sum_{j=1}^{N} |\phi_j|. \tag{113}$$

Hence,

$$\|R_n w_n\|_\infty \le \max_j |\beta_j| \max_{x \in D} \sum_{j=1}^{N} |\phi_j| = \max_j |\beta_j| \left\| \sum_{j=1}^{N} |\phi_j| \right\|_\infty. \tag{114}$$

Now from (111)

$$|\beta_j| \le \max_k |E_k| \sum_{j=1}^{N} |\Phi_{jk}^{-1}|, \tag{115}$$

so that

$$\max_j |\beta_j| \le \varepsilon_1 \max_j \left(\sum_{k=1}^{N} |\Phi_{jk}^{-1}| \right) = \varepsilon_1 \|\Phi^{-1}\|_\infty, \tag{116}$$

where $\varepsilon_1 = \max_k |E_k|$ and $\|\Phi^{-1}\|_\infty$ is the usual infinity norm of Φ^{-1} induced by the infinity norm on R^N. Thus

$$\|R_n w_n\|_\infty \le \varepsilon_1 \|\Phi^{-1}\|_\infty \left\| \sum_{j=1}^{N} |\phi_j| \right\|_\infty. \tag{117}$$

To estimate $\|R_n\|_n$ we use the linearity of the error functional E_k to get

$$E_k(k\phi_k w_n) = \sum_{j=1}^{N} b_j E_{kj}(k\phi_k \phi_j), \tag{118}$$

where $E_{kj}(k\phi_k \phi_j)$ is the quadrature error in evaluating $\langle K\phi_j, \phi_k \rangle$ by $Q_M \times Q_M(k\phi_k \phi_j)$. Thus

$$|E_k| \leq (\max_j |b_j|) N\varepsilon_2, \tag{119}$$

where $\varepsilon_2 = \max_{j,k} |E_{jk}|$.

Now it follows from Eq. 3.2 of Ref. 8 that

$$\max_j |b_j| \leq c \|\Phi^{-1}\|_\infty^{1/2} \|w_n\|_\infty,$$

which gives

$$\|R_n w_n\|_\infty \leq cN\varepsilon_2 \|\Phi^{-1}\|_\infty^{3/2} \|w_n\|_\infty. \tag{120}$$

Thus

$$\|R_n\|_n \leq cN\varepsilon_2 \|\Phi^{-1}\|_\infty^{3/2}. \tag{121}$$

Using (117) and (121) we can now prove a general convergence theorem for the discrete Galerkin method given by (101)-(102).

Theorem 4.1. See Ref. 8. Suppose that $k(x, y)$ and $f(x)$ in Eq. (2) are continuous and that (2) is solved by the discrete Galerkin method (101)-(102). Suppose that $\{X_n\}$ and the partitions $\{D_j\}$ have the following properties:

(i) $\{P_n\}$ are uniformly bounded, $n \geq 1$, and $\bigcup_{n \geq 1} X_n$ is dense in $X = R(D)$.

(ii) The basis $\{\phi_1, \ldots, \phi_N\}$ for X_n satisfies $\|\sum_{j=1}^{N} |\phi_j|\|_\infty \leq c$, where c is a constant not depending on N.

(iii) If $g \in C^r(D)$, $r \geq 1$, then

$$\inf_{s_n \in X_n} \|g - s_n\|_\infty \leq Ch^{r^*}, \qquad r \geq 1, \tag{122}$$

where $r^* = \min(r, l)$, $l - 1$, $l \geq 1$, is the highest degree of the polynomials in X_n, and h is the maximum of the diameters of $\{D_j\}$.

(iv) $\|\Phi^{-1}\|_\infty \leq ch^{-\gamma}$, $\gamma \geq 0$; \hfill (123)

Perturbed Projection Methods

$$\text{(v)} \quad |E(f\phi_k)| \leq ch^{\gamma+\delta}, \delta > 0, \tag{124}$$

where $E(f\phi_k)$ is the integration error in evaluating $\langle f, \phi_k \rangle$ by $Q_M(f\phi_k)$, $k = 1, 2, \ldots, N$.

$$\text{(vi)} \quad E_1 = \max_j |E(ku_n\phi_j)| \leq ch^{\gamma+\xi}, \xi > 0; \tag{125}$$

(vii) $\varepsilon_2 \leq ch^{\mu+3\gamma/2}$, $\mu > m$; and

$$\text{(viii)} \quad N \leq ch^{-m}. \tag{126}$$

If $h \to 0$ as $n \to \infty$, then for all n sufficiently large the discrete Galerkin approximation v_n exists, and converges uniformly to u. Moreover, if $u \in C^r(D)$, $r \geq 1$, then

$$\|u - v_n\|_\infty \leq ch^\theta, \quad h \to 0, \tag{127}$$

where $\theta = \min(l, r, \delta, \xi)$.

Proof. We merely have to verify the conditions of Theorem 3.1 and then use the error estimate (107).

From the conditions on k, f, $\{P_n\}$, and $\{X_n\}$, it follows from Theorem 2.3 that u_n, the Galerkin approximation, exists for all n sufficiently large and, $\|u - u_n\|_\infty \to 0$, $n \to \infty$. From (121)(iv), (vii), and (viii)

$$\|R_n\|_n \leq cN\varepsilon_2\|\Phi^{-1}\|^{3/2} \leq ch^{-m}c_1 h^{\mu+3\gamma/2}c_2 h^{-3\gamma/2} = ch^{\mu-m} \to 0, \quad n \to \infty,$$

because $\mu > m$ and $h \to 0$, $n \to \infty$.

Using a calculation analogous to that for bounding $\|R_n w_n\|_\infty$, one finds that

$$\|r_n\|_\infty \leq \max_j |E(f\phi_j)| \|\Phi^{-1}\|_\infty \left\| \sum_{j=1}^N |\phi_j| \right\|_\infty \leq ch^{\gamma+\delta}h^\gamma = ch^\delta \to 0, \quad n \to \infty. \tag{128}$$

Thus $\|R_n\|_n \to 0$ and $\|r_n\|_\infty \to 0$, $n \to \infty$, so the conditions of Theorem 3.1 are satisfied and $\|u - v_n\|_\infty \to 0$.

If $u \in C^r(D)$, $r \geq 1$, it follows from (122) that $\|u - u_n\|_\infty \leq ch^{r^*}$. Also

$$\|R_n u_n\|_\infty \leq \varepsilon_1 \|\Phi^{-1}\|_\infty \left\| \sum_{j=1}^N |\phi_j| \right\|_\infty \leq ch^{-\gamma}h^{\gamma+\xi} = ch^\xi, \tag{129}$$

so that (107), (122), (128), and (129) yield

$$\|u - v_n\|_\infty \le c(h^{r^*} + h^\delta + h^\xi) \le ch^\theta, \qquad h \to 0. \qquad \square$$

From (127) we see that for v_n to be consistent with u_n it is sufficient to have $\delta, \xi \ge r^*$.

Example 4.1. To see that there are situations where all of the conditions of Theorem 4.1 hold, let us examine the problem of solving Eq. (2) using piecewise constant approximations as in Example 2.4.

Assume now that $k(x, y)$ and $f(x)$ are in $C^1(D)$ and that the boundary ∂D of D is also C^1. Then the solution $u \in C^1(D)$ and a standard argument using Taylor's theorem show that $\|u - u_n\|_\infty \le ch$ (Refs. 24-25). Since the basis element ϕ_k is the characteristic function of D_k, $\Phi = \text{diag}[\text{meas}(D_k)]$, so that $\|\Phi^{-1}\|_\infty = 1/\min_k \text{meas}(D_k) \le ch^{-m}$ as follows from assumption (viii) of Theorem 4.1. Thus $\gamma = m$ in (iv). In addition, if we use the rules

$$\int_{D_k} f\, dx \simeq f(x_k)\, \text{meas}(D_k), \qquad x_k \in D_k \qquad (130)$$

and

$$\int_{D_j \times D_k} f(x, y)\, dx\, dy \simeq f(x_j, x_k)\, \text{meas}(D_j)\, \text{meas}(D_k), \qquad (131)$$

$E(f\phi_k)$ and $E_{kj}(k\phi_k\phi_j)$ are the quadrature errors in numerically integrating $f(x)$ over D_k and $k(x, y)$ over $D_j \times D_k$. Then using Taylor's theorem again gives (Ref. 25) $|E(f\phi_k)| \le ch^{m+1}$, and $|E_{kj}(k\phi_k\phi_j)| \le ch^{2m+1}$, and $|E_k(k\phi_k u_n)| = |\sum_{j=1}^N a_j E_{kj}(k\phi_k\phi_j)|$ where $a_j = \langle \phi_j, u_n\rangle/\text{meas}(D_j)$. Hence

$$\left|\sum_{j=1}^N a_j E_{kj}(k\phi_k\phi_j)\right| \le \max_{k,j}|E_{kj}(k\phi_k\phi_j)| \sum_{j=1}^N |a_j|$$

$$\le c\|u_n\|_\infty N h^{2m+1} \le ch^{-m} \cdot h^{2m+1} = ch^{m+1},$$

because $\{\|u_n\|_\infty\}$ are uniformly bounded.

As $\{\phi_k\}_{k=1}^N$ form a partition of unity, $\|\sum_{k=1}^N |\phi_k|\|_\infty = 1$, and the conditions of Theorem 4.1 are satisfied if $\gamma = m$, $\delta = \xi = 1$, $\mu = \frac{1}{2}m + 1$, and the partition $\{D_j\}$ is chosen so that (viii) holds. Since m must be less than μ, this gives $m = 1$, which agrees with a particular case of Chandler's result (Refs. 4, 15), but apparently rules out any multidimensional cases.

However, if x_k is chosen to be the centroid of D_k, then the rules (130)-(131)—which we call the compound centroidal rules—have precision one, and the errors $|E(f\phi_k)|$, E_1, and E_2 are of orders $O(h^{m+2})$, $O(h^{2m+2})$, and $O(h^{m+2})$ if $f \in C^2(D)$, $k \in C^2(D \times D)$, and $u \in C^2(D)$. In this case $\gamma = m$, $\delta = \xi = 2$, and $\mu = \frac{1}{2}m + 2$, so that the condition $\mu > m$ gives $m \le 3$. Thus using a quadrature rule of precision one gives consistency if the space has dimension at most three. □

An interesting question is whether one needs to increase the precision of the quadrature rule Q_M as m increases. As has been shown in Ref. 4, this is not necessary, and can be proved using the nonperturbation analysis of the discrete Galerkin equation (105), to be discussed in Section 5. At present the reason for this discrepancy is not entirely clear, but probably arises from an overestimate in the bound on $\|R_n\|_n$ (Refs. 3, 8). In fact, in this situation, using the integration rules (130)-(131), the discrete Galerkin method is equivalent to the collocation method of Example 2.5. The analysis in Ref. 4 then shows $\|u - v_n\|_\infty \le ch$, so that v_n is consistent with u_n independent of the dimension m and the location of the collocation point x_k in D_k.

We now consider the consistency problem for the discrete Sloan iterate

$$v_n^s = f + K_n v_n, \qquad (132)$$

and in order to keep the discussion from getting overly complicated, we impose conditions on k and f which are somewhat stronger than those in Ref. 8. For practical problems, the differences are probably of little consequence. Again our main purpose is to illustrate how the perturbation formalism can be directly used to establish sufficient conditions for the superconvergence of v_n^s.

Assume that $f(x) \in C^r(D)$, $k(x, y) \in C^r(D \times D)$, $r \ge l \ge 1$, and the boundary of D is smooth enough so that $u \in C^r(D)$ and the solution $g_x(y)$ of the adjoint equation

$$g_x(y) = k(x, y) + \int_D k(s, y) g_x(s)\, ds$$

is in $C^r(D \times D)$.

Theorem 4.2. Assume that $f(x)$, $k(x, y)$, $u(x)$, and $g_x(y)$ satisfy the conditions given above. In addition, suppose conditions (i)-(iv), (vii)-(viii) of Theorem 4.1 hold, and $h \to 0$ as $n \to \infty$. Assume that
 (i) $\sup_{x \in D} |E(\langle k_x, v_n \rangle)| \le ch^\rho$, $\rho > 0$,
 (ii) $\|u_n - v_n\|_\infty \le ch^\lambda$, $\lambda > 0$,
 (iii) $\sup_{x \in D} |E(\langle P_n g_x, f \rangle)| \le ch^\eta$, $\eta > 0$,
 (iv) $\sup_{x \in D} |E(\langle P_n g_x, K v_n \rangle)| \le ch^\tau$, $\tau > 0$,

where (i), (iii), and (iv) are integration errors. Then,

$$\|u - v_n^s\|_\infty \leq ch^\sigma, \qquad h \to 0,$$

where $\sigma = \min(2l, \rho, \eta, \tau)$.

Proof. Using (78) with $H = I$, and $\delta_n = (K_n - K)v_n$,

$$\|u - v_n^s\| \leq \|u - u_n^s\|_\infty + \|K(u_n - v_n)\|_\infty + \|(K_n - K)v_n\|_\infty. \quad (133)$$

From Theorem 3.2 of Chapter 2 and our assumptions, $\|u - u_n^s\| \leq ch^{2r^*} = ch^{2l}$ and $\|(K - K_n)v_n\|_\infty \leq ch^\rho$, since

$$\|(K - K_n)v_n\|_\infty = \sup_{x \in D} |(K - K_n)v_n| = \sup_{x \in D} |E(\langle k_x, v_n \rangle)| \leq ch^\rho.$$

To bound $\|K(u_n - v_n)\|_\infty$ observe that (we use a duality argument analogous to that in Section 4 of Chapter 2)

$$Ku_n - Kv_n = \langle k_x, u_n - v_n \rangle$$

$$= \langle (I - K^*)g_x, (u_n - v_n) \rangle = \langle g_x, (I - K)(u_n - v_n) \rangle$$

$$= \langle P_n g_x, (I - K)(u_n - v_n) \rangle$$

$$+ \langle (I - P_n)g_x, (I - K)(u_n - v_n) \rangle. \quad (134)$$

Since $P_n^2 = P_n$ and $P_n^* = P_n$,

$$\langle P_n g_x, (I - K)(u_n - v_n) \rangle = \langle P_n g_x, P_n(I - K)(u_n - v_n) \rangle.$$

However, $P_n(I - K)u_n = u_n - P_n K u_n = P_n f$, and $P_n(I - K)v_n = v_n - K_n P_n v_n = R_n v_n + P_n f + r_n$, so that the first term in the last line of (134) becomes

$$\langle P_n g_x, -R_n v_n - r_n \rangle = -\langle P_n g_x, R_n v_n \rangle - \langle P_n g_x, r_n \rangle$$

$$= -\langle P_n g_x, (\hat{P}_n K_n - P_n K)v_n \rangle - \langle P_n g_x, \hat{P}_n f - P_n f \rangle$$

$$= \langle P_n g_x, (K - \hat{P}_n K_n)v_n \rangle + \langle P_n g_x, f - \hat{P}_n f \rangle$$

$$= E(\langle P_n g_x, Kv_n \rangle) + E(\langle P_n g_x, f \rangle).$$

Hence

$$|\langle P_n g_x, (I - K)(u_n - v_n)\rangle| \le c(h^\tau + h^\eta) \le ch^\sigma. \qquad (135)$$

Finally we estimate the second term in (134). Thus

$$\langle (I - P_n)g_x, (I - K)(u_n - v_n)\rangle$$

$$= \langle (I - P_n)g_x, (I - P_n K)(u_n - v_n)\rangle$$

$$+ \langle (I - P_n)g_x, (P_n K - K)(u_n - v_n)\rangle$$

$$= \langle (I - P_n)g_x, (P_n K - K)(u_n - v_n)\rangle,$$

because $(I - P_n K)(u_n - v_n) \in X_n$ is orthogonal to $(I - P_n)g_x \in X_n^\perp$. Hence

$$|\langle (I - P_n)g_x, (I - K)(u_n - v_n)\rangle| \le \|(I - P_n)g_x\|_1 \|(K - P_n K)(u_n - v_n)\|_\infty$$

$$\le \|(I - P_n)g_x\|_1 \|K - P_n K\| \|u_n - v_n\|_\infty$$

$$\le ch^l \cdot c_1 h^l \cdot c_2 h^\lambda \le ch^{2l+\lambda} \le ch^\sigma. \qquad (136)$$

Using (135)-(136) in (133) gives $\|u - v_n^s\|_\infty \le ch^\sigma$. □

Example 4.2. Here we show that the Sloan iterate v_n^s of the piecewise polynomial approximation v_n in Example 4.1 is consistent with the unperturbed Sloan iterate, provided the centroidal rule is used to evaluate the integrals, all the data f, k, u, and g_x are C^2, and $m \le 3$.

Since $l = 1$, $2l = 2$, it suffices to show that $(\rho, \eta, \tau) \ge 2$. Arguing as in Example 4.1 the integration errors in (i), (iii)-(iv) of Theorem 4.2 are $O(h^2)$. Thus $\rho = \eta = \tau = 2$ and $\sigma = \min(2, 2, 2, 2, 2 + \lambda)$ where $\lambda = 1$ because $\|u_n - v_n\|_\infty \le \|u - u_n\|_\infty + \|u - v_n\|_\infty \le ch$ from Example 4.1.

For $m \le 3$, this result agrees with that in Application 2 of Ref. 4. However, the restriction on the dimension m is unnecessary. In fact, in this case, the iterated discrete Galerkin approximation is equivalent to a Nyström approximation using the compound centroidal rule (Ref. 4). The error is then known to be of the order of the integration error in evaluating Ku (Ref. 4), and this is $O(h^2)$, independent of m, when the data are smooth.

5. Direct Analysis of Galerkin's Method with Quadrature Errors

Because the perturbation analysis of the discrete Galerkin method does not appear to yield optimal error estimates, Atkinson and Bogomolny in Ref. 4 considered a direct analysis of (105) which in certain cases appears to give sharper results than those currently obtainable from the methods of Sections 3 and 4. We now present a discussion of their approach.

Again we only consider the case where D is the closure of an open bounded subset of R^m, $m \geq 1$, assume that the data $f(x)$, $k(x, y)$ are at least continuous, and the boundary is smooth enough so that u is continuous as well. For simplicity we work in $C(D)$, the space of continuous functions on D, with the sup norm, the inner product (97), and consider Galerkin methods defined by a sequence of subspaces $\{X_n\}$ whose corresponding orthogonal projections are uniformly bounded in $X = C(D)$. Then, as before, it follows from Theorems 2.2 and 2.3 that the Galerkin approximations $\{u_n\}$ exist for $n \geq n_0$ and converge uniformly to u (in contrast to our previous analysis, this fact will *not* be needed in what follows).

Assume now that the inner products $\{\langle \phi_k, f \rangle\}$, $\{\langle K\phi_j, \phi_k \rangle\}$, and $\{\langle \phi_k, \phi_j \rangle\}$ in (96) are approximated by the quadrature rules Q_M and $Q_M \times Q_M$ in (99)-(100). Then the discrete Galerkin approximation v_n is given by

$$v_n = \sum_{j=1}^{N} b_j \phi_j, \tag{137}$$

where $\{b_j\}_{j=1}^{N}$ satisfy

$$\sum_{j=1}^{N} \left[\sum_{p=1}^{M(n)} w_p \phi_j(x_p) \phi_k(x_p) - \sum_{m=1}^{M(n)} \sum_{p=1}^{M(n)} w_m w_p k(x_p, y_m) \phi_j(x_p) \phi_k(y_m) \right] b_j$$

$$= \sum_{p=1}^{M} w_p f(x_p) \phi_k(x_p), \quad k = 1, 2, \ldots, N. \tag{138}$$

To develop the convergence analysis of Atkinson and Bogomolny some additional conditions need to be imposed on the quadrature rules $\{Q_M\}$. They are:

(i) $Q_M(f) \to \int_D f(x) \, dx$, $n \to \infty$, $\forall f \in C(D)$;
(ii) the number of quadrature points $M(n) \geq N = \dim(X_n)$;
(iii) $w_p > 0$, $p = 1, 2, \ldots, M$;
(iv) there exists a subset S_N of N points of the quadrature nodes $\{x_p\}_{p=1}^{M(n)}$ such that $\{\phi_k\}_{k=1}^{N}$ is unisolvent with respect to S_N; i.e., the $N \times N$ matrix $\psi = [\phi_k(x_p)]$, $x_p \in S_N$, $k = 1, 2, \ldots, N$, is invertible.

From (ii)–(iv) it follows that

$$\langle f, g \rangle_n = \sum_{p=1}^{M(n)} w_p f(x_p) g(x_p) \tag{139}$$

defines an inner product on X_n, so that

$$\hat{\Phi} = [\langle \phi_j, \phi_k \rangle_n]_{(j,k)=1}^{N}$$

is symmetric and positive definite, and so is invertible (Ref. 4) ($\hat{\Phi}$ may be regarded as an approximation of Φ—see Ref. 4 for further comments on this point). Letting

$$\pi_n u = \sum_{j=1}^{N} \sum_{k=1}^{N} \hat{\Phi}_{jk}^{-1} \langle u, \phi_k \rangle_n \phi_j, \qquad u \in C(D), \tag{140}$$

it can be shown that π_n defines a projection from $C(D)$ to X_n, and the discrete Galerkin equations can be written as

$$v_n = \pi_n K_n v_n + \pi_n f, \tag{141}$$

where K_n is given by (103) (Ref. 4). Moreover, if

$$v_n^s = f + K_n v_n \tag{142}$$

is the discrete Sloan iterate, then

$$\pi_n v_n^s = \pi_n f + \pi_n K_n v_n = v_n, \tag{143}$$

so that v_n^s satisfies

$$v_n^s = f + K_n \pi_n v_n^s. \tag{144}$$

We will base our analysis on v_n^s, since the continuity of f and k implies that v_n^s is continuous, so that one can discuss piecewise polynomial (possibly discontinuous) approximations without having to consider more complicated spaces than $C(D)$ (Ref. 4).

If $\{\pi_n\}$ are uniformly bounded in $C(D)$, then it is shown in Ref. 4 that $\{K_n \pi_n\}$ is a pointwise convergent collectively compact set of operators on $C(D)$, so that for $n \geq n_0$ $(I - K_n \pi_n)^{-1}$ exists, and $\{(I - K_n \pi_n)^{-1}\}$, $n \geq n_0$, are uniformly bounded. Consequently, for $n \geq n_0$,

$$v_n^s = (I - K_n \pi_n)^{-1} f, \tag{145}$$

so that

$$u - v_n^s = (I - K)^{-1} f - (I - K_n \pi_n)^{-1} f$$
$$= (I - K_n \pi_n)^{-1}(Ku - K_n \pi_n u) \tag{146}$$

and
$$\|u - v_n^s\|_\infty \le c\|Ku - K_n\pi_n u\|_\infty \to 0,$$
since $K_n\pi_n$ converges pointwise to K. Thus v_n^s converges uniformly to u. From (143)
$$u - v_n = u - \pi_n v_n^s = u - \pi_n u + \pi_n(u - v_n^s). \tag{147}$$

By assumption $\bigcup_{n\ge 1} X_n$ is dense in X, so the uniform boundedness of $\{\pi_n\}$ implies that $\pi_n u \to u$, $\forall u \in C(D)$. Hence
$$\|u - v_n\|_\infty \le \|u - \pi_n u\|_\infty + c\|u - v_n^s\|_\infty, \tag{148}$$

and $v_n \to u$.

From (146) and (147) we see that establishing convergence rates for both v_n and v_n^s depends on bounding $\|\pi_n\|$ and estimating $\|Ku - K_n\pi_n u\|_\infty$. As for the unperturbed Galerkin method and the analysis in Theorem 4.2 this depends on duality-type arguments (Ref. 4). To see this, write

$$Ku - K_n\pi_n u = Ku - K_n u + K_n(I - \pi_n)u$$
$$= Ku - K_n u + \langle k_x, (I - \pi_n)u\rangle_n, \tag{149}$$

where the last step in (149) follows from the definitions of K_n and the inner product $\langle \cdot, \cdot \rangle_n$.

As shown in Ref. 4, π_n is self-adjoint with respect to $\langle \cdot, \cdot \rangle_n$ so that $\langle k_x, (I - \pi_n)u\rangle_n = \langle (I - \pi_n)k_x, (I - \pi_n)u\rangle_n$ and it is this last identity which gives the superconvergence properties of v_n^s. We now use (149) to derive a number of the convergence results in Ref. 4, some of which improve on the results of Chandler (Refs. 3, 8) and those for the piecewise constant approximations given in Examples 4.1 and 4.2. Complete details and further interesting consequences of (147) and (149) can be found in Ref. 4.

Example 5.1. With the same setup as in Examples 4.1 and 4.2 we show that using the centroidal integration rule leads to consistency of v_n with u_n and v_n^s with u_n^s. We begin by estimating $\|\pi_n\|$ (we could—and will—use the general bound in Ref. 4. However, since the calculation there is somewhat complicated, it is useful to do at least one ad-hoc computation).

From Eq. (4.20) of Ref. 4,
$$\|\pi_n\| \le \|\hat{\Phi}^{-1}\|_\infty \left[\sup_{x\in D} \sum_{k=1}^N |\phi_k(x)|\right]\left[\max_{1\le k \le N} \sum_{p=1}^M w_p|\phi_k(t_p)|\right] \tag{150}$$

[this follows in a straightforward fashion from (140) (Ref. 4)]. Because $\{\phi_k\}_{k=1}^N$ are characteristic functions, and Q_M has precision 1, $\sum_{k=1}^N |\phi_k(x)| = 1$, $\sum_{p=1}^M w_p |\phi_k(t_p)| = \int_D \phi_k(x)\,dx = \text{meas}(D_k)$, and $\|\pi_n\| \leq \|\hat{\Phi}^{-1}\|_\infty \max_k \text{meas}(D_k)$. Again using the fact that Q_M has precision one, the elements $\langle \phi_j, \phi_k \rangle_n$ of $\hat{\Phi}$ are the integrals $\int_D \phi_j(x)\phi_k(x)\,dx = \int_{D_j} \phi_k(x)\,dx = \delta_{jk} \text{meas}(D_k)$, so that $\|\hat{\Phi}^{-1}\|_\infty = 1/\min_k \text{meas}(D_k)$. Thus $\|\pi_n\| \leq \max_k \text{meas}(D_k)/\min_k \text{meas}(D_k) \leq c$ and $\{\pi_n\}$ are uniformly bounded. Thus all the conditions for (149) to hold are met and [see Eqs. (5.15)-(5.16) of Ref. 4]

$$\|u - v_n^s\|_\infty \leq \|Ku - K_n u\|_\infty + c(\sup_{x \in D} \|(I - \pi_n)k_x\|_\infty \|u - \pi_n u\|_\infty). \quad (151)$$

Assume now that $k \in C^2(D \times D)$ and $u \in C^2(D)$, then $\|(K - K_n)u\|_\infty \leq ch^2$, because $Ku - K_n u$ is the numerical integration error in evaluating Ku by $K_n u$. Also

$$\|u - \pi_n u\|_\infty \leq c \inf_{s_n \in X_n} \|u - s_n\|_\infty \leq ch$$

and $\|(I - \pi_n)k_x\|_\infty \leq ch$ as well. Thus $\|u - v_n^s\|_\infty \leq ch^2$, which shows that v_n^s is consistent with u_n^s (Ref. 4).

From Ref. 25 $\|u - u_n\|_\infty \leq ch$, showing that v_n is consistent with u_n, with the same result holding for an arbitrary integration rule (130). These results improve those in Section 4, since they are true for all $m \geq 1$.

Example 5.2. In Ref. 15, Chandler considered the problem of solving (13) by Galerkin's method using polynomial splines of degree $l - 1$, $l \geq 1$, and continuity $-1 \leq r \leq l - 2$ ($r = -1$ corresponds to discontinuous polynomials). A rule of precision d was chosen on each interval $[t_k, t_{k+1})$, $k = 0, \ldots, n - 1$ (the notation is that of Example 2.2), and the corresponding compound rule Q_M was used to evaluate $\langle f, \phi_k \rangle$. The product rule $Q_M \times Q_M$ was used to evaluate $\langle K\phi_j, \phi_k \rangle$ and d was chosen to be at least $2l - 2$, so that $\hat{\Phi} = \Phi$ in this case. Under the conditions that $k \in C^{2l}(D \times D)$ and $f \in C^{2l}(D)$, it was shown in Ref. 15 that $\|u - v_n\|_\infty \leq ch^l$ and $\|u - v_n^s\|_\infty \leq ch^{2l}$ if in addition $d = 2l - 1$ (Gaussian integration).

In Ref. 4 it was shown that if $r = -1, 0$ and discontinuous polynomials were used as a basis for $r = -1$ and B-splines as a basis for $r = 0$, then $\|\pi_n\| \leq c$. In this case, using an integration rule of precision at least $l - 1$ satisfying conditions (i)-(iv) gives $\|u - v_n\|_\infty \leq ch^l$, which improves on the results of Chandler (Ref. 15) and Joe (Ref. 8). Again one needs $d = 2l - 1$, Gaussian integration, to obtain the full order of superconvergence $O(h^{2l})$ for v_n^s, although a rule having $d > l - 1$ will still yield faster convergence than for u_n or v_n.

Although these results are interesting theoretically, for smooth data it was shown in Ref. 4 [and follows from (144)—because $K_n \pi_n = K_n$] that using v_n^s with Gaussian integration is equivalent to the Nyström method with the compound Gauss rule. It then follows from (144) (or standard error estimates) that $\|u - v_n^s\|_\infty \leq ch^{2l}$.

6. Collocation Methods with Quadrature Errors

In this section we consider the analysis of collocation methods for solving Eq. (2) where the collocation matrix is evaluated by some type of quadrature method. Here, as for Galerkin's method, we will be primarily concerned with giving sufficient conditions for the discrete collocation method to be consistent with the exact one. We will consider the use of both the perturbed projection method approach, and the analogue of the Atkinson–Bogomolny direct analysis. Again, we shall see that the Atkinson–Bogomolny approach appears to give sharper results than perturbation arguments. In addition, we will relate some of our results to an apparently different technique which has been used by Brunner (Refs. 5-6, 36) to obtain many interesting superconvergence results for collocation methods for solving Volterra equations.

Although Sloan in Chapter 2 indicates that these results have a somewhat different flavor than those in the Fredholm case, they can be derived by utilizing the standard error estimates for (u_n, u_n^s, v_n, v_n^s) in a slightly different fashion than is customary in the Fredholm case. This leads, for instance, to a somewhat different proof of the superconvergence of u_n^s than that given in Theorem 5.2 of Chapter 2.

6.1. Collocation Method. To solve Eq. (2) by the collocation method in a Banach space X, choose a sequence of subspaces $\{X_n\}$, $n \geq 1$, where $\bigcup_{n \geq 1} X_n$ is dense in X, and let $\{\phi_j\}_{j=1}^N$ be a basis for X_n. If $x_k \in D$, $k = 1, 2, \ldots, N$, are distinct points in D, assume that $\psi = [\phi_j(x_k)]_{(j,k)=1}^N$ is invertible for all N. Letting

$$u_n = \sum_{j=1}^N a_j \phi_j(x), \tag{152}$$

denote the approximation to u, the coefficients $\{a_j\}_{j=1}^N$ are determined by solving

$$\sum_{j=1}^N a_j \phi_j(x_k) = \sum_{j=1}^N a_j K \phi_j(x_k) + f(x_k), \qquad k = 1, 2, \ldots, N \tag{153}$$

[of course we must choose X so that the values $\{\phi_j(x_k)\}$, $\{f(x_k)\}$ are well defined (Ref. 4)].

If $P_n: X \to X_n$ is defined by

$$P_n u = \sum_{j=1}^{N} \sum_{k=1}^{N} \psi_{jk}^{-1} u(x_k) \phi_j(x), \tag{154}$$

where ψ_{jk}^{-1} is the jkth element of ψ^{-1}, then P_n is a projection (Ref. 2). Keeping with our assumptions in Section 2, we assume that $\{P_n\}$ are uniformly bounded (although much of our analysis can go through without this assumption), and then u_n is given by the projection method

$$u_n = P_n K u_n + P_n f.$$

If K is compact, and $P_n g \to g$, $\forall g \in X$, it follows from Theorem 2.2 that $u_n \to u$, $\forall f \in X$.

As for Galerkin's method, the integrals $\{K\phi_j(x_k)\}$ usually have to be obtained numerically, particularly in the important case where $k(x, y)$ is not continuous (Ref. 1). To allow for this possibility, we assume that $K\phi_j(x)$ is given by a quadrature rule of the form

$$K\phi_j(x) \simeq \sum_{p=1}^{M(n)} w_p(x) \phi_j(y_p) = Q_M(\phi_j), \tag{155}$$

$y_p \in D$, $p = 1, 2, \ldots, M(n)$ and $\{w_p(x)\}_{p=1}^{M(n)}$ do not depend on $\{\phi_j\}$, but generally depend on K. This allows us to consider not only rules of the form (155) for continuous integrands, but product quadrature methods for nonsmooth kernels as well.

For the remainder of this section we shall assume that $X = C(D)$ or $X = R(D)$ with the sup norm, and that $\{w_p(x)\}_{p=1}^{M(n)}$ are continuous functions of x. Moreover, in this situation we assume that the rule (155) is defined for all $u \in X$ and that $Q_M(u) \to Ku$, $n \to \infty$, $\forall u \in X$.

If $K\phi_j$, $j = 1, 2, \ldots, N$, is replaced by (155) in (153), then u is approximated by

$$v_n = \sum_{j=1}^{N} b_j \phi_j, \tag{156}$$

where $\{b_j\}_{j=1}^{N}$ satisfy

$$\sum_{j=1}^{N} b_j \phi_j(x_k) = \sum_{j=1}^{N} \left[\sum_{p=1}^{M(n)} w_p(x_k) \phi_j(y_p) \right] b_j + f(x_k), \quad k = 1, 2, \ldots, N. \tag{157}$$

It is customary to refer to (156)-(157) as a discrete collocation method. Defining $K_n: X \to X$ by

$$K_n u(x) = \sum_{p=1}^{M} w_p(x) u(y_p), \qquad (158)$$

a straightforward calculation shows that v_n satisfies

$$v_n = P_n K_n v_n + P_n f. \qquad (159)$$

Letting $R_n = P_n K_n - P_n K$,

$$v_n = P_n K v_n + R_n v_n + P_n f, \qquad (160)$$

so that v_n defines a perturbed projection method with $r_n = 0$. From Theorem 3.1, v_n will be consistent with u_n if $\|R_n\|_n \to 0$ and $\|R_n u_n\|_\infty = O(\|u - u_n\|_\infty)$. For this the analysis follows along the lines of Section 4.

First observe that if $w_n \in X_n$ is arbitrary, then

$$\|R_n w_n\|_\infty = \|P_n(K - K_n) w_n\|_\infty \leq \|P_n\| \|(K - K_n) w_n\|_\infty, \qquad (161)$$

since $\{P_n\}$ are uniformly bounded. However, if $w_n = \sum_{j=1}^{N} c_j \phi_j$,

$$(K - K_n) w_n = \sum_{j=1}^{N} c_j \left[\int_D k(x, y) \phi_j(y) \, dy - \sum_{p=1}^{M} w_p(x) \phi_j(y_p) \right]$$

$$= \sum_{j=1}^{N} c_j E(k_x \phi_j) \equiv \sum_{j=1}^{N} c_j e_j(x),$$

where $E(k_x \phi_j)$ is the error in evaluating $k_x \phi_j$ by Q_M. Thus

$$|(K - K_n) w_n| \leq N \max_j |c_j| \max_j |e_j(x)|, \qquad (162)$$

so that

$$\|(K - K_n) w_n\|_\infty \leq \max_j |c_j| NE, \qquad (163)$$

where

$$E = \max_{x \in D} \max_j |e_j(x)| < \infty. \qquad (164)$$

Perturbed Projection Methods

Using Eq. (3.2) of Ref. 8 again,

$$\max_j |c_j| \leq c \|\Phi^{-1}\|_\infty^{1/2} \|w_n\|_\infty, \tag{165}$$

which finally gives

$$\|(K - K_n)w_n\|_\infty \leq cNE \|\Phi^{-1}\|_\infty^{1/2} \|w_n\|_\infty, \tag{166}$$

where again Φ is the Grammian of $\{\phi_j\}_{j=1}^N$. Hence

$$\|R_n\|_n \leq cNE \|\Phi^{-1}\|_\infty^{1/2}, \tag{167}$$

so that v_n converges to u provided that $cNE \|\Phi^{-1}\|_\infty^{1/2} \to 0$. If this is the case, we have the error estimate

$$\|u - v_n\|_\infty \leq \|u - u_n\|_\infty + C \|R_n u_n\|_\infty. \tag{168}$$

Example 6.1. We consider the piecewise polynomial collocation method discussed in Example 2.5 and assume that the conditions on the boundary of D are the same as in Example 4.1 and that the kernel is C^2. Then, as pointed out in Refs. 24–25, u_n converges uniformly to u and v_n will converge if $\|R_n\|_n \to 0$.

From Example 4.1 $\|\Phi^{-1}\|_\infty^{1/2} \leq ch^{-m/2}$, and assuming that the partition of D is such that $N \leq ch^{-m}$, $\|R_n\|_n \leq ch^{-3m/2} E$. If the integrals $K\phi_j$ are evaluated by the centroidal rule (recalling that ϕ_j are characteristic functions), $E(k_x \phi_j)$ is the error in integrating k_x over D_j. From Example 4.1 this gives $E \leq ch^{m+2}$, because Q_M has precision one. Thus $\|R_n\|_n \leq ch^{-3m/2} \cdot h^{m+2} = ch^{2-m/2}$, so that v_n converges if $m \leq 3$. If $u \in C^1(D)$, then $\|u - u_n\|_\infty \leq ch$ (Ref. 24), and $\|R_n u_n\|_\infty \leq ch^{2-m/2}$, so that v_n is consistent with u_n if $m \leq 2$.

It remains to be seen if the restriction on the dimension m can be removed, and whether an integration rule of precision zero is sufficient to yield consistency. For this we turn to an analysis of (159) similar to that of Section 5.

6.2. Direct Analysis of the Discrete Collocation Method. From our assumptions on Q_M, it follows that the operators $\{K_n\}$ are compact (they have finite rank) and $K_n u \to Ku$, $\forall u \in X$. Assuming that $\{K_n\}$ is a collectively compact sequence of operators (Ref. 30), for all n sufficiently large, $n \geq n_0$,

$(I - P_nK_n)^{-1}$ exists and $\{\|(I - P_nK_n)^{-1}\|\}$ are uniformly bounded. If $v_n^s = K_nu_n + f$ is the Sloan iterate of v_n, then arguing as for the discrete Galerkin method, $P_nv_n^s = v_n$, so that v_n^s satisfies

$$v_n^s = K_nP_nv_n^s + f, \tag{169}$$

and

$$u - v_n^s = (I - K_nP_n)^{-1}(Ku - K_nP_nu). \tag{170}$$

Thus $v_n^s \to u$, because $K_nP_nu \to Ku$. Hence, as for (147),

$$u - v_n = (u - P_nu) + P_n(u - v_n^s), \tag{171}$$

and

$$\|u - v_n\|_\infty \le \|u - P_nu\|_\infty + c\|u - v_n^s\|_\infty, \tag{172}$$

so that v_n converges uniformly to u. To get convergence rates, we must analyze $\|u - v_n^s\|_\infty$ and this requires examination of $\|Ku - K_nP_nu\|_\infty$.

Now

$$Ku - K_nP_nu = Ku - K_nu + K_n(I - P_n)u, \tag{173}$$

so that

$$\|Ku - K_nP_nu\|_\infty \le \|Ku - K_nu\|_\infty + c\|(I - P_n)u\|_\infty, \tag{174}$$

since by the principle of uniform boundedness $\{K_n\}$ are uniformly bounded. Thus

$$\|u - v_n\|_\infty \le c\|u - P_nu\|_\infty + \|Ku - K_nu\|_\infty, \tag{175}$$

and it follows from (175) that $\{v_n\}$ will be consistent with $\{u_n\}$ if $\|Ku - K_nu\|_\infty = O(\|u - P_nu\|_\infty)$. This will certainly be the case for the collocation method in Example 6.1 if $u \in C^1(D)$, $k(x, y) \in C^1(D \times D)$, and the integration rule (130) is used to evaluate $\{K\phi_j\}_{j=1}^N$. This follows because $\|u - P_nu\|_\infty \le ch$, and $\|Ku - K_nu\|_\infty \le ch$, which can be obtained by estimating the integration error.

One can now use (173) to determine the possible superconvergence of v_n^s. However, the duality argument leading to (149) cannot be used here because $\{P_n\}$ are not self-adjoint.

Rather than consider a general analysis, let us reconsider the piecewise polynomial collocation method for solving the one-dimensional equation $u(x) = \int_a^b k(x, t)u(t) \, dt + f(x)$, considered by Sloan in Chapter 2.

Here we assume that $k \in C^{2l}$, and $f \in C^{2l}$, and assume that piecewise polynomials of degree $\leq l - 1$, $l \geq 1$, are used to approximate u. If the collocation points $x_{j,k} \in (t_k, t_{k+1})$, $j = 1, 2, \ldots, l$, $k = 0, \ldots, n - 1$, are the Gauss points, and Q_M is the compound Gauss rule $[M = nl = \dim(X_n)]$ relative to the partition $a = t_0 < t_1 < \cdots < t_n = b$ of $[a, b]$, then

$$K_n u = \sum_{k=0}^{n-1} \sum_{j=1}^{l} w_{jk} k(x, x_{j,k}) u(x_{j,k}), \qquad (176)$$

and

$$K_n(I - P_n)u = K_n u - \sum_{k=0}^{n-1} \sum_{j=1}^{l} w_{jk} k(x, x_{j,k}) P_n u(x_{j,k}). \qquad (177)$$

But $P_n u(x_{j,k}) = u(x_{j,k})$, because P_n is an interpolatory projection, so that $K_n P_n = K_n u$ and $K_n(I - P_n)u = 0$. Thus (170) and (177) give $u - v_n^s = (I - K_n P_n)^{-1}(Ku - K_n u_n)$, so that $\|u - v_n^s\|_\infty \leq c\|Ku - K_n u\|_\infty \leq ch^{2l}$, because $u \in C^{2l}$ and Q_M has precision $2l - 1$ in this case (Ref. 23). From Theorem 5.2 of Chapter 2 $\|u - u_n^s\|_\infty \leq ch^{2l}$, so that v_n^s is consistent with u_n^s.

This result has been obtained by a number of authors using other arguments (Refs. 6–7, 17), and is not surprising since the discrete iterated collocation method is equivalent to the Nyström method in this case. In fact it is easily shown that for any discrete collocation method (156)–(157) where the collocation points $\{x_k\}_{k=1}^N$ and the integration points $\{y_p\}_{p=1}^N$ are the same, the iterated discrete collocation method is a product quadrature method based on Q_M (Ref. 2). In fact, as in (144), $K_n P_n v_n^s = K_n v_n^s$, so that v_n^s satisfies $v_n^s = K_n v_n^s + f$, the Nyström equations.

Using this result one finds that v_n^s, the Sloan iterate of v_n in Example 6.1, satisfies $\|u - v_n^s\|_\infty \leq ch^2$ if $k(x, y) \in C^2$ and the centroidal rule (130) is used to evaluate $\{K\phi_j\}$. This shows that v_n^s is consistent with u_n^s provided that $u \in C^2(D)$ (Ref. 25).

Recalling the argument in Chapter 2, it appears that the superconvergence of v_n^s is "easier" to prove than u_n^s. However, when the data are smooth, the superconvergence of u_n^s can also be obtained by estimating quadrature errors. This result was obtained by Brunner in Ref. 6 where an "ad-hoc" argument apparently was also used. However, re-examination of the basic error estimate (41) for $u - u_n^s$ shows that his results can be obtained from the general functional analytic framework given here and in Chapter 2.

Using $u_n^s = (I - KP_n)^{-1}f$ and $u = (I - K)^{-1}f$, we found in Section 3 that $u - u_n^s = (I - KP_n)^{-1}(Ku - KP_n u)$. For Galerkin's method, as shown in detail in Chapter 2, this allows one to exploit the self-adjointness of $\{P_n\}$ to obtain superconvergence results. Since collocation projections are not self-adjoint relative to standard inner products, manipulating $Ku - KP_n u$ to get correct error bounds becomes tricky. However, by observing that

$$u - u_n^s = (I - K)^{-1}(Ku_n^s - KP_n u_n^s)$$
$$= (I - K)^{-1}K(u_n^s - P_n u_n^s)$$
$$= (I - K)^{-1}K\alpha_n,$$

we get

$$u - u_n^s = \int_a^b R(x,t)\alpha_n(t)\,dt, \qquad (178)$$

where $R(x,t)$ is the resolvent kernel of $k(x,t)$ and $\alpha_n(t) = (u_n^s - P_n u_n^s)(t)$. Since $P_n u_n^s = u_n$, $u_n^s - P_n u_n^s = u_n^s - u_n = f + Ku_n - u_n = \delta_n$, which Brunner refers to as the defect (Ref. 6) (it is the negative of the usual residual). Thus (178) is the formula given in Ref. 6 and it may be exploited to give superconvergence results for iterated collocation. Before doing this, we note that the forgoing result is quite general, so that if $k(x,t)$ is a Volterra kernel (and so discontinuous on $[a,b] \times [a,b]$), then

$$u - u_n^s = \int_a^x R(x,t)\delta_n(t)\,dt, \qquad (179)$$

which forms the basis for Brunner's superconvergence results for Volterra equations obtained in Ref. 6. We shall return to this shortly.

To see how (179) can be used to get superconvergence results for piecewise polynomial collocation, assume as before that $f \in C^{2l}$ and $k \in C^{2l}$. Then $\delta_n(t)$ is C^{2l} on each of the open intervals (t_k, t_{k+1}), $k = 0, \ldots, n-1$, and evaluating (178) by compound Gaussian quadrature gives

$$\int_a^b R(x,t)\delta_n(t)\,dt = \sum_{k=0}^{n-1} \sum_{j=1}^{l} w_{jk} R(x, x_{k,j})\delta_n(x_{k,j}) + \sum_{k=0}^{n-1} E_k(x), \qquad (180)$$

where $\{E_k(x)\}$ are the integration error terms from each subinterval (t_k, t_{k+1}). Since the residual, by definition, is zero at the collocation points, the first term in (180) vanishes, and $\int_a^b R(x,t)\delta_n(t)\,dt = \sum_{k=0}^{n-1} E_k(x)$. Since Gaussian quadrature has precision $2l - 1$, $|E_k| \le c_k h^{2l+1}$, $k = 0, 1, 2, \ldots, n$ (Ref. 6), where the constants $\{c_k\}$ are bounded in terms of the $2l$th derivatives of $R(x,t)\delta_n(t)|_{t \in (t_k, t_{k+1})}$. Since $\delta_n = f + Ku_n - u_n$, it suffices to show that the first $l - 1$ derivatives of $u_n|_{(t_k, t_{k+1})}$, $k = 0, \ldots, n-1$ are uniformly bounded. This may be done as in Ref. 36 so that $\{c_k\}$ do not depend on n, and $\|u - u_n^s\|_\infty \le ch^{2l}$, showing again that iterated collocation gives the full order of superconvergence if the Gauss points are used as collocation points.

If other collocation points are used, then (180) shows that if k and f are smooth, the convergence rate of u_n^s depends on the maximal precision of the compound integration rule that one can obtain using the collocation points as nodes. For instance, if one uses the Lobatto points rather than the Gauss points, then $|E_k(x)| \le ch^{2l-2}$, and superconvergence occurs if $l \ge 3$. If this is the case, since the nodes $\{t_k\}$ are collocation points, u_n is continuous as well as u_n^s (Refs. 5-7).

Since $P_n u_n^s = u_n$, these results show that we get local superconvergence of u_n at the collocation points if they are the Gauss points, and local superconvergence at the mesh points if the Lobatto points are used (Ref. 6). From our previous results, this property also holds for v_n, provided that k and f are smooth and the corresponding Gauss or Lobatto rules are used in (176). This agrees with Brunner's results in Ref. 6.

When k is not smooth, then one generally cannot expect to have global superconvergence of u_n^s or v_n^s. However, in certain circumstances local superconvergence can be obtained. This has been shown by Brunner in Ref. 6 for Volterra kernels, and by Chatelin and Lebbar for Green's function kernels in Ref. 17. As a final application of Brunner's formula we rederive some of the results in Ref. 6 for Volterra equations. Many other interesting results along these lines can be found in Ref. 5 and the references quoted there.

6.3. Superconvergence of Collocation for Volterra Equations.

Since a Volterra equation may be regarded as a particular case of a Fredholm equation, one can consider solving

$$u(x) = \int_a^x k(x,t)u(t)\,dt = f(x), \qquad a \le x \le b, \tag{181}$$

by the piecewise collocation method of the previous subsection. If $k(x,t) \in C^{2l}(S)$, $S = \{a \le t \le x \le b\}$ and $f(x) \in C^{2l}[a,b]$, it follows from Theorems 2.2 and 2.3 that u_n exists for all n sufficiently large and $\|u - u_n\|_\infty \le ch^l$ (see Ref. 36 for another proof of this result). Then, for $n \ge n_0$, $u_n^s = f + Ku_n$ is well defined. In this case the analysis of Chapter 2 will not hold, and as shown in Ref. 6 u_n^s does not generally exhibit global superconvergence. However, if the collocation points are the Gauss points, then u_n^s is locally superconvergent on $\{t_k\}_{k=1}^n$, the mesh points, i.e., $|u(t_k) - u_n^s(t_k)| \le ch^{2l}$, $k = 1, 2, \ldots, n$.

To prove this use (179) to give

$$u(t_k) - u_n^s(t_k) = \int_a^{t_k} R(t_k, t)\delta_n(t)\,dt, \tag{182}$$

and evaluating the integral in (182) by compound Gauss quadrature as in (180), and using the fact that $\delta_n(t)$ vanishes at the collocation points, shows that $|u(t_k) - u_n^s(t_k)| \leq ch^{2l}$, $k = 1, 2, \ldots, n$.

One can also prove local superconvergence results for u_n using an argument similar to that in (180). In fact, as shown in Ref. 6,

$$u(x) - u_n(x) = \delta_n(x) + \int_a^x R(x, t)\delta_n(t)\, dt, \tag{183}$$

so that

$$u(t_k) - u_n(t_k) = \delta_n(t_k) + \int_a^{t_k} R(t_k, t)\delta_n(t)\, dt. \tag{184}$$

If the collocation points are the Lobatto points, for instance, then $\{t_k\}_{k=1}^n$ are collocation points, and $\delta_n(t_k) = 0$, $k = 1, 2, \ldots, n$. Thus

$$u(t_k) - u_n(t_k) = \int_a^{t_k} R(t_k, t)\delta_n(t)\, dt, \tag{185}$$

and discretizing the integral in (185) by compound Lobatto quadrature shows that $|u(t_k) - u_n(t_k)| \leq ch^{2l-2}$ (Ref. 6), since Lobatto quadrature has precision $2l - 3$ if l quadrature points are used. Hence, local superconvergence occurs if $l \geq 3$. If the Radau points are used as collocation points, then the same argument shows that $|u(t_k) - u_n(t_k)| \leq ch^{2l-1}$. However, full local superconvergence, $|u(t_k) - u_n(t_k)| = O(h^{2l})$, cannot be obtained [this would require the Gauss points as collocation points, but these are not mesh points, so that $\delta_n(t_k) \neq 0$ in (184)].

As in the Fredholm case, the collocation integrals $\{K\phi_j\}$ usually have to be obtained by numerical integration (Ref. 6). Since $k(x, t)$ is generally not even continuous on $[a, b]$, the recursive structure of the collocation method needs to be exploited in order to obtain appropriate quadrature rules (Ref. 6). If this is done as in Ref. 6 and $k(x, t) \in C^2 \in (S)$, then the full rates of global convergence and local superconvergence can be retained for v_n and v_n^s (in the following τ_k, $k = 1, 2, \ldots, n$, denote the collocation points). These facts are proved in Ref. 6 by a perturbation analysis, which can be obtained from our general analysis in Section 3. Here, the interpolatory nature of $\{P_n\}$ is crucial, since this permits a local, rather than a global error analysis, as is required for Galerkin's method. For brevity we consider only an outline of the analysis for v_n and v_n^s. For complete details we refer the reader to Ref. 6.

Since v_n is a piecewise polynomial, it suffices to estimate the errors $u_n(\tau_k) - v_n(\tau_k)$, as the error $\|u_n - v_n\|_\infty$ can then be obtained by interpolation (Ref. 6).

Now, if K_n represents the approximation to K when the integrals in (159) are evaluated as in Ref. 6, then v_n, the discrete collocation approximation, satisfies

$$v_n = P_n K_n v_n + P_n f,$$

so that

$$v_n(\tau_k) = (K_n v_n)(\tau_k) + f(\tau_k), \qquad k = 1, 2, \ldots, nl,$$

because P_n is an interpolatory projection. Since $u_n = P_n K u_n + P_n f$, $u_n(\tau_k) = K u_n(\tau_k) + f(\tau_k)$, and discretizing $K u_n(\tau_k)$ as in Ref. 6, one finds that $K u_n(\tau_k) = K_n u_n(\tau_k) + E_k$, $k = 1, 2, \ldots, nl$, where E_k is the numerical integration error. Thus

$$u_n(\tau_k) - v_n(\tau_k) = K_n(u_n - v_n)(\tau_k) + E_k, \qquad k = 1, 2, \ldots, nl, \qquad (186)$$

and letting $\eta_k = u_n(\tau_k) - v_n(\tau_k)$, $k = 1, 2, \ldots, nl$, and using the formula for K_n given in Ref. 6, it can be shown that $\{\eta_k\}_{k=1}^{nl}$ satisfy linear equations

$$\eta_k = \sum_{j=1}^{nl} M_{kj} \eta_j + E_k, \qquad k = 1, 2, \ldots, nl. \qquad (187)$$

Using Banach's lemma, Brunner showed in Ref. 6 that $[\{\delta_{kj} - M_{kj}\}] = \mathbf{M}$ has an inverse for all n sufficiently large (h sufficiently small), and $\|\mathbf{M}^{-1}\|_\infty$ is uniformly bounded for all $n \geq n_0$. Thus if $\boldsymbol{\eta} = (\eta_1, \ldots, \eta_{nl})$, $\|\boldsymbol{\eta}\|_1 \leq c\|\mathbf{E}\|_1$, $n \geq n_0$, and since $\|\mathbf{E}\|_1 \leq ch^{2l}$ if Gaussian quadrature is used, one finds that $\|u - v_n\|_\infty \leq c_1 h^l + c_2 h^{2l} \leq ch^l$, $h \to 0$ (Ref. 6). Thus the quadrature scheme proposed in Ref. 6 produces a discrete collocation approximation v_n consistent with u_n.

This result has been extended by Brunner in Ref. 36 to deal with weakly singular Volterra equations

$$u(x) = \int_a^x \frac{k(x,t) u(t)\, dt}{(x-t)^\alpha} + f(x), \qquad (188)$$

where $k(x, t)$ and $f(x)$ are suitably differentiable and $0 < \alpha < 1$. Here the solution $u(x)$ is singular near $x = 0$, and graded meshes need to be employed to obtain optimal rates of convergence for u_n and suitable product quadratures are required to retain these optimal convergence rates for v_n.

Returning to the continuous case, it can be shown that v_n^s is locally superconvergent on $\{t_k\}_{k=1}^n$ if the Gauss points are used for collocation. In fact

$$u(t_k) - v_n^s(t_k) = u(t_k) - u_n^s(t_k) + u_n^s(t_k) - v_n^s(t_k).$$

By our previous calculations

$$|u(t_k) - u_n^s(t_k)| \le ch^{2l}, \qquad k = 1, 2, \ldots, n,$$

so it suffices to obtain the same bound for $u_n^s(t_k) - v_n^s(t_k)$.

Now $u_n^s(t_k) = f(t_k) + Ku_n(t_k)$ and $v_n^s(t_k) = f(t_k) + K_n v_n(t_k)$ so that

$$(u_n^s - v_n^s)(t_k) = Ku_n(t_k) - K_n v_n(t_k)$$

$$= K_n u_n(t_k) - K_n v_n(t_k) + E_k = K_n(u_n - v_n)(t_k) + E_k.$$

Using the expression for K_n given in Ref. 6, the bounds for $|E_k|$, and $|u_n(t_k) - v_n(t_k)| \le ch^{2l}$, which is derivable from (186) by interpolation (Ref. 6),

$$|u(t_k) - v_n^s(t_k)| \le ch^{2l}, \qquad k = 1, 2, \ldots, n,$$

and $v_n^s(t_k)$ is consistent with $u_n^s(t_k)$, $k = 1, 2, \ldots, n$. Similar results can be obtained from Fredholm equations with Green's function kernels (Refs. 6, 17).

7. Galerkin's Method for Equations with Positive-Definite Dominant Parts

So far in our treatment of projection methods for Eq. (1), we have assumed that $P_n H u_n = H u_n$, that is, P_n projects onto the range of HX_n. Although this condition is quite natural in the treatment of Galerkin and collocation methods for equations of the second kind where $H = I$, and for a number of polynomial approximation methods for Cauchy singular and some other singular equations as well (Refs. 18, 37), it does not occur when solving such singular equations by piecewise and trigonometric polynomials (Refs. 11-14). In these cases $P_n H u_n \ne H u_n$ and the analysis becomes more difficult because the straightforward conversion of $P_n H u_n + P_n K u_n = P_n f$ to the equivalent equation $H u_n + P_n K u_n = P_n f$ is not possible.

Perturbed Projection Methods

However, in many problems, particularly those arising from boundary integral equations used to solve elliptic partial differential equations, H has a positive-definiteness property which allows one to invert $P_n H P_n$ and then the analysis, at least for Galerkin's method, proceeds pretty much along the lines as before. The arguments used are well known for finite-element approximations to elliptic partial differential equations (Ref. 38), and the basic theorem, Cea's lemma (Ref. 11), has been generalized by Wendland and his colleagues to analyze Galerkin and collocation methods for quite general classes of boundary integral equations. Here we present an abstract version of these convergence analysis for Galerkin's method applied to operators $A = H + K$ where H is positive definite and K is compact. We will then specialize to treat trigonometric approximations for an equation of the first kind arising in potential theory (Ref. 14). Piecewise polynomial approximations can be handled in a similar fashion. Further examples and extensions can be found in Refs. 11–13 and in Chapter 1 by Atkinson.

Now let X, Y, and Z be real Hilbert spaces (this is done for convenience only) with $X \subseteq Y \subseteq Z$ with the inclusions being compact. Let $\langle \cdot, \cdot \rangle_Y$ denote the inner product on Y, and assume that X and Z are in duality with respect to $\langle \cdot, \cdot \rangle_Y$. That is, X and Z are dual spaces, and $\langle x, z \rangle_Y \leq \|x\|_X \|z\|_Z$, $x \in X$, $z \in Z$. Also $\|\cdot\|_X \geq \|\cdot\|_Y \geq \|\cdot\|_Z$. As before, we assume that $H: Z \to X$, $A = H + K: Z \to X$ are bounded, and have bounded inverses. $K: Z \to X$ is compact. In addition, we assume that H is positive definite [sometimes referred to as coercive (Ref. 11)] in the sense that $\langle Hx, x \rangle_Y \geq c\|x\|_Z^2$, $\forall x \in Z$, $c > 0$ (if $X = Y = Z$, then H is positive definite in the usual sense).

To approximate the solution to $Au = f$ by Galerkin's method, let $\{Y_n\} \subseteq Y$ be a sequence of subspaces such that $\bigcup_{n \geq 1} Y_n$ is dense in Y. If $\{\phi_k\}_{k=1}^N$ is a basis of Y_n, then $u \simeq u_n = \sum_{j=1}^N a_j \phi_j$ where $\{a_j\}_{j=1}^N$ are determined by solving

$$\sum_{j=1}^N a_j \langle A\phi_j, \phi_k \rangle_Y = \langle f, \phi_k \rangle_Y, \qquad k = 1, 2, \ldots, N. \tag{189}$$

Now let P_n be the orthogonal projection of Y onto Y_n with respect to $\langle \cdot, \cdot \rangle_Y$, and assume (this is crucial to the theory) that P_n can be extended to Z so that $\|x - P_n x\|_Z \to 0$, $\forall x \in Z$. By the principle of uniform boundedness $\{\|P_n\|_{Z \to Z}\}$ are uniformly bounded, and by duality $\{\|P_n^*\|_{X \to X}\}$ are uniformly bounded as well (Ref. 11) (for convenience we will not make any distinction between P_n and P_n^*—both will be written as P_n). With these definitions the Galerkin equations (189) can be written as

$$P_n H P_n u_n + P_n K u_n = P_n f. \tag{190}$$

Since we do not assume that $P_n H u_n = P_n H P_n u_n = H u_n$ as in Section 2, the convergence analysis of $\{u_n\}$ is somewhat more complicated than before. In the finite-element literature, when X, Y, and Z are Sobolev spaces, the convergence of $\{u_n\}$ is given by Cea's lemma. Here, for the convenience of the reader, we give an abstract version of this important theorem.

Theorem 7.1. Under the given assumptions on H, K, and $\{P_n\}$, $A_n = P_n H P_n + P_n K : Y_n \to Y_n$ has an inverse for all n sufficiently large, $n \geq n_0$, and the induced norms $\|A_n^{-1}\|_n$ are uniformly bounded. Thus for $n \geq n_0$, $u_n = A_n^{-1} P_n f$ exists, and $G_n = A_n^{-1} P_n A$ is bounded from Z to Y_n. Moreover,

$$\|u - u_n\|_Z \leq c \inf_{s_n \in Y_n} \|u - s_n\|_Z, \tag{191}$$

so that by our assumptions on $\{P_n\}$, u_n converges to u in Z.

Proof. We begin by showing that $H_n = P_n H P_n : Y_n \to Y_n$ has an inverse H_n^{-1} and $\{\|H_n^{-1}\|_n\}$ are uniformly bounded.

Since Y_n is finite dimensional, it suffices to show that if $y_n \in Y_n$ and $H_n y_n = 0$, then $y_n = 0$. Now

$$0 = \langle H_n y_n, y_n \rangle_Y = \langle P_n H P_n y_n, y_n \rangle_Y$$

$$= \langle H P_n y_n, P_n y_n \rangle_Y \geq c \| P_n y_n \|_Z^2 = c \| y_n \|_Z^2,$$

so that $\| y_n \|_Z = 0$. Thus $y_n = 0$ so that H_n is one-to-one and hence is invertible.

To bound $\|H_n^{-1}\|_n$ we begin with the result $\langle H_n y_n, y_n \rangle_Y \geq c \|y_n\|_Z^2$ proved above, and let $y_n = H_n^{-1} x_n$. Then $\langle x_n, H_n^{-1} x_n \rangle_Y \geq c \|H_n^{-1} x_n\|_Z^2$. Using the duality between X and Z, $\langle x_n, H_n^{-1} x_n \rangle_Y \leq \|x_n\|_X \|H_n^{-1} x_n\|_Z$, so that

$$\|H_n^{-1} x_n\|_Z \leq \|x_n\|_X / c, \quad \forall x_n \in Y_n. \tag{192}$$

By the definition of the induced norm, $\|H_n^{-1}\|_n \leq 1/c$, $c > 0$, $n \geq 1$.

Now $A_n = H_n + P_n K = H_n(I + H_n^{-1} P_n K)$, and the approximation properties of P_n and the uniform boundedness of $\{H_n^{-1}\}$ show that $H_n^{-1} P_n K z \to H^{-1} K z$, $\forall z \in Z$. Since K is compact, the convergence is uniform, and for all n sufficiently large, $n \geq n_1$, $(I + H_n^{-1} P_n K)|_{Y_n}$ has a uniformly bounded inverse. Thus for $n > n_2 = \max(1, n_1)$, $A_n^{-1}|_{Y_n} = (I + H_n P_n K)^{-1} H_n^{-1}|_{Y_n}$ exists and (190) is uniquely solvable. Moreover, $\{\|A_n^{-1}\|_n\}$ are uniformly bounded.

Letting $G_n = A_n^{-1} P_n A$, $\{\|G_n\|\}$ are uniformly bounded, and if $y_n \in Y_n$, $G_n y_n = A_n^{-1} P_n A y_n = A_n^{-1} P_n A P_n y_n = y_n$ and G_n projects onto Y_n. Thus, $u_n = A_n^{-1} P_n f = A_n^{-1} P_n A u = G_n u$ so that

$$u - u_n = u - G_n u = (u - s_n) - G_n(u - s_n), \qquad (193)$$

where $s_n \in Y_n$ is arbitrary. Taking norms in (193), using the uniform boundedness of $\{\|G_n\|\}$, gives $\|u - u_n\|_Z \le c \|u - s_n\|_Z$ and (191) follows. □

Example 7.1. See Ref. 14. As an example of Theorem 7.1 we consider the integral equation

$$\int_C \log(\|x - y\|) v(y) \, d\sigma(y) = f(x), \qquad (x, y) \in C, \qquad (194)$$

where C is a simply connected C^∞ curve in the plane and, $\|\ \|$ denotes the Euclidean norm of vectors in R^2. Equation (194) arises in the solution of the Dirichlet problem for Laplace's equation by single-layer potentials (Ref. 13).

If C is parametrized by $\alpha(t) = (x(t), y(t))$, where $(x(t), y(t))$, $0 \le t \le 1$ are 1-periodic functions, then (194) can be written as

$$Au = \int_0^1 L(t, \tau) u(\tau) \, d\tau = f(t), \qquad 0 \le t \le 1, \qquad (195)$$

where $u(\tau) = v(x(\tau), y(\tau))$, $f(t) = g(x(t), y(t))$, and

$$L(t, \tau) = \log\{[x(t) - x(\tau)]^2 + [y(t) - y(\tau)]^2\}^{1/2} r(\tau),$$

where $r(\tau) = \{[dx(\tau)/d\tau]^2 + [dy(\tau)/d\tau]^2\}^{1/2}$.

To solve (194) we consider the Galerkin method using trigonometric polynomials developed in Ref. 14. To do this, and to analyze the convergence of the method, it is convenient to introduce appropriate Hilbert spaces of 1-periodic functions.

For s real, let H^s be the Sobolev space of 1-periodic functions on R obtained by completing the 1-periodic C^∞ functions with respect to the norm

$$\|f\|_s = \left[|\hat{f}(0)|^2 + \sum_{\substack{-\infty < k < \infty \\ k \ne 0}} |\hat{f}(k)|^2 |2\pi k|^{2s} \right]^{1/2},$$

where

$$\hat{f}(k) = \int_0^1 \exp(-2\pi i k t) f(t) \, dt, \qquad i^2 = -1,$$

is the kth Fourier coefficient of $f(t)$. The inner product of $f \in H^s$ and $g \in H^s$ is given by

$$\langle f, g \rangle_s = \hat{f}(0)\overline{\hat{g}(0)} + \sum_{\substack{-\infty < k < \infty \\ k \neq 0}} \hat{f}(k)\overline{\hat{g}(k)}|2\pi k|^{2s},$$

where the overbar denotes complex conjugation.

For (195) it is well known that A can be defined as a bounded operator from $H^{s-1/2}$ to $H^{s+1/2}$ for $s \geq 0$. In particular if $s = 0$, then A is bounded from $H^{-1/2}$ to $H^{1/2}$. In the terminology of pseudodifferential operators, A has order -1 (Ref. 11). Moreover if we choose $X = H^{1/2}$, $Y = H^0 \equiv L_2$, and $Z = H^{-1/2}$, then A can be decomposed as $A = H + K$ where H is positive definite and K is compact (Ref. 13).

Now let $Y_n = \text{span}\{e^{2\pi i k t} = \phi_k\}_{k=-n}^{n}$ and approximate u by u_n where $\{a_j\}_{j=-n}^{n}$ are determined by solving

$$\langle A u_n, \phi_k \rangle_0 = \langle f, \phi_k \rangle_0, \qquad k = -n, \ldots, n. \tag{196}$$

In this case the projection P_n is defined by

$$P_n f = \sum_{k=-n}^{n} \hat{f}(k)\phi_k, \tag{197}$$

because $\{\phi_k\}$ are orthogonal in L_2. Thus (196) is equivalent to the operator equation

$$P_n A u_n = P_n f, \tag{198}$$

where we assume that A^{-1} exists [this depends on the curve C (Ref. 39)]. Since $P_n f \to f$, $n \to \infty$, $\forall f \in H^s$ (Ref. 14), one can show that all the conditions of Theorem 7.1 are met (Ref. 14). Consequently u_n converges to u in $H^{-1/2}$ and by (191)

$$\|u - u_n\|_{-1/2} \leq c \inf_{s_n \in Y_n} \|u - s_n\|_{-1/2}. \tag{199}$$

If u is sufficiently smooth, i.e., it belongs to H^s for sufficiently large $s > 0$, then one can use (199) to obtain convergence rates in various Sobolev norms. For this we need the following approximation and inverse properties of $\{\phi_k\}$ (Ref. 14).

(i) If $-\infty \leq r \leq s$ and $f \in H^s$, then

$$\|f - P_n f\|_r \leq c n^{r-s} \|u\|_s. \tag{200}$$

(ii) If $g \in \bigcup_{n \geq 1} Y_n$ and $r \leq s$, then

$$\|g\|_s \leq cn^{s-r}\|g\|_r. \tag{201}$$

From (199)-(201) it follows that if $t \geq -1/2$ and $t \leq s$, then

$$\|u - u_n\|_s \leq cn^{t-s}\|u\|_s. \tag{202}$$

To get (202), observe from (199) and (200) that

$$\|u - u_n\|_{-1/2} \leq c\|u - P_n u\|_{-1/2} \leq cn^{-1/2-s}\|u\|_s. \tag{203}$$

By the triangle inequality,

$$\|u - u_n\|_t \leq \|u - P_n u\|_t + \|u_n - P_n u\|_t$$

and from the inverse assumption (201)

$$\|u_n - P_n u\|_t \leq cn^{t+1/2}\|u_n - P_n u\|_{-1/2},$$

so that

$$\|u - u_n\|_t \leq \|u - P_n u\|_t + c_1 n^{t+1/2}\|u_n - P_n u\|_{-1/2}$$

$$\leq c_2 n^{t-s}\|u\|_s + c_1 n^{t+1/2}(\|u - u_n\|_{-1/2} + \|u - P_n u\|_{-1/2}). \tag{204}$$

However, from (200) and (203), $\|u - u_n\|_{-1/2} \leq c_3 n^{-1/2-s}\|u\|_s$, $\|u - P_n u\|_{-1/2} \leq c_4 n^{-1/2-s}\|u\|_s$ and substituting these into (204) gives

$$\|u - u_n\|_t \leq c_1 n^{t-s}\|u\|_s + c_5 n^{t+1/2} \cdot n^{-1/2-s}\|u\|_s = c_6 n^{t-s}\|u\|_s,$$

as required.

In particular if $t = 0$, then $\|u - u_n\|_0 \leq Cn^{-s}\|u\|_s$, so that if $u \in C^\infty$, u_n converges in L_2 faster than any power of s. For analytic solutions this can be improved to exponential convergence (Ref. 14).

Similar results can be obtained for spline approximations to u and for quite general classes of boundary integral equations. The proofs follow essentially the same pattern as for (202). Further examples may be found in Ref. 13, Chapter 1 of this book, and in the numerous references cited there.

As we have discussed in some detail already for equations of the second kind, the inner products in (196) usually have to be computed numerically. For boundary integral equations the splitting $A = H + K$ usually requires one to use different rules to evaluate $\langle H\phi_k, \phi_j \rangle_Y$ and $\langle K\phi_k, \phi_j \rangle_Y$ (Refs. 12, 14). In some instances, when using trigonometric approximations, $\{\langle H\phi_k, \phi_j \rangle_Y\}$ can be obtained analytically. In Ref. 14 an integration rule permitting the use of the fast Fourier transform is developed for evaluating $\langle K\phi_k, \phi_j \rangle$, leading to an efficient algorithm for trigonometric approximations [since these rules can be quite involved, we refer the reader to the original papers (Refs. 12, 14) for full details].

At any rate, once this has been done one ultimately obtains an approximation $v_n \in Y_n$ to u satisfying an equation of the form

$$\hat{A}_n v_n = f_n, \qquad f_n \in Y_n, \tag{205}$$

the perturbed version of (198). It is now important to be able to analyze the convergence of $\{v_n\}$. Usually one wishes to prove this convergence in Y (L_2 when Sobolev spaces are used), and this requires that we obtain bounds on $\|A_n^{-1}\|$ in the norm of Y_n rather than the norm induced by using X and Z. Rather than get further involved we will merely examine the convergence in Z ($H^{-1/2}$ in Example 7.1) and leave the more technical details for the interested reader to consult in Ref. 14, for example.

Suppose now that $\hat{A}_n$ can be considered as an operator from $Z \to X$, that $\|R_n\|_n = \|A_n - \hat{A}_n\|_n \to 0$ and $\|\hat{f}_n - P_n f\|_X \to 0$, $n \to \infty$, then under the assumptions of Theorem 7.1, $\|u - v_n\|_Z \to 0$, $n \to \infty$.

Since $\hat{A}_n = R_n + A_n = R_n + P_n A P_n = A_n(I + A_n^{-1} R_n)$, and $\{\|A_n^{-1}\|_n\}$ are uniformly bounded in n, Banach's lemma applied to $I + A_n^{-1} R_n$ shows that it has an inverse for all $n \geq n_0$ so that $\hat{A}_n^{-1}$ exists for all n sufficiently large. Also an easy calculation shows that

$$\hat{A}_n(u_n - v_n) = (\hat{A}_n - A_n)u_n + P_n f - \hat{f}_n, \tag{206}$$

so that for $n \geq n_0$,

$$u_n - v_n = \hat{A}_n^{-1}(\hat{A}_n - A_n)u_n + \hat{A}_n^{-1}(P_n f - \hat{f}_n), \tag{207}$$

and taking norms in (207) gives

$$\|u_n - v_n\|_Z \leq c[\|R_n\|_n + \|P_n f - \hat{f}_n\|_X] \to 0, \qquad n \to \infty.$$

8. Conclusions

We have developed a theory for proving the convergence of various perturbed projection methods and their variants for a class of operator equations which includes Fredholm, Volterra, and many Cauchy singular equations. The theory has been applied to rederive a number of recent convergence theorems for discrete Galerkin and collocation methods in a unified fashion. These results have then been compared to some nonperturbative ones of Atkinson and Bogomolny to illustrate the limitations of perturbation theory when the perturbations give rise to integration errors. Some suggestions are made for further work which should be done to fully understand the discrepancies between the two approaches.

References

1. BAKER, C. T. H., *The Numerical Treatment of Integral Equations*, Oxford University Press, Oxford, England, 1977.
2. ATKINSON, K. E., *A Survey of Numerical Methods for the Solution of Fredholm Integral Equations of the Second Kind*, Society for Industrial and Applied Mathematics, Philadelphia, Pennsylvania, 1976.
3. SPENCE, A., AND THOMAS, K., *On Superconvergence Properties of Galerkin's Method for Compact Operator Equations*, IMA Journal of Numerical Analysis, Vol. 3, pp. 253-271, 1983.
4. ATKINSON, K. E., AND BOGOMOLNY, A., *The Discrete Galerkin Method for Integral Equations*, Mathematics of Computation, Vol. 38, pp. 595-616, 1987.
5. BRUNNER, H., *The Application of Variation of Constants Formulas in the Numerical Analysis of Integral and Integrodifferential Equations*, Utilitas Mathematica, Vol. 19, pp. 255-290, 1981.
6. BRUNNER, H., *Iterated Collocation Methods and Their Discretizations*, SIAM Journal on Numerical Analysis, Vol. 21, pp. 1132-1145, 1984.
7. JOE, S., *Discrete Collocation Methods for Second Kind Fredholm Integral Equations*, SIAM Journal on Numerical Analysis, Vol. 22, pp. 1167-1177, 1985.
8. JOE, S., *Discrete Galerkin Methods for Fredholm Equations of the Second Kind*, IMA Journal of Numerical Analysis, Vol. 7, pp. 149-164, 1987.
9. MIEL, G., *On the Galerkin and Collocation Methods for a Cauchy Singular Integral Equation*, Technical Report No. 81, Department of Mathematics, Arizona State University, 1983.
10. GOLBERG, M. A., *The Perturbed Galerkin Method for Cauchy Singular Integral Equations*, Applied Mathematics and Computation, Vol. 26, pp. 1-33, 1988.
11. WENDLAND, W. L., *Asymptotic Accuracy and Convergence*, Progress in Boundary Element Methods, Edited by C. A. Brebbia, Pentech Press, London, England, pp. 289-313, 1981.
12. HSIAO, G. C., KOPP, P., AND WENDLAND, W. L., *A Galerkin Collocation Method for Some Integral Equations of the First Kind*, Computing, Vol. 25, pp. 89-130, 1980.
13. HSIAO, G. C., KOPP, P., AND WENDLAND, W. L., *Some Applications of a Galerkin Collocation Method for Integral Equations of the First Kind*, Mathematical Methods in Applied Science, Vol. 6, pp. 280-325, 1984.

14. LAMP, U., SCHLEICHER, K. T., AND WENDLAND, W. L., *The Fast Fourier Transform and the Numerical Solution of One-Dimensional Boundary Integral Equations*, Numerische Mathematik, Vol. 47, pp. 1-24, 1985.
15. CHANDLER, G. A., *Superconvergence of Numerical Solutions to Second Kind Integral Equations*, Australian National University, PhD Thesis, 1979.
16. FAATH, J., *Diskrete Galerkin-änlichte Verfahren zur Losung von Operatorgleichungen*, Kaiserslautern University, PhD Thesis, 1986.
17. CHATELIN, F., AND LEBBAR, R., *Superconvergence Results for the Iterated Projection Method Applied to a Fredholm Integral Equation of the Second Kind and the Corresponding Eigenvalue Problem*, Journal of Integral Equations, Vol. 6, pp. 71-91, 1984.
18. GOLBERG, M. A., *Introduction to the Numerical Solution of Cauchy Singular Integral Equations*, Chapter 5, this Volume.
19. GOLBERG, M. A., *Numerical Solution of Cauchy Singular Integral Equations with Constant Coefficients*, Journal of Integral Equations, Vol. 9, pp. 127-151, 1985.
20. JUNGHANNS, P., AND SILBERMANN, B., *Zur Theorie der Naherungsverfahren fur Singulare Integralgleichungen auf Intervallen*, Mathematische Nachrichten, Vol. 103, pp. 199-244, 1981.
21. DELVES, L. M., *A Fast Method for the Solution of Fredholm Integral Equations*, IMA Journal, Vol. 20, pp. 173-182, 1977.
22. DELVES, L. M., ABD-ELAL, L. M., AND HENDRY, J. A., *A Fast Galerkin Algorithm for Singular Integral Equations*, IMA Journal, Vol. 23, pp. 139-166, 1979.
23. SZËGO, G., *Orthogonal Polynomials*, American Mathematical Society, Providence, Rhode Island, 1975.
24. GRAHAM, I. G., *Collocation Methods for Two Dimensional Weakly Singular Integral Equations*, Journal of the Australian Mathematical Society, Series B, Vol. 22, pp. 456-475, 1981.
25. GRAHAM, I. G., *Numerical Methods for Multidimensional Integral Equations*, Department of Mathematics Technical Report 12-1983, University of Melbourne, 1983.
26. BREBBIA, C. A., Editor, *Topics in Boundary Elements Research, Vol. 3, Computational Aspects*, Springer-Verlag, New York, New York, 1987.
27. SLOAN, I. H., NOUSSAIR, E., AND BURN, B. J., *Projection Methods for Equations of the Second Kind*, Journal of Mathematical Analysis and Applications, Vol. 69, pp. 84-103, 1979.
28. DOUGLAS, J., DUPONT, T., AND WAHLBIN, L., *The Stability in L^∞ of the L^2-Projection into Finite Element Function Spaces*, Numerische Mathematik, Vol. 23, pp. 193-197, 1975.
29. FROMME, J., GOLBERG, M., AND WERTH, J., *Two Dimensional Aerodynamic Interference Effects on Oscillating Airfoils with Flaps in Ventilated Subsonic Wind Tunnels*, NASA Contractor Report No. 3210, Washington, DC, 1979.
30. ANSELONE, P., *Collectively Compact Operator Approximation Theory*, Prentice-Hall, Englewood Cliffs, New Jersey, 1971.

31. SLOAN, I. H., *Four Variants of the Galerkin Method for Integral Equations of the Second Kind*, IMA Journal of Numerical Analysis, Vol. 4, pp. 9-17, 1984.
32. SLOAN, I. H., *Superconvergence*, Chapter 2, this Volume.
33. FRIEDMAN, M. B., AND DELLWO, D. R., *Accelerated Projection Methods*, Journal of Computational Physics, Vol. 45, pp. 108-126, 1982.
34. PRÖSSDORF, S., *On Approximation Methods for the Solution of One-Dimensional Singular Integral Equations*, Applicable Analysis, Vol. 7, pp. 259-270, 1977.
35. MIEL, G., *Perturbed Projection Methods for Split Equations of the First Kind*, Technical Report No. 80, Department of Mathematics, Arizona State University, 1983.
36. BRUNNER, H., AND VANDERHOUWEN, P. J., *The Numerical Solution of Volterra Equations*, North-Holland, Amsterdam, Holland, 1986.
37. GOLBERG, M. A., *The Convergence of Several Algorithms for Solving Integral Equations with Finite Part Integrals, II*, Journal of Integral Equations, Vol. 9, pp. 259-270, 1985.
38. STRANG, G., AND FIX, G. J., *An Analysis of the Finite Element Method*, Prentice-Hall, Englewood Cliffs, New Jersey, 1973.
39. JASWON, M. A., AND SYMM, G. T., *Integral Equation Methods in Potential Theory and Elastostatics*, Academic Press, London, England, 1977.

4

Numerical Solution on Parallel Processors of Two-Point Boundary-Value Problems of Astrodynamics

G. MIEL

Abstract. We describe and analyze an experimental technique for numerically solving a class of two-point boundary-value problems typical of applications in astrodynamics. A prototype of this class of problems is that of finding the trajectory of a spacecraft passing between two given positions in a specified time. The numerical technique deals with the equivalent Hammerstein integral equation rather than the differential problem. It consists of discretizing each Fredholm integral equation that results from quasi-linearization by means of a perturbed Galerkin method based on Legendre polynomials and Gaussian quadrature. To reduce complexity, this is implemented by repeated use of an equivalent and more efficient Nyström method based on the Gaussian rule. The last approximant is then transformed to the desired polynomial expansion via a simple matrix-vector multiplication. The integral equation approach yields several benefits, among them robust parallelization for use on multiprocessor systems. Relevant analytical and physical concepts are described along with parallel algorithms.

1. Introduction

Given a vector-valued function $f:[0, T] \times D \to \mathbb{R}^m$, $D \subset \mathbb{R}^m$, and two constant vectors r_0, $r_T \in \mathbb{R}^m$, we consider the boundary-value problem involving a second-order system of nonlinear differential equations,

$$\ddot{r}(t) = f(t, r(t)), \qquad (1)$$

G. MIEL • Hughes Research Laboratories, Malibu, California 90265. This paper is dedicated to the memory of Dr. Franklin-Merrel Wolff who sought harmony and whose Stone House still offers welcome refuge.

in which a solution $r:[0, T] \to \mathbb{R}^m$ is required to satisfy the end conditions

$$r(0) = r_0, \qquad r(T) = r_T. \tag{2}$$

Here the dot denotes differentiation with respect to t, so that $\ddot{r}(t)$ is the second derivative. A prototype of this class of problem is that of computing the trajectory of a space vehicle that passes from an initial position r_0 to a final position r_T in a specified time T. In this case, the dimension of the problem is $m = 3$, $r(t)$ is the position vector of the spacecraft at time t, and $f(t, r(t))$ is a force vector per unit mass.

The dimension of the two-point boundary value problem (1)–(2) is allowed to be $m > 3$ in order to account for models that represent simultaneously the motions of several spacecraft or models that formulate the optimization of a trajectory into a problem in higher dimension (an example of the latter case is given in Section 7.2 in the sequel).

The two-point boundary value problem (1)–(2) is equivalent to the *Hammerstein integral equation*

$$r(t) = \int_0^T \gamma(t, s) f(s, r(s)) \, ds + w(t), \tag{3}$$

where $w:[0, T] \to \mathbb{R}^m$ is given by

$$w(t) = t(r_T - r_0)/T + r_0, \tag{4}$$

and where $\gamma:[0, T] \times [0, T] \to \mathbb{R}$ defined by

$$\gamma(t, s) = \begin{cases} -(T-s)t/T, & t \le s, \\ -(T-t)s/T, & s \le t, \end{cases} \tag{5}$$

is called a *Green's function*.

The aim of this chapter is to present a numerical technique for approximately solving the integral equation (3). Consideration of this integral equation instead of the differential problem (1)–(2) provides several benefits. Numerical techniques operating on the differential equation may suffer numerical instabilities that are avoided by the integral approach (Refs. 1–2). The discretization, shooting, and parallel shooting methods applied on the differential problem all require the solution of large systems of nonlinear equations (Ref. 3, p. 4, and Ref. 4). However, nonlinear solvers may lack robustness and their parallelization is little understood at this time. The technique proposed here involves only the solution of linear systems. It also offers the convenience of an orthogonal polynomial expansion with fast convergence when the function f is smooth.

There is yet another advantage to the integral equation approach, this time more of a theoretical nature. Criteria for the existence and uniqueness of a solution to the boundary value problem (1)-(2) are nearly always obtained by analyzing the corresponding integral equation (3). Moreover, such criteria are often of a constructive nature and as such they provide in the abstract a method for approximating the solution. It is satisfying to correlate analytical thought with computational practice. Our numerical technique relies on discretizing the *Newton-Kantorovich method*, also called *quasi-linearization*. In the idealized case, the convergence of that method follows from hypotheses that also imply the local well-posedness of the integral equation.

Our technique consists of discretizing Fredholm integral equations that result from quasi-linearization by use of a method based on an n-term Legendre polynomial expansion. In concept, each linear integral equation is approximately solved by a perturbed Galerkin method in which the n-point Gauss-Legendre quadrature rule is used to approximate integral transforms and inner products. In practice, in order to reduce the computational workload, a more efficient Nyström method is repeatedly applied, each time involving the solution of an $mn \times mn$ system of linear equations, to get the values of the polynomial expansion at the n nodes of the quadrature rule. The last approximation is then transformed via a simple matrix-vector multiplication to the desired polynomial expansion. In this process, we extend to systems of integral equations results for the scalar case presented earlier by the author (Ref. 5).

The approach of discretizing a sequence of linearized problems, as compared to that of directly discretizing the nonlinear problem, yields more robust software and better control of numerical stability. Moreover, parallel algorithms for solving systems of linear equations are better understood at this time than those for solving systems of nonlinear equations.

The approximation of the solution by a Legendre polynomial expansion works well when the true solution is smooth, which is the case for the type of problems considered here. The approximate solution is stored conveniently and economically as m vectors, each representing the n coefficients of a Legendre expansion of degree $n - 1$. This approximate solution is thus an explicit function of the independent variable; it can be conveniently manipulated algebraically, differentiated, or integrated. The approximate solution is easily computed for any value of the independent variable, using a three-term recurrence relation, eliminating the need for interpolation. Moreover, the error in the representation can be gauged by inspection of high-order terms of the expansion.

While presenting an experimental technique for numerically solving two-point boundary value problems of type (1)-(2), we endeavor at the same time to illustrate and relate diverse notions touching on the numerical analysis. There is considerable synergy among the analytical, computational, and physical concepts at hand. The Fréchet differentiability of the Hammerstein operator relies on the premise that the differential equations of motion are regular. The issue of well-posedness of the corresponding nonlinear equation (3) connects theoretical criteria with constructive methods of approximation. The use of a Hilbert product space leads naturally to parallel algorithms based on partitioning the computation into dimensional components of the problem. The method of patched conics in astrodynamics is a manifestation of the concept of transforming an ill-posed problem into several well-posed problems. In turn, the repetitive nature of this process brings about the desirability of parallel processing in order to reduce computing time.

The outline of the chapter is as follows. The following section defines our notation and describes mathematical preliminaries. Section 3 summarizes relevant concepts of astrodynamics in order to illustrate the type of two-point boundary value problems considered here. The following section deals with the issue of well-posedness of the integral equation (3). In Section 5, we analyze the perturbed Galerkin method, based on Legendre polynomial expansions and the Gauss–Legendre quadrature rule, for systems of linear integral equations. Parallel algorithms, based on a coarse-grain partition corresponding to the dimension of the problem, are described in Section 6. The next section illustrates the basic numerical technique on two prototype problems. Finally, Section 8 describes areas for future research.

2. Mathematical Preliminaries

Our notation is as follows. The ith component function of a vector-valued function $r:[0, T] \to \mathbb{R}^m$ is denoted by $r_i(t)$. Thus Eq. (3) is the vector form of a system of m nonlinear integral equations

$$r_i(t) = \int_0^T \gamma(t, s) f_i(s, r_1(s), \ldots, r_m(s)) \, ds + w_i(t), \qquad 1 \le i \le m. \qquad (6)$$

The space of $m \times m$ real matrices is denoted by $\mathbb{R}^{m \times m}$. A vector in $\mathbb{R}^m$ is denoted by $[r_i]$ and a matrix in $\mathbb{R}^{m \times m}$ by $[k_{ij}]$ with the letters i, j always denoting the row and column subscripts, respectively, and the understanding that $1 \le i, j \le m$. Similarly, $[\xi_\lambda]$ and $[\gamma_{\lambda\mu}]$ denote respectively a vector in $\mathbb{R}^n$ and a matrix in $\mathbb{R}^{n \times n}$ with the understanding that the indices always satisfy $1 \le \lambda, \mu \le n$. Likewise, the indices in the vector $[x_\rho]$ and matrix $[a_{\rho\sigma}]$ are assumed to always satisfy $1 \le \rho, \sigma \le mn$. A diagonal matrix whose

(i, i)th element is d_i is denoted by diag$[d_i]$. We shall use capital script letters to denote linear spaces. In particular, $\mathscr{E}'$ denotes the Hilbert space

$$L^2[0, T] = \left\{x:[0, T] \to \mathbb{R} \,\bigg|\, \int_0^T |x(t)|^2 \, dt < \infty\right\},$$

with inner product

$$\langle x, y \rangle' = \int_0^T x(t)y(t) \, dt,$$

and $\mathscr{P}_m$ denotes the space of polynomials of degree $\leq m$. All linear spaces are over the real field. As background references on the functional analytic topics used in the sequel, we refer the reader to Krasnosel'skii *et al.* (Ref. 6), Noble (Ref. 7), and Rall (Ref. 8). The symbol □ indicates the end of a proof.

2.1. The Integral Equation. In order to show that the boundary value problem (1)-(2) and the integral equation (3) have the same solution set, we use two formulas:

$$\frac{d}{dt}\int_a^t F(t, s) \, ds = \int_a^t \frac{\partial F}{\partial t}(t, s) \, ds + F(t, t), \tag{7}$$

$$\int_a^t \int_a^\xi g(s) \, ds \, d\xi = \int_a^t (t - s)g(s) \, ds, \tag{8}$$

where a is a constant and the functions $F(t, s)$, $g(s)$ satisfy appropriate smoothness conditions. Relation (7) is a special case of Leibnitz's formula (Ref. 9, p. 22). It can be established directly by use of the definition of the derivative and a mean value theorem for integrals. Relation (8) is easily derived by integrating by parts on its left side.

First, we show that a solution of the integral equation must necessarily be a solution of the boundary value problem. Assume that $r(t)$ solves (3). Since $\gamma(0, s) = \gamma(T, s) = 0$, the boundary conditions are satisfied. By definition of $\gamma(t, s)$,

$$r(t) = \int_0^t -\frac{s}{T}(T - t)f(s, r(s)) \, ds$$

$$+ \int_T^t \frac{t}{T}(T - s)f(s, r(s)) \, ds + w(t). \tag{9}$$

Use (7) to get

$$\dot{r}(t) = \int_0^t \frac{s}{T} f(s, r(s)) \, ds + \int_t^T \frac{1}{T}(T-s) f(s, r(s)) \, ds + \dot{w}(t). \quad (10)$$

Differentiate once more, using the Fundamental Theorem of Calculus, to show that $\ddot{r}(t) = f(t, r(t))$.

Conversely, we now demonstrate that a solution of the boundary value problem also solves the integral equation. Integrate twice the differential equation to get

$$r(t) = \int_0^t \int_0^\xi f(s, r(s)) \, ds \, d\xi + c_1 t + c_2, \quad (11)$$

where c_i are constants. Use formula (8) and the boundary conditions to find that

$$r(t) = \int_0^t (t-s) f(s, r(s)) \, ds - \frac{t}{T} \int_0^T (T-s) f(s, r(s)) \, ds + w(s). \quad (12)$$

In the second integral, partition the interval of integration into the subintervals $[0, t]$ and $[t, T]$, and then simplify to conclude that $r(t)$ satisfies (3).

2.2. The Legendre Polynomials. The normalized Legendre polynomials in the Hilbert space $\mathscr{E}' = L^2[0, T]$ are given by

$$\varphi_n(t) = \sqrt{\frac{2n-1}{T}} P_{n-1}\left(\frac{2t-T}{T}\right), \quad n \geq 1, \quad (13)$$

where $P_m \in \mathscr{P}_m$ denotes the Legendre polynomial over the interval $[-1, 1]$ satisfying the condition $P_m(1) = 1$. See Abramovitz and Stegun (Ref. 10). Figure 1 shows the graphs of the first few polynomials φ_n over the unit interval $[0, 1]$. The following properties hold:

$$\langle \varphi_m, \varphi_n \rangle' = \delta_{mn}, \quad (14)$$

$$\varphi_n(T-t) = (-1)^{n-1} \varphi_n(t), \quad (15)$$

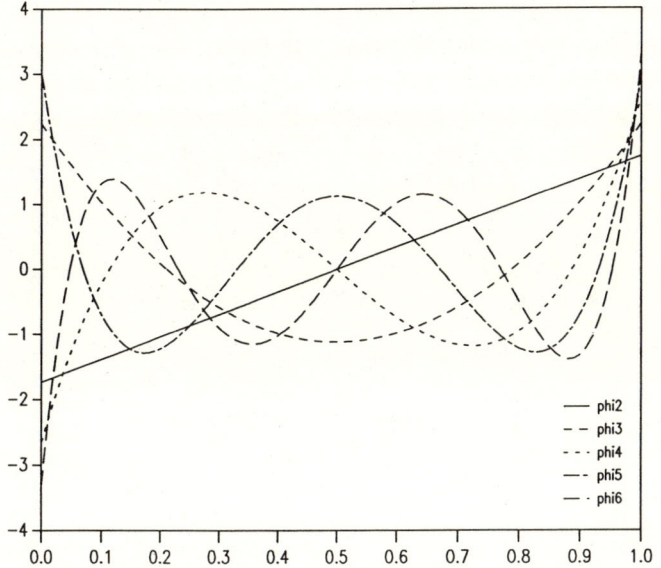

Fig. 1. The normal Legendre polynomials $\varphi_2, \ldots, \varphi_6$ over the interval $[0, 1]$.

$$|\varphi_n(t)| \leq \sqrt{\frac{2n-1}{T}}. \tag{16}$$

The three-term recurrence relation

$$\varphi_1(t) = \sqrt{\frac{1}{T}}, \tag{17a}$$

$$\varphi_2(t) = \sqrt{\frac{3}{T^3}}(2t - T), \tag{17b}$$

$$\varphi_n(t) = \frac{2\sqrt{(2n-3)(2n-1)}}{(n-1)T}\left(t - \frac{T}{2}\right)\varphi_{n-1}(t)$$

$$- \frac{n-2}{n-1}\sqrt{\frac{2n-1}{2n-5}}\,\varphi_{n-2}(t), \tag{18}$$

will be used in Sections 6.4 and 6.5 in order to formulate certain algorithms. We have L^2-convergence,

$$\left\| u - \sum_{j=1}^{n} \langle u, \varphi_j \rangle' \varphi_j \right\| \to 0, \quad \text{as } n \to \infty, \tag{19}$$

for any $u \in \mathscr{E}'$. If u is continuous with a piecewise continuous derivative u' then the Legendre expansion converges uniformly and absolutely on the open interval $(0, T)$.

The n-point Gauss–Legendre quadrature rule is given by

$$\int_0^T u(t)\, dt = \sum_{j=1}^n a_j u(s_j) + e_n(u), \tag{20}$$

where $s_1, \ldots, s_n$ are the n roots of φ_{n+1} and the error functional satisfies $e_n(u) = 0$ for every $u \in \mathscr{P}_{2n-1}$. See Stroud and Secrest (Ref. 11).

The formula

$$\langle v, w \rangle_n = \sum_{j=1}^n a_j v(s_j) w(s_j) \tag{21}$$

approximates the inner product $\langle v, w \rangle'$ with an error $e_n(vw)$. We see that $\langle \cdot, \cdot \rangle_n$ is a pseudo-inner product which coincides with the inner product $\langle \cdot, \cdot \rangle'$ on the subspace $\mathscr{P}_{n-1} \subset \mathscr{E}'$. The orthogonal projection of u on $\mathscr{P}_{n-1}$ is given by

$$\sum_{j=1}^n \langle u, \varphi_j \rangle' \varphi_j, \tag{22}$$

while the expression

$$\sum_{j=1}^n \langle u, \varphi_j \rangle_n \varphi_j \tag{23}$$

defines the collocation projection that maps u to the polynomial in $\mathscr{P}_{n-1}$ which interpolates u at the n nodes s_j.

2.3. The Hilbert Product Space.

An orthonormal basis $\{\varphi_n\}_{n=1}^\infty$ for $\mathscr{E}'$ consists of the normalized Legendre polynomials $\varphi_n \in \mathscr{P}_{n-1}$ defined above. An orthonormal basis for the subspace $\mathscr{P}_{n-1}$ is given by $\{\varphi_1, \ldots, \varphi_n\}$.

Let $\mathscr{E}$ denote the m-fold direct product of the space $\mathscr{E}'$, namely, the set consisting of ordered m-tuples

$$x = (x_1, \ldots, x_m), \qquad x_i \in \mathscr{E}'. \tag{24}$$

For $x, y \in \mathscr{E}$ and $a \in \mathbb{R}$, define addition and scalar multiplication,

$$x + y = (x_1 + y_1, \ldots, x_m + y_m), \tag{25a}$$

$$ax = (ax_1, \ldots, ax_m), \tag{25b}$$

and also an inner product

$$\langle x, y \rangle = \sum_{j=1}^{m} \langle x_j, y_j \rangle'. \tag{26}$$

The product space $\mathscr{E}$ is then a Hilbert space. The norm of $r \in \mathscr{E}$ is given by

$$\|r\| = \sqrt{\int_0^T (|r_1(t)|^2 + \cdots + |r_m(t)|^2)\, dt}. \tag{27}$$

For notational simplicity, we shall also use the symbol $\|v\|$ to denote the Euclidean norm of a vector $v \in \mathbb{R}^m$. With this convention, we can write (27) as

$$\|r\| = \sqrt{\int_0^T \|r(t)\|^2\, dt}, \tag{28}$$

since $r(t) \in \mathbb{R}^m$ for every $t \in [0, T]$.

Next, consider a subspace of $\mathscr{E}$,

$$\mathscr{E}^n = \{(x_1, \ldots, x_m) \mid x_i \in \mathscr{P}_{n-1}\}, \tag{29}$$

consisting of ordered m-tuples of polynomial of degree $\leq n - 1$. For $1 \leq i \leq m$, let e_i denote the m-tuple with unity in its ith entry and zero remaining entries. An orthonormal basis for $\mathscr{E}^n$ is given by

$$e_i \varphi_\lambda, \quad 1 \leq i \leq m, \quad 1 \leq \lambda \leq n. \tag{30}$$

We see that $\mathscr{E}^n$ is an mn-dimensional subspace of $\mathscr{E}$.

For computational purposes, we order the basis elements (30) as follows:

$$\Phi_\rho = e_i \varphi_\lambda, \tag{31}$$

$$\rho = (i-1)n + \lambda, \tag{32a}$$

$$i = (\rho - 1)\,\text{div}\,n + 1, \tag{32b}$$

$$\lambda = (\rho - 1)\,\text{mod}\,n + 1, \tag{32c}$$

$$1 \leq \rho \leq mn, \tag{33a}$$

$$1 \leq i \leq m, \tag{33b}$$

$$1 \leq \lambda \leq n, \tag{33c}$$

The orthogonal projection $P^n : \mathscr{E} \to \mathscr{E}^n$ is given by

$$P^n x = \sum_{\rho=1}^{mn} \langle x, \Phi_\rho \rangle. \tag{34}$$

Observe that

$$P^n x = \sum_{i=1}^{m} \left(\sum_{\lambda=1}^{n} \langle x_i, \varphi_\lambda \rangle' \varphi_\lambda \right) e_i, \tag{35}$$

where the expression in parentheses is the orthogonal projection of x_i on $\mathscr{P}_{n-1}$.

2.4. Operator Equations.
Given a mapping

$$k : [0, T] \times [0, T] \to \mathbb{R}^{m \times m}, \tag{36}$$

consider the vector integral transform

$$K : \mathscr{E} \to \mathscr{E}, \qquad Kx(t) = \int_0^T k(t, s) x(s) \, ds. \tag{37}$$

If

$$\alpha^2 \equiv \int_0^T \int_0^T \|k(t, s)\|_F^2 \, dt \, ds < \infty, \tag{38}$$

where $\|\cdot\|_F$ is the Frobenius matrix norm, then K is a bounded linear operator, called a *Fredholm operator*, with

$$\|K\| \leq \alpha. \tag{39}$$

In this case, for a given $v \in \mathscr{E}$, the linear operator equation

$$u = Ku + v \tag{40}$$

is called a *Fredholm equation* (of the second kind) on the product space $\mathscr{E}$.

A method for approximately solving the operator equation (40) is the so-called *Galerkin method*, in which one seeks an approximate solution $u^n \in \mathscr{E}^n$ such that

$$u^n = P^n K u^n + P^n v. \tag{41}$$

Since the integral operator K and the inner product $\langle \cdot, \cdot \rangle$ needed in defining P^n usually cannot be evaluated exactly, the Galerkin method cannot be carried out as it stands. In the sequel, we shall use a *perturbed Galerkin method* in which the n-point Gaussian quadrature rule (20) is used to approximate K and $\langle \cdot, \cdot \rangle$. In effect, the approximate operator equation (41) will be replaced by another approximate operator equation

$$(I - P^n K - R^n) u^n = P^n v + r^n, \tag{42}$$

where $R^n : \mathscr{E}^n \to \mathscr{E}^n$ and $r^n \in \mathscr{E}^n$ represent perturbations due to quadrature errors.

In order to analyze the numerical solution of the two-point boundary value problem (1)–(2), we view the equivalent integral equation (3) as an operator equation on the product space $\mathscr{E}$,

$$r = Hr + w, \tag{43}$$

where

$$H : \mathscr{E} \to \mathscr{E}, \qquad Hx(t) = \int_0^T \gamma(t, s) f(s, x(s)) \, ds. \tag{44}$$

The nonlinear operator H is called a *Hammerstein operator* and the operator equation is called a *Hammerstein equation* on the product space $\mathscr{E}$.

A strategy for finding an approximate solution to this nonlinear operator equation is to generate a sequence of well-posed Fredholm equations on $\mathscr{E}$,

$$r^{(\nu)} = K^{(\nu)} r^{(\nu)} + w^{(\nu)}, \qquad \nu \geq 1, \tag{45}$$

with the hope that the sequence of solutions $\{r^{(\nu)}\}$ will converge to a solution of Eq. (43). The process by which the nonlinear operator equation is transformed into a sequence of linear operator equations is described next.

2.5. The Newton–Kantorovich Method. A nonlinear operator F on the product space $\mathscr{E}$ is said to be *Fréchet differentiable* at $r^0 \in \mathscr{E}$ if there exists a bounded linear operator D such that

$$\lim_{\|h\| \to 0} \frac{\|F(r^0 + h) - Fr^0 - Dh\|}{\|h\|} = 0. \tag{46}$$

If such an operator D exists, it is unique. This operator, which is denoted by $F'(r^0)$, is called the *Fréchet derivative* of F at r^0.

An *affine operator* on the space $\mathscr{E}$ is an operator $L: \mathscr{E} \to \mathscr{E}$ of the form

$$Lr = Ar - b, \tag{47}$$

where A is a bounded linear operator on $\mathscr{E}$ and b is a fixed element of $\mathscr{E}$. When it exists, the Fréchet derivative $F'(r^0)$ of a nonlinear operator F can be used to approximate F locally by the affine operator

$$Lr = F'(r^0)(r - r^0) + Fr^0, \tag{48}$$

$$A = F'(r^0), \tag{49a}$$

$$b = Ar^0 - Fr^0. \tag{49b}$$

The process of replacing F by (48) is analogous in the scalar case to approximating a smooth function by its tangent line at a given point. Under proper conditions, an approximate solution to the nonlinear operator equation $Fr = 0$ can be obtained by solving the corresponding affine operator $Lr = 0$, namely, by solving the linear equation $Ar = b$.

The *Newton-Kantorovich method* for approximately solving $Fr = 0$ consists of successive applications of the above linearization process. After choosing an initial estimate $r^{(0)}$, the iterate $r^{(\nu)}$ is obtained by solving the affine operator equation

$$L^{(\nu)}r \equiv F'(r^{(\nu-1)})(r - r^{(\nu-1)}) + Fr^{(\nu-1)} = 0, \qquad \nu \geq 1. \tag{50}$$

This iteration is analogous in the scalar case to Newton's well-known method for approximating a root of a function.

The Newton method was first extended to operator equations by Kantorovich (Ref. 12). Bellman (Ref. 13) independently presented the method for nonlinear two-point boundary value problems. The method was further investigated by Bellman and Kalaba (Ref. 14). A modern perspective on the Newton-Kantorovich and related methods can be found, for example, in Refs. 15-16. Refinements in error bounds for such methods were recently reported (Refs. 17-18).

If the m^2 partial derivatives $\partial f_i/\partial r_j$ exist and are continuous over an appropriate region in $[0, T] \times \mathbb{R}^m$, then the Fréchet derivative of the Hammerstein operator H exists and is given by

$$H'(r^0)x(t) = \int_0^T \gamma(t, s) \frac{\partial f}{\partial r}(s, r^0(s))x(s)\, ds, \tag{51}$$

where

$$\frac{\partial f}{\partial r}(s, r) = \left[\frac{\partial f_i}{\partial r_j}(s, r)\right] \in \mathbb{R}^{m \times m} \tag{52}$$

is the Jacobian of f. The Hammerstein equation (43) can be rewritten as $Fr = 0$, where $Fr = r - Hr - w$. The Fréchet derivative at r^0 of F is $F'(r^0) = I - H'(r^0)$.

It follows that the Newton-Kantorovich iterate $r^{(\nu)}$ is the solution of the affine operator equation

$$r = H'(r^{(\nu-1)})(r - r^{(\nu-1)}) + Hr^{(\nu-1)} + w. \tag{53}$$

Equations (53) are Fredholm equations corresponding to (45) with

$$K^{(\nu)} = H'(r^{(\nu-1)}), \tag{54a}$$

$$w^{(\nu)} = -K^{(\nu)}r^{(\nu-1)} + Hr^{(\nu-1)} + w. \tag{54b}$$

In our numerical technique, rather than solving (53) we solve instead the Fredholm equation

$$u = K^{(\nu)}u + v^{(\nu)}, \tag{55}$$

$$v^{(\nu)} = Hr^{(\nu-1)} - r^{(\nu-1)} + w, \tag{56}$$

and then set $r^{(\nu)} = r^{(\nu-1)} + u$. Moreover, each Fredholm equation (55) will be approximately solved by using the perturbed Galerkin method mentioned earlier.

The operator equation (53) has the form

$$r(t) = \int_0^T \gamma(t, s) \frac{\partial f}{\partial r}(s, r^{(\nu-1)}(s))[r(s) - r^{(\nu-1)}(s)] \, ds$$

$$+ \int_0^T \gamma(t, s) f(s, r^{(\nu-1)}(s)) \, ds + w(s). \tag{57}$$

One can intuitively interpret the integral equation (57) as follows. Approximate $f(t, r)$ by the first two terms of a Taylor expansion at $r^{(\nu-1)}$:

$$f(t, r) \approx f(t, r^{(\nu-1)}) + \frac{\partial f}{\partial r}(t, r^{(\nu-1)}) \cdot (r - r^{(\nu-1)}) \equiv g(t, r). \tag{58}$$

Then the two-point boundary value problem

$$\ddot{r}(t) = g(t, r(t)), \tag{59}$$

$$r(0) = r_0, \tag{60a}$$

$$r(T) = r_T, \tag{60b}$$

is a linear problem that approximates the nonlinear problem (1)-(2). Moreover, the integral equation equivalent to the linear problem (59)-(60) is precisely the integral equation (57).

Throughout the remainder of this paper, we assume that the m^2 partial derivatives $\partial f_i/\partial r_j$ exist and are continuous over some appropriate region $[0, T] \times D$, $D \subset \mathbb{R}^m$. This assumption means that we are able to consider theoretical criteria for the well-posedness of the two-point boundary value problem (1)-(2) based on the Fréchet differentiability of the Hammerstein operator H. The assumption also allows us to construct approximation schemes based on quasi-linearization. Our numerical technique consists of discretizing the Newton-Kantorovich iterates via the use of the perturbed Galerkin method.

From an engineering point of view, the smoothness of the Hammerstein operator means that the differential equations of motion are regular. This is the case for motion resulting from a Newtonian gravitational field except at collision. We therefore suppose that during the time interval $[0, T]$ there are no collisions, and that accelerations due to nongravitational effects are smooth. The proximity of a singularity at $t_0 \in [0, T]$ may cause rapid oscillations in some of the orbital elements in a time interval about t_0. Hence, near-singular conditions may create numerical instabilities in the approximate solution. In order to avoid tedious issues on stability, we also assume in this chapter that near-singularities do not occur during the time interval $[0, T]$.

3. Equations of Motion

We encapsulate relevant notions of astrodynamics in order to illustrate the type of problems that we are dealing with. As general references on the topic, we point to the texts by Battin and Escobal (Refs. 20-23).

The motion of a space vehicle relative to a predominant celestial body is given by the differential equation (1), with $m = 3$, and a force vector per unit mass given by

$$f(t, r) = g(r) + g^M(r) + h(t, r), \tag{61}$$

$$g(r) = -\frac{Gm_0 r}{\|r\|^3}, \tag{62a}$$

$$g^M(r) = -\sum_{j=1}^{M} Gm_j \left(\frac{r - r^j}{\|r - r^j\|^3} + \frac{r^j}{\|r^j\|^3} \right), \tag{62b}$$

where G is *Newton's universal gravitational constant*, $g(r)$ is the gravitational acceleration due to the predominant body, assumed to have mass m_0, $g^M(r)$ is the perturbative acceleration due to the gravitational pull of M additional celestial bodies, assumed to have masses m_j, and $h(t, r)$ represents remaining perturbative accelerations due to, for example, aspherical effects of the predominant body or nongravitational forces such as solar radiation, thrust, etc. The norms in the functions g and g^M are Euclidean norms. A three-dimensional Cartesian coordinate system is used with origin at the center of mass of the predominant celestial body, with $r(t)$ the position vector at time t of the space vehicle and $r^j(t)$ the position vector at t of the center of mass of the jth perturbing celestial body.

The above differential equation represents the motion of a negligible mass influenced by $1 + M$ masses whose motion is known *a priori* and not influenced by the negligible mass. In this context, the differential equation represents a restricted $(2 + M)$-body problem. When $h \equiv 0$, the function $f = g + g^M$ is the Newtonian gravitational field resulting from the assumption that the $1 + M$ celestial bodies are spheres with a continuous and spherically symmetric distribution of matter.

In the *heliocentric* (Sun-centered) coordinate system, the $r_1 r_2$-plane is taken as the *ecliptic* (the plane of the Earth's orbit), with the r_1-axis directed toward a point among the stars known as the *vernal equinox*, and with the r_3-axis oriented so that the angular motion of the Earth on the $r_1 r_2$-plane is positive. In a *geocentric* (Earth-centered) coordinate system, the $r_1 r_2$-plane is taken as the equatorial plane of the Earth, with the r_1-axis pointing toward the vernal equinox, and the positive r_3-axis through the Earth's north pole. Since the Earth's orbit and its axis of rotation are perturbed by various torques, specific corrections and conventions are taken in order to ascertain that heliocentric and geocentric coordinates are well defined. We refer to aforementioned texts on astrodynamics for precise definitions. All that matters for our purposes is that the Cartesian system in use be regarded as fixed.

The value of Newton's universal constant of gravitation depends solely on the choice of units, for example,

$$G = 6.672 \times 10^{-11} \text{ m}^3/(\text{kg} \cdot \text{s}^2). \tag{63}$$

The constant $\mu = Gm_0$ is called the gravitational constant of the central celestial body. The corresponding gravitational field is then represented as

$$g(r) + g^M(r) = -\mu \frac{r}{\|r\|^3} - \sum_{j=1}^{M} \mu m_j' \left(\frac{r - r^j}{\|r - r^j\|^3} + \frac{r^j}{\|r^j\|^3} \right), \qquad (64)$$

where $m_j' = m_j/m_0$ are normalized masses relative to the mass of the central body. The gravitational constants, and other data, of solar planets are listed in Table 1.

For interplanetary trajectories, it is customary to use *astronomical units* in which one unit of length (0.149597870×10^9 km) approximates the semi-major axis of the Earth's elliptical orbit, one unit of mass is the mass of the Sun (0.19891×10^{31} kg), and one unit of time is one day (86,400 s). When heliocentric coordinates are used with these units, the constant

$$k = \sqrt{\mu} = 0.01720209895 \text{ (au)}^{3/2}/\text{day} \qquad (65)$$

is called the *Gaussian gravitational constant*.

The position vectors r^j of the M perturbing masses m_j, needed in the evaluation of the function $g^M(r)$, can be obtained by numerical integration of differential equations that describe the motion of these bodies, or, more accurately, if these bodies are in the solar system, by interpolation of data in ephemeris tapes produced by the Jet Propulsion Laboratory (JPL) of Pasadena, California (Refs. 24-25).

Table 1. Planetary Data

	Gravitational constant (km^3/s^2)	Radius of influence (10^6 km)	Inverse mass (Sun = 1)	Mean distance from Sun (10^6 km)	Period (Earth = 1)
Sun	0.1327154×10^{12}	∞	1	—	—
Mercury	0.2168553×10^5	0.11178	6,023,600	57.9	0.24
Venus	0.3247695×10^6	0.61696	408,523	108.1	0.62
Earth	0.3986032×10^6	0.92482	332,946	149.7	1.00
Mars	0.4297780×10^5	0.57763	3,098,710	227.7	1.88
Jupiter	0.1267106×10^9	48.141	1,047	777.8	11.86
Saturn	0.3791870×10^8	54.774	3,498	1,426.1	29.46
Uranus	0.5803292×10^7	51.755	22,869	2,869.2	84.02
Neptune	0.7026072×10^7	86.952	19,314	4,495.8	164.79
Pluto	0.3317886×10^6	35.812	3,000,000	5,914.4	248.43

3.1. Keplerian Motion.
This is the case with $M = 0$, $g^M \equiv 0$, and $h \equiv 0$. The solution of the resulting differential equation, one of the quintessential problems of celestial mechanics, is called the *two-body problem*. The motion is that of a negligible mass influenced solely by the gravitational potential of a spherically symmetric central body. This simple model is used to obtain a first approximation to the motion of a near-Earth spacecraft. The two-body problem is one of the few problems in astrodynamics that is analytically soluble. It is readily shown that the motion takes place in a fixed plane and that the trajectory is a conic section with one focus at the origin (Kepler's First Law). The orientation in space of the orbital plane needs to be specified by three angles defined relative to the three-dimensional Cartesian frame.

The two-point boundary value problem corresponding to Keplerian motion,

$$\ddot{r} = -\mu r / \|r\|^3, \tag{66a}$$

$$r(0) = r_0, \tag{66b}$$

$$r(T) = r_T, \tag{66c}$$

is called *Lambert's problem*. If $r = (r_1, r_2)$ denotes the position vector of the spacecraft, relative to a planar Cartesian frame in the orbiting plane, then the equivalent integral equations are

$$r_i(t) = \int_0^T \gamma(t, s) r_i(s) R(s)^{-3} \, ds$$

$$+ t(r_{T_i} - r_{0_i}) T^{-1} + r_{0_i}, \quad i = 1, 2, \tag{67}$$

where $R = [r_1^2 + r_2^2]^{1/2}$.

3.2. Aspherical Gravitational Potential.
A differential equation that models the gravitational effects of an aspherical central mass m_0 is given by

$$\ddot{r} = \nabla \Phi \equiv g(r) + \nabla \Phi_N, \tag{68}$$

where ∇ denotes the gradient operation and the potential (or negative potential, depending on one's convention) is given by

$$\Phi = \frac{\mu}{\|r\|} + \Psi_N, \tag{69}$$

$$\Psi_N = \mu \sum_{n=2}^{N} \frac{a^n}{\|r\|^{n+1}} \sum_{j=0}^{n} [C_{nj} \cos j\lambda + S_{nj} \sin j\lambda] P_{nj}(\sin \phi), \qquad (70)$$

where N is the degree and order of the potential model, a is the equatorial radius of the central body, C_{nj} and S_{nj} are constants called harmonic coefficients, and P_{nj} is the Legendre function of degree n and order j. Note that P_{n0} is the Legendre polynomial of degree n. If $\Psi_N \equiv 0$, the motion is Keplerian. Harmonic coefficients corresponding to the Earth's gravitational field have been extensively tabulated (Ref. 26). In this case, the coefficient C_{20}, which represents the Earth's flattening at the poles, is about three orders of magnitude larger than any other harmonic coefficient. The main effect of C_{20} is to produce periodic and secular variations in perigee and node of near-Earth satellites.

3.3. Earth Satellite Orbits.

The motion of an artificial satellite orbiting the Earth can be modeled by a differential equation of the form

$$\ddot{r} = g(r) + \nabla \Psi_8 + g^2(r) + h(t, r), \qquad (71)$$

where r is the geocentric position vector of the satellite, $g(r)$ is the acceleration due to the central part of the Earth's gravitational potential, Ψ_8 is the noncentral part of that potential, approximated to degree and order $N = 8$, $g^2(r)$ is the perturbative acceleration due to the Moon and Sun, and $h(t, r)$ is the acceleration due to other forces. Here μ is the gravitational constant of the Earth, r^1 and r^2 are the geocentric position vectors of the Moon and Sun, obtained for a given time t by quadrature or interpolation from a JPL ephemeris tape.

The function $h(t, r)$ consists of a sum of accelerations due to such forces as Earth tides, ocean tides, direct solar radiation, or solar radiation reflected from the Earth (*albedo*). For example, the acceleration due to direct solar radiation can be modeled as

$$h^1(t, r) = \nu c \|r^2\|^2 \cdot \frac{r - r^2}{\|r - r^2\|^2}, \qquad (72)$$

where norms are Euclidean norms; r^2 is the position vector of the Sun; c is a constant that depends on the area-to-mass ratio and reflective properties of the satellite; the constant ν is an eclipse factor, with $\nu = 0$ when the satellite is in the Earth's shadows, $\nu = 1$ when it is in sunlight, and $0 < \nu < 1$ if it is in the penumbra. See Ref. 27 for a description of $h^1(t, r)$ and other models that are usually included in the function $h(t, r)$.

Table 2. Effects of Perturbative Forces on GPS Satellites

Source	Perturbing acceleration (m/s^2)	Resulting deviations (m)	
		3-hour arc	2-day arc
Asphericity of Earth:			
C_{20} harmonic	5×10^{-5}	$\simeq 2{,}000$	$\simeq 14{,}000$
other harmonics	3×10^{-7}	5–80	100–1500
Gravitational effects g^2 of Moon and Sun	5×10^{-6}	5–150	1000–3000
Solar radiation h^1	1×10^{-7}	5–10	100–800
Albedo	1×10^{-9}	—	1.0–1.5
Ocean tides	1×10^{-9}	—	0.0–2.0
Earth tides	1×10^{-9}	—	0.5–1.0

3.4. Perturbations of GPS Orbits. The Global Positioning System (GPS) was designed to provide real-time positioning, for military and civilian navigation and survey purposes, using signals transmitted by a fleet of satellites. There are currently seven usable GPS satellites in orbit. When completed in the 1990s, GPS will consist of over 20 satellites in six orbital planes, each with roughly a 12-hour orbit, and altitude of about 20,200 km. The configuration will allow for simultaneous visibility of at least four satellites almost anywhere on Earth at any time of day.

Rizos and Stolz (Ref. 27) studied the effects of perturbative forces on GPS orbits by recording the differences between computed orbits with and without perturbative accelerations for both short arcs (a few hours) and long arcs (up to 2 days). The results are summarized in Table 2. Note that GPS satellites are at such high altitudes that atmospheric drag needs not be modeled.

4. The Well-Posedness Issue

The numerical solution of a boundary value problem (1)–(2) with low dimension m is not a large-scale computation and the speedup that results from multiprocessor computing is not needed. On the other hand, parallelization of a numerical technique may be beneficial, or required, when the dimension m is large or the problem needs to be solved repeatedly very quickly. An example is the determination of the trajectories in $\mathbb{R}^3$ of a constellation of N missiles scheduled to reach specified targets at time T. This situation, which can be modeled as a problem of dimension $m = 3N$, could very well benefit from multiprocessor computing.

The desirability of parallel processing, in the case a problem needs to be solved repeatedly, touches on the issue of existence and uniqueness of a solution. Boundary value problems of type (1)–(2), particularly for large values of T, are frequently ill-posed, namely, a solution does not exist or there are multiple solutions. In the latter case, such solutions can be densely packed: in every neighborhood of a trajectory there are other trajectories. Experiments showing this type of behavior, for the restricted three-body problem, were recorded by Gallagher and Perlin (Ref. 1). A possible strategy for dealing with an ill-posed boundary value problem is to transform the problem into a finite sequence of related well-posed problems. We illustrate this concept on a very simple problem.

4.1. Example. The scalar ($m = 1$) boundary value problem,

$$\ddot{r}(t) = -r(t), \tag{73a}$$

$$r(\alpha) = a, \tag{73b}$$

$$r(\beta) = b, \tag{73c}$$

has a unique solution if and only if

$$\det \begin{vmatrix} \sin \alpha & \cos \alpha \\ \sin \beta & \cos \beta \end{vmatrix} \neq 0. \tag{74}$$

Consider the ill-posed problem with the boundary conditions

$$r(0) = -1, \tag{75a}$$

$$r(\pi) = b. \tag{75b}$$

If $b = 1$ there is an infinite number of solutions. If $b \neq 1$ there is no solution. In the case $b = 1$, we can select one solution in the infinite solution set by breaking down the ill-posed problem into two well-posed problems: we choose $t_1 \in (0, \pi)$ and $r_1 \in \mathbb{R}$ and take the problems with respective boundary conditions

$$r(0) = -1, \tag{76a}$$

$$r(t_1) = r_1, \tag{76b}$$

and

$$r(t_1) = r_1, \tag{77a}$$

$$r(\pi) = 1. \tag{77b}$$

Generally, for an arbitrary value of b, we partition the interval $[0, \pi]$ into $N \geq 2$ subintervals

$$[t_{i-1}, t_i], \quad 1 \leq i \leq N, \tag{78}$$

$$0 = t_0 < t_1 < \cdots < t_N = \pi, \tag{79}$$

we let $r_0 = 1$ and $r_N = b$, and we pick values $r_1, \ldots, r_{N-1} \in \mathbb{R}$. The N problems with boundary conditions

$$r(t_{i-1}) = r_{i-1}, \quad r(t_i) = r_i, \quad 1 \leq i \leq N, \tag{80}$$

are each well-posed. The resulting solutions can be patched to yield a continuous function over $[0, \pi]$ which satisfies the boundary conditions (75) and which "sectionally" satisfies the differential equation.

4.2. Method of Patched Conics. This technique, based on the concept illustrated above, is used in the analysis of interplanetary trajectories. The motion of a spacecraft is partitioned into contiguous segments, each of which is assumed to be a Keplerian two-body orbit. The endpoints of a segment neighboring a planet are the entry and exit points of the spacecraft as it passes through a fictitious sphere called the planet's *sphere of influence*. The center of this sphere is at the center of mass of the planet and its radius is chosen so that the gravitational attraction of the planet and that of the Sun are nearly equal on its surface. Inside the sphere, the motion of the spacecraft is assumed to be a planetocentric Keplerian orbit, while upon exit from the sphere, the motion is assumed to become a heliocentric Keplerian orbit.

It can be shown that the radius of the sphere of influence is given by

$$r = R(m/m_0)^{2/5}, \tag{81}$$

where m is the mass of the planet, m_0 is the mass of the Sun, and R is the distance between the Sun and the planet (Ref. 20). These radii for the planets of the solar system are shown in Table 1.

Consider a flyby trajectory to N planets, from an initial position $r_0 = r_0^{(1)}$ at time $0 = t_0^{(1)}$ inside the sphere of influence of the first planet, to a final position $r_T = r_1^{(N)}$ at time $T = t_1^{(N)}$ inside the sphere of influence of the Nth planet. Since the corresponding two-point boundary value problem is likely to be ill-posed, we consider instead $2N - 1$ Lambert problems

$$\ddot{r} = g^{(i)}(r), \qquad r(t_0^{(i)}) = r_0^{(i)}, \qquad r(t_1^{(i)}) = r_1^{(i)}, \qquad 1 \leq i \leq N, \quad (82)$$

$$\ddot{r} = g(r), \qquad r(t_1^{(i)}) = r_0^{(i)}, \qquad r(t_0^{(i+1)}) = r_0^{(i+1)}, \qquad 1 \leq i \leq N - 1, \quad (83)$$

where $t_0^{(i)}$ and $t_1^{(i)}$ are respectively the time of entry into and the time of exit from the sphere of influence of the ith planet, and $r_j^{(i)}$ is the desired position of the space vehicle at time $t_j^{(i)}$. Here $r(t)$ is the position vector of the spacecraft in heliocentric coordinates, $g^{(i)}$ is its gravitational acceleration due to the ith planet, and g is its gravitational acceleration due to the Sun. Assuming that each Lambert problem is well-posed, the $2N - 1$ conic segments are computed individually and then patched together to get a candidate trajectory for the flyby mission.

4.3. Grand Tour of Voyager 2. Figure 2 illustrates the four-planet flyby of NASA's space probe Voyager 2. During a span of more than 8 years, Voyager has transmitted invaluable data and spectacular photographs

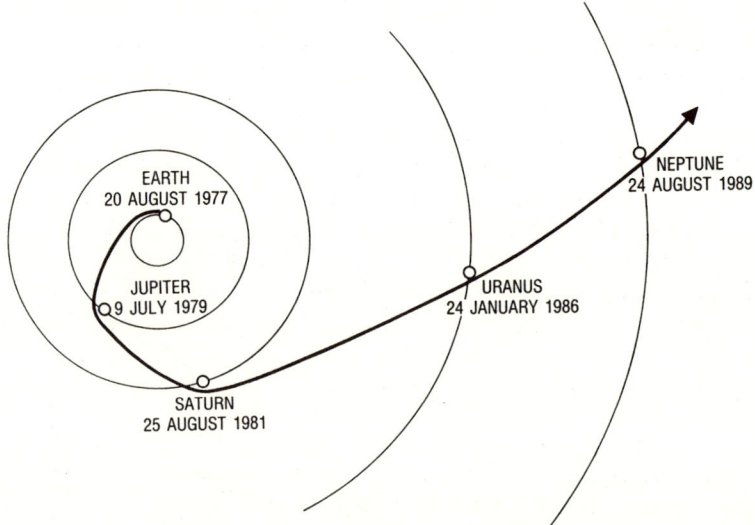

Fig. 2. Grand tour to the outer planets: flyby trajectory of Voyager 2.

of Jupiter and its moon Io, Saturnian rings, and Uranus and its moon Miranda. The spacecraft was scheduled to rendezvous with Neptune on August 24, 1989. The method of patched conics can be applied to get an approximation of the flyby trajectory, with hyperbolic segment inside the spheres of influence of the planets, and heliocentric elliptic segments in between. The hyperbolic segments represent the gravitational slingshots which helped Voyager in its journey. Without the gravitational boosts due to Jupiter and Saturn, the trip to Neptune would have taken more than 30 years instead of 12 years.

Neptune's orbit has a period of about 165 years. However, since the planet has not been observed for one complete orbit since its discovery in the 1840s, its period is not known accurately. To compound the uncertainty, an accurate estimate of its mass is also not known. Hence, navigators at JPL, who have been aiming Voyager at a point 4300 km above Neptune's north pole, had to use somewhat fuzzy data in their computations. The resulting uncertainty is problematical since the mission calls for using Neptune's gravitational pull to help swing the probe behind the moon Triton. This maneuver, which is needed for certain measurements, requires that the spacecraft be properly positioned at a critical time. The method of patched conics can be used to approximate alternate trajectories as a function of the fuzzy astrodynamical constants.

During the actual flyby, Voyager will look for Neptunian moons in addition to the two already identified from Earth, with a view of using data on the motion of such moons to pin down accurate estimates of Neptune's mass and its center of mass.

The numerical technique under study in this paper requires that the target problem be well-posed. We have indicated that ill-posed problems can sometimes be transformed into a finite sequence of well-posed problems, each of which can therefore be handled by our numerical technique. The issue remains, however, as to how one can test for existence and uniqueness of a solution to a given problem (1)-(2). Theoretical criteria, obtained by analysis of the equivalent integral equation (3), are based on hypotheses that the corresponding integral transform is smooth, contractive, or positive (Refs. 7, 28).

In what follows, we consider two theoretical criteria for the existence and uniqueness of a solution to the Hammerstein equation (43) on the product space $\mathscr{E}$. Our first result uses the Banach contraction theorem while the second result uses the Kantorovich theorem for Newton's method. Proofs of these underlying functional analytic theorems can be found in various references (e.g., Refs. 6, 8, 15). It is interesting to observe that the exercise of testing for the well-posedness of Eq. (3) provides a marriage of theory and practice, in the sense that the two theorems below are constructive in nature, namely, they each yield—in principle at least—a method for approximating the solution. The first method is *Picard iteration* while the second method is *Newton-Kantorovich iteration* described earlier in Section 2.5.

4.4. Picard Iteration. In the following theorem, we use the notation and definitions introduced in Section 2.

Theorem 4.1. Assume that the Hammerstein operator H is Fréchet differentiable on a closed and convex set $\mathcal{D} \subset \mathcal{E}$. If

$$\beta = \sup_{r \in \mathcal{D}} \|H'(r)\| < 1 \tag{84}$$

and

$$H\mathcal{D} \subset \mathcal{D}, \tag{85}$$

then the operator equation $r = Hr + w$ has a unique solution $r \in \mathcal{D}$ and for any $r^{(0)} \in \mathcal{D}$, the Picard iterates

$$r^{(\nu)} = Hr^{(\nu-1)} + w, \quad \nu \geq 1, \tag{86}$$

converge to r.

Proof. A mean value theorem for Fréchet derivatives states that

$$\|Hx - Hy\| \leq \sup_{0 \leq \lambda \leq 1} \|H'(x + \lambda(y - x))\| \cdot \|x - y\|, \quad x, y \in \mathcal{D}. \tag{87}$$

Consequently,

$$\|Hx - Hy\| \leq \beta \|x - y\|, \quad x, y \in \mathcal{D}, \tag{88}$$

that is, the operator H is contractive on $\mathcal{D}$. The result follows from the Banach contraction principle. □

One usually attempts to show that the operator H is contractive either on the entire space or on a closed ball in the space. In particular, the theorem holds when

$$\mathcal{D} = \bar{\mathcal{B}}(r^{(0)}, \alpha) \equiv \{r \in \mathcal{E} \mid \|r - r^{(0)}\| \leq \alpha\}, \tag{89}$$

and condition (85) is replaced by

$$\|Hr^{(0)} - r^{(0)}\| \le \alpha(1 - \beta). \tag{90}$$

Indeed, the ball $\bar{\mathscr{B}}(r^{(0)}, \alpha)$ is closed and convex and condition (90) implies condition (85).

The norm of the Fréchet derivative of H satisfies

$$\|H'(r)\| \le \int_0^T \int_0^T |\gamma(t, s)| \cdot \left\|\frac{\partial f}{\partial r}(s, r(s))\right\|_F dt\, ds. \tag{91}$$

We also have

$$\int_0^T \int_0^T |\gamma(t, s)|\, dt\, ds = T^3/12. \tag{92}$$

Hence, if

$$\max_{0 \le s \le T} \left\|\frac{\partial f}{\partial r}(s, r(s))\right\|_F \le J, \quad r \in \bar{\mathscr{B}}(r^{(0)}, \alpha), \tag{93}$$

and

$$\beta = JT^3/12 < 1, \quad \left\|\int_0^T \gamma(t, s)f(s, r^{(0)}(s))\, ds - r^{(0)}(s)\right\| \le \alpha(1 - \beta), \tag{94}$$

then the boundary value problem (1)-(2) has a unique solution in the ball $\bar{\mathscr{B}}(r^{(0)}, \alpha)$, and moreover, the Picard iterates

$$r^{(\nu)}(t) = \int_0^T \gamma(t, s)f(s, r^{(\nu-1)}(s))\, ds + w(s) \tag{95}$$

converge to that solution. The iterative method (95) suffers from two major drawbacks that prevent it from being implementable as it stands: the rate of convergence is slow and the integrals need to be approximated numerically.

4.5. Example: Perturbed Keplerian Motion. Consider the motion of a spacecraft defined by

$$f(t, r) = g(r) + h(t, r), \tag{96}$$

$$g(r) = -\mu r/R^3, \tag{97a}$$

$$R = [r_1^2 + r_2^2 + r_3^2]^{1/2}, \tag{97b}$$

where $h(t, r)$ denotes a smooth acceleration due to perturbing forces. Define

$$J = \max_{0 \le s \le T} \left(\left\| \frac{\partial g}{\partial r}(r(s)) \right\|_F + \left\| \frac{\partial h}{\partial r}(s, r(s)) \right\|_F \right), \quad r \in \bar{\mathscr{B}}(r^{(0)}, \alpha). \tag{98}$$

We can test for well-posedness of the boundary value problem on the ball $\bar{\mathscr{B}}(r^{(0)}, \alpha)$ by applying criteria (94).

The case for Lambert's problem, when $h \equiv 0$, is particularly simple. We can then use planar Cartesian coordinates in the plane of motion, thereby reducing the dimension of the problem to $m = 2$. We find that

$$\left\| \frac{\partial g}{\partial r} \right\|_F = \mu[(3r_1^2 R^{-5} - R^{-3})^2 + 2(3r_1 r_2 R^{-5}) + (3r_2^2 R^{-5} - R^{-3})^2]^{1/2}$$

$$= \mu\sqrt{5}/R^3. \tag{99}$$

We can thus take

$$J = \sqrt{5}\mu/\delta^3, \tag{100}$$

where

$$\delta = \min\{\|r(t)\| \,|\, r \in \bar{\mathscr{B}}(r^{(0)}, \alpha), t \in [0, T]\}. \tag{101}$$

It follows that if

$$T < a\delta \tag{102}$$

and

$$\int_0^T (|u_1(t)|^2 + |u_2(t)|^2) \, dt \le \alpha^2 (1 - \beta)^2, \tag{103}$$

where

$$a^3 = 12/\mu\sqrt{5}, \quad \beta = (T/a\delta)^3, \tag{104}$$

and

$$u_i(t) = \mu \int_0^T \gamma(t, s) r_i^{(0)}(s) [|r_1^{(0)}(s)|^2 + |r_2^{(0)}(s)|^2]^{-3/2} \, ds + r_i^{(0)}(t), \tag{105}$$

then Lambert's problem has a unique solution in the ball $\bar{\mathscr{B}}(r^{(0)}, \alpha)$.

4.6. Newton–Kantorovich Iteration.
In the following theorem, $\mathcal{B}(a, \alpha)$ denotes the interior of the closed ball $\bar{\mathcal{B}}(a, \alpha)$.

Theorem 4.2. Assume that the Hammerstein operator H is Fréchet differentiable on an open and convex set $\mathcal{D} \in \mathcal{E}$, that the operator $G = I - H'(r^{(0)})$ is invertible for some $r^{(0)} \in \mathcal{D}$, and that

$$\|G^{-1}(H'(x) - H'(y))\| \le K\|x - y\|, \qquad x, y \in \mathcal{D}, \tag{106}$$

$$\|G^{-1}(Hr^{(0)} - r^{(0)} + w)\| \le a, \tag{107}$$

$$\mathcal{B}(r^{(0)}, t^*) \subset \mathcal{D}, \tag{108a}$$

$$t^* = (1 - \sqrt{1-h})/K, \tag{108b}$$

$$h = 2Ka \le 1. \tag{108c}$$

Then the Newton–Kantorovich iterates

$$r^{(\nu+1)} = r^{(\nu)} + [I - H'(r^{(\nu)})]^{-1}(Hr^{(\nu)} - r^{(\nu)} + w), \qquad \nu \ge 0, \tag{109}$$

exist, remain in $\mathcal{B}(r^{(0)}, t^*)$, and converge to a solution r of the operator equation $r = Hr + w$. This solution is unique in $\mathcal{B}(r^{(0)}, t^{**}) \cap \mathcal{D}$, $t^{**} = (1 + \sqrt{1-h})/K$, if $h < 1$ and in $\bar{\mathcal{B}}(r^{(0)}, t^{**})$ if $h = 1$.

Proof. Apply directly the Kantorovich theorem on the nonlinear operator equation $Fr = 0$, $Fr = r - Hr - w$. □

In both Theorems 4.1 and 4.2, it is possible to include error bounds and rates of convergence of the respective iterates. A unifying concept based on majorization by scalar sequences allows one to derive simultaneously such results in both cases (Ref. 15). Such estimates, however, are of limited practical interest since neither the pertinent constants nor the iterates themselves can be computed accurately.

The numerical methods studied by Feagin (Ref. 29) may be viewed as discretizations based on Chebyshev polynomials of the Picard and Newton–Kantorovich iterates. The basic numerical method proposed in the following section may be viewed as a discretization, resulting from a perturbed Galerkin method based on Legendre polynomials, of the Newton–Kantorovich iteration.

5. Perturbed Galerkin Method

Our aim is to solve numerically the Fredholm equations (55) obtained by quasi-linearization of the Hammerstein equation (43). In this section, we describe and analyze the perturbed Galerkin method that will be used to do so. In the process, we extend results for the scalar case $m = 1$ (Ref. 5) to systems of $m \geq 1$ linear integral equations.

Consider a system

$$u_i(t) = \int_0^T \sum_{j=1}^m k_{ij}(t, s) u_j(s) \, ds + v_i(t), \qquad 1 \leq i \leq m, \qquad (110)$$

in which, given m^2 kernel functions $k_{ij}(t, s)$ and m functions $v_i(t)$, the goal is to find m functions $u_i(t)$. Letting $u, v : [0, T] \to \mathbb{R}^m$ and $k : [0, T] \times [0, T] \to \mathbb{R}^{m \times m}$ be defined by

$$u(t) = [u_i(t)], \qquad (111a)$$

$$v(t) = [v_i(t)], \qquad (111b)$$

$$k(s, t) = [k_{ij}(t, s)], \qquad (111c)$$

the system is concisely represented as an operator equation on the product space $\mathscr{E}$,

$$u = Ku + v, \qquad (112)$$

where $K : \mathscr{E} \to \mathscr{E}$ is defined by

$$Kx(t) = \int_0^T k(t, s) x(s) \, ds. \qquad (113)$$

We assume throughout that

$$\int_0^T \int_0^T \|k(t, s)\|_F^2 \, dt \, ds < \infty, \qquad (114)$$

so that the integral transform K is a compact operator on $\mathscr{E}$.

We first need to describe the *Nyström method* for numerically solving Eq. (112). Using the n-point Gaussian quadrature rule (20), we replace the integral transform K by the operator

$$K^n x(t) = \sum_{\mu=1}^n a_\mu k(t, s_\mu) x(s_\mu). \qquad (115)$$

The goal is to discretize the approximate operator equation

$$(I - K^n)u^N(t) = v(t) \qquad (116)$$

into a system of mn linear equations in the mn unknowns

$$\eta_{j\mu} = u_j^N(s_\mu), \qquad \begin{cases} j = 1, \ldots, m, \\ \mu = 1, \ldots, n. \end{cases} \qquad (117)$$

Since

$$K^n x(t) = \sum_{i=1}^{m} \left(\sum_{\mu=1}^{n} a_\mu \sum_{j=1}^{m} k_{ij}(t, s_\mu) x_j(s_\mu) \right) e_i \qquad (118)$$

and

$$u^N(t) = \sum_{i=1}^{m} u_i^N(t) e_i, \qquad v(s) = \sum_{i=1}^{m} v_i(s) e_i, \qquad (119)$$

the operator equation (116) is equivalent to the m conditions

$$\left(u_i(t) - \sum_{\mu=1}^{n} a_\mu \sum_{j=1}^{m} k_{ij}(t, s_\mu) \eta_{j\mu} \right) e_i = v_i(t) e_i, \qquad 1 \le i \le m. \qquad (120)$$

Now set $t = s_\lambda$, $1 \le \lambda \le n$, in each of these conditions to get the mn linear equations

$$\eta_{i\lambda} - \sum_{j=1}^{m} \sum_{\mu=1}^{n} a_\mu k_{ij}(s_\lambda, s_\mu) \eta_{j\mu} = v_i(s_\lambda), \qquad 1 \le i \le m, \quad 1 \le \lambda \le n, \qquad (121)$$

in the mn unknowns (117).

A matrix representation $Ax = b$ of this linear system depends on the order used to form the components of the unknown vector $x = [x_\rho] \in \mathbb{R}^{mn}$. We choose the ordering given by

$$\eta_{j\mu} = x_{(j-1)n+\mu}. \qquad (122)$$

We thus have

$$x = (\eta_{11}, \ldots, \eta_{1n}; \eta_{21}, \ldots, \eta_{2n}; \ldots; \eta_{m1}, \ldots, \eta_{mn}), \qquad (123)$$

with

$$x_\rho = \eta_{j\mu}, \qquad \begin{cases} j = (\rho - 1) \text{ div } n + 1, \\ \mu = (\rho - 1) \text{ mod } n + 1, \end{cases} \quad \rho = 1, \ldots, mn. \qquad (124)$$

The ith n-tuple $\eta_i = (\eta_{i1}, \ldots, \eta_{in})$ in x consists of the values of the ith component function $u_i^N(t)$ at the nodes $s_1, \ldots, s_n$ in that order.

In the *Galerkin method*, we approximate a solution of Eq. (112) by an element $u^n \in \mathscr{E}^n$,

$$u^n = \sum_{j=1}^{m} \sum_{\mu=1}^{n} \xi_{j\mu} \phi_\mu e_j, \tag{125}$$

such that the residual $r^n = u^n - Ku^n - v$ is orthogonal to the subspace $\mathscr{E}^n$. We thus get mn linear equations

$$\langle r^n, \phi_{i\lambda} e_i \rangle = 0, \quad \begin{cases} i = 1, \ldots, m, \\ \lambda = 1, \ldots, n, \end{cases} \tag{126}$$

in the mn unknowns $\xi_{j\mu}$. A matrix representation of this linear system depends on the ordering of the basis elements $\phi_\mu e_j$. We choose the ordering described earlier in Section 2.3,

$$\varphi_\mu e_j = \Phi_{(j-1)n+\mu}. \tag{127}$$

The corresponding unknown vector $y = [y_\rho] \in \mathbb{R}^{mn}$ is

$$y = (\xi_{11}, \ldots, \xi_{1n}; \xi_{21}, \ldots, \xi_{2n}; \ldots; \xi_{m1}, \ldots, \xi_{mn}), \tag{128}$$

with

$$y_\rho = \xi_{j\mu}, \quad \begin{cases} j = (\rho - 1) \operatorname{div} n + 1, \\ \mu = (\rho - 1) \bmod n + 1, \end{cases} \rho = 1, \ldots, mn. \tag{129}$$

The ith n-tuple $\xi_i = (\xi_{i1}, \ldots, \xi_{in})$ in y consists of the Fourier coefficients $\xi_{i\lambda} = \langle u_i^n, \varphi_\lambda \rangle$, of the ith component function of u^n. With this ordering, the linear system

$$[\langle \Phi_\rho, \Phi_\sigma - K\Phi_\sigma \rangle][y_\rho] = [\langle \Phi_\rho, v \rangle] \tag{130}$$

represents the mn conditions (126).

The Galerkin method offers the convenience of a polynomial approximation $u^n \in \mathscr{E}^n$, and for problems with well-behaved functions k and v, the advantages of fast convergence of u^n to u and of superconvergence of certain functionals (Refs. 30–35). However, since the inner product $\langle \cdot, \cdot \rangle$ and the integral transform K nearly always need to be approximated with quadrature rules, the method in practice cannot be implemented exactly.

We are thus led to consider a *perturbed Galerkin method* in which we approximate as before K by K^n and, in addition, we use the n-point Gaussian quadrature rule to approximate the inner product $\langle x, y \rangle$ by

$$\langle x, y \rangle^n = \sum_{j=1}^{m} \sum_{\mu=1}^{n} a_\mu x_j(s_\mu) y_j(s_\mu). \tag{131}$$

The $mn \times mn$ linear system corresponding to the perturbed Galerkin method is

$$[\langle \Phi_\rho, \Phi_\sigma - K^n \Phi_\sigma \rangle^n][y_\rho] = [\langle \Phi_\rho, v \rangle^n]. \tag{132}$$

Since the n-point Gaussian rule is exact whenever $x_j \cdot y_j \in \mathscr{P}_{2n-1}$, $\langle x, y \rangle^n$ is a pseudo-inner product that coincides with the inner product $\langle x, y \rangle$ on the subspace $\mathscr{E}^n$.

5.1. Equivalence. While the perturbed Galerkin method preserves advantageous characteristics of the Galerkin method, system (132) is a very costly setup. Fortunately, it turns out that system (132) is equivalent, in a sense to be made precise in the theorem below, to the simple and efficient Nyström method described earlier.

Theorem 5.1. With the ordering of the components of the vectors $x, y \in \mathbb{R}^{mn}$ given by (124) and (129), the linear equations resulting from the Nyström and perturbed Galerkin methods are given respectively by

$$(I - B)x = b \tag{133}$$

and

$$S(I - B)Ty = Sb, \tag{134}$$

where

$$b = (v_{11}, \ldots, v_{1n}; v_{21}, \ldots, v_{2n}; \ldots; v_{m1}, \ldots, v_{mn}), \quad v_{j\mu} = v_j(s_\mu), \tag{135}$$

$$B = \begin{bmatrix} K_{11} & K_{12} & \cdots & K_{1m} \\ K_{21} & K_{22} & \cdots & K_{2m} \\ & & \cdots & \\ K_{m1} & K_{m2} & \cdots & K_{mm} \end{bmatrix} \in \mathbb{R}^{mn \times mn}, \tag{136}$$

in which the submatrix K_{ij} is given by

$$K_{ij} = [a_\mu k_{ij}(s_\lambda, s_\mu)] \in \mathbb{R}^{n \times n}, \tag{137}$$

and where S, T are block diagonal matrices

$$S = \begin{bmatrix} S' & & & \\ & S' & & \\ & & \ddots & \\ & & & S' \end{bmatrix}, \quad T = \begin{bmatrix} T' & & & \\ & T' & & \\ & & \ddots & \\ & & & T' \end{bmatrix} \in \mathbb{R}^{mn \times mn}, \quad (138)$$

respectively consisting of m copies of the submatrices

$$S' = [a_\mu \varphi_\lambda(s_\mu)], \quad T' = [\varphi_\mu(s_\lambda)] \in \mathbb{R}^{n \times n}. \quad (139)$$

Moreover, the matrices S, T are invertible with $T^{-1} = S$.

Proof. With the ordering (124), the linear system (121) becomes

$$x_{(i-1)n+\lambda} - \sum_{j=1}^{m} \sum_{\mu=1}^{n} a_\mu k_{ij}(s_\lambda, s_\mu) x_{(j-1)n+\mu} = v_{i\lambda}. \quad (140)$$

The matrix representation of this system is (133). Use the definitions (115) and (131) of K^n and $\langle \cdot, \cdot \rangle^n$ to derive (134). Since the polynomials $\varphi_1, \ldots, \varphi_n$ constitute a Haar system over the interval $[0, T]$, the submatrix T' is invertible. The matrix $U' = \text{diag}[a_\lambda]$ is invertible because the weights a_λ are strictly positive. Thus $S' = T'^t U'$ is also invertible. Moreover, we have

$$S'T' = \left[\sum_{i=1}^{n} a_i \phi_\lambda(s_i) \phi_\mu(s_i) \right] = [\langle \phi_\lambda, \phi_\mu \rangle'] = I \in \mathbb{R}^{n \times n}, \quad (141)$$

where the second equality holds because the n-point quadrature rule is exact for polynomials of degree $\leq 2n - 1$. Thus $T'^{-1} = S'$ and consequently $T^{-1} = S$. □

Linear system (133) corresponding to the Nyström method is considerably easier to set up than system (134) corresponding to the perturbed Galerkin method. Given a solution x to system (133), we find a corresponding solution y to system (134) by computing

$$y = Sx. \quad (142)$$

Due to the ordering of the components of x, y and the block diagonal structure of S, the matrix-vector multiplication (142) in dimension mn corresponds to m independent matrix-vector operations in dimension n:

$$\xi_j = S' \eta_j, \quad 1 \leq j \leq m. \quad (143)$$

Recall here that $\xi_j = [\xi_{j\lambda}]$, $\eta_j = [\eta_{j\lambda}] \in \mathbb{R}^n$. The independence in (143) can be used advantageously in a parallel computing environment.

The approximant

$$\begin{bmatrix} u_1^n \\ u_2^n \\ \vdots \\ u_n^n \end{bmatrix} = \begin{bmatrix} \xi_{11} \\ \xi_{21} \\ \vdots \\ \xi_{n1} \end{bmatrix} \varphi_1 + \begin{bmatrix} \xi_{12} \\ \xi_{22} \\ \vdots \\ \xi_{n2} \end{bmatrix} \varphi_2 + \cdots + \begin{bmatrix} \xi_{1n} \\ \xi_{2n} \\ \vdots \\ \xi_{nn} \end{bmatrix} \varphi_n, \qquad (144)$$

satisfies the mn interpolation conditions

$$u_j^n(s_\mu) = \eta_{j\mu}. \qquad (145)$$

Thus the function $u^n \in \mathscr{E}^n$ is an alternate interpolant to the usual Nyström interpolant.

While the Nyström method (133) is represented by the approximate operator equation (116), the perturbed Galerkin method (134) is represented by the approximate operator equation

$$(I - P^n K - R^n)u^n = P_\nu^n + r^n, \qquad u^n \in \mathscr{E}^n, \qquad (146)$$

where P^n is the orthogonal projection of the space $\mathscr{E}$ onto the subspace $\mathscr{E}^n$, $R^n: \mathscr{E}^n \to \mathscr{E}^n$ is a linear operator, and r^n is an element of $\mathscr{E}^n$. The inclusion in (146) emphasizes that an approximate solution is sought in the finite-dimensional subspace $\mathscr{E}^n$. The operator R^n and the element r^n may be viewed as perturbations due to the quadrature rule in approximating the integral transform and the inner products. With $R^n = 0$ and $r^n = 0$, Eq. (146) represents the exact Galerkin method.

5.2. Analytic Principle. Our next goal is to establish under general conditions the least-squares convergence of the approximants u^n to a true solution of the Fredholm equation. In order to do so, we temporarily generalize the notation. Let $\mathscr{E}$ be an arbitrary Hilbert space. Denote as usual the space of bounded linear operators $L: \mathscr{E} \to \mathscr{E}$ by $\mathscr{L}(\mathscr{E})$. Consider

$\mathscr{E}^n$, a closed subspace of $\mathscr{E}$,
$P^n: \mathscr{E} \to \mathscr{E}^n$, the orthogonal projection of $\mathscr{E}$ onto $\mathscr{E}^n$,
$R^n \in \mathscr{L}(\mathscr{E}^n)$, a perturbation operator,
$r^n \in q^n$, a perturbation element.

Consider a generic operator equation (112) where $K \in \mathscr{L}(\mathscr{E})$ and $v \in \mathscr{E}$ are given and a corresponding approximate operator equation (146). The following lemma describes the functional analytic principle (Refs. 6, 36, 37) underlying our convergence analysis.

Lemma 5.1. Assume that K is a compact operator and that $L = I - K$ has an inverse in $\mathscr{L}(\mathscr{E})$. If the orthogonal projections P^n converge strongly to the identity operator and if

$$\|R^n\| \to 0 \quad \text{and} \quad \|r^n\| \to 0, \tag{147}$$

as $n \to \infty$, then for sufficiently large n, $L^n = I - P^n K - R^n$ has an inverse in $\mathscr{L}(\mathscr{E}^n)$, the inverse operators L_n^{-1} are uniformly bounded, and the approximate solutions $u^n = L_n^{-1}(P^n v + r^n)$ converge to the exact solution $u = L^{-1}v$ with a rate

$$\|u - u^n\| \leq c b^n, \tag{148}$$

where $b^n = \max(\|u - P^n u\|, \|R^n\|, \|r^n\|)$ and c is a constant independent of n.

Proof. See Theorem 1 with $J = I$ in Ref. 37, p. 270. □

The lemma shows that the well-posedness of the Fredholm equation (112) implies the well-posedness, for large enough n, of the approximate equation (146). Moreover, the perturbed Galerkin method is stable, in the sense that the inverse operators L_n^{-1} are uniformly bounded; see Ref. 38 and Ref. 39, Theorem 3, p. 60.

5.3. Convergence Result. We now return to the original notation and we specialize the lemma to the integral equation (112) and the perturbed Galerkin method engendered by linear system (134).

Theorem 5.2. If the vector-valued functions k and v are Riemann integrable on $[0, T] \times [0, T]$ and $[0, T]$, respectively, and the operator $I - K$ has a bounded inverse on the product Hilbert space $\mathscr{E}$ then, for sufficiently large n, the linear system (134) has a unique solution, and as $n \to \infty$, the inverses of the coefficient matrices are uniformly bounded and the approximants u^n defined by (144) converge in $\mathscr{E}$ to the true solution $u = (I - K)^{-1}v$.

Proof. An argument for the scalar case $m = 1$ can be found in Ref. 5, Theorem 4.2. The extension to $m > 1$ involves technicalities on product spaces. In outline, one uses the natural isomorphism between

$$(\xi_{11}\varphi_1 + \cdots + \xi_{1n}\varphi_n)e_1 + \cdots + (\xi_{m1}\varphi_1 + \cdots + \xi_{mn}\varphi_n)e_m \in \mathscr{E}^n \tag{149}$$

and

$$(\xi_{11}, \ldots, \xi_{1n}; \ldots; \xi_{m1}, \ldots, \xi_{mn}) \in \mathbb{R}^{mn}, \tag{150}$$

to establish the correspondence between the linear system (132) and the approximate operator equation (146). The perturbations R^n and r^n are expressed in terms of the error functional of the n-point Gaussian quadrature rule. One shows that $R^n \to 0$ and $r^n \to 0$, with the desired conclusions then following directly from Lemma 5.1. □

Under smoothness conditions on the functions k and v, one can readily establish uniform convergence of the approximants u^n to u (Refs. 33, 40).

6. Parallel Algorithms

A procedural description of the Newton–Kantorovich iteration (55)–(56) is given by:

   Pick an initial r.
   For $\nu = 1, 2, \ldots, N$ do:
Step A_ν: $K := H'(r)$,
Step B_ν: $v := Hr - r + w$,
Step C_ν: Solve $u = Ku + v$,
Step D_ν: $r := r + u$,
Step E_ν: If $\|u\| < \varepsilon$ then exit.
   Exit.

The algorithmic format of this description allows us to avoid messy indices in the sequel. Here N and ε are input parameters. The iteration is performed a maximum number of N times or until the improvement u in an iteration has a norm within a prescribed tolerance ε.

In practice, the procedure cannot be carried out as it stands since the operators H and K cannot be evaluated exactly. In our implementation, the element r remains in a finite-dimensional subspace $\mathscr{E}^n$, for a fixed integer n, and the Fredholm equation in Step C_ν is each time approximately solved by using the corresponding perturbed Galerkin method. To reduce computational complexity, the approximate solution is obtained by using the equivalent Nyström method, and only the final approximant is transformed to a Legendre polynomial expansion.

We shall use the vectors $x, y, z \in \mathbb{R}^{mn}$ with

$$z = (r_1(s_1), \ldots, r_1(s_n); \ldots; r_m(s_1), \ldots, r_m(s_n)), \tag{151}$$

$$x = (u_1(s_1), \ldots, u_1(s_n); \ldots; u_m(s_1), \ldots, u_m(s_n)), \quad (152)$$

$$y = (\xi_{11}, \ldots, \xi_{1n}; \ldots; \xi_{m1}, \ldots, \xi_{mn}). \quad (153)$$

The correspondence among components is given by

$$z_\rho = r_j(s_\mu), \quad x_\rho = u_j(s_\mu) = \eta_{j\mu}, \quad y_\rho = \xi_{j\mu} = \langle r_j, \varphi_\mu \rangle', \quad (154)$$

$$\rho = (j-1)n + \mu, \quad j = (\rho - 1) \text{ div } n + 1,$$

$$\mu = (\rho - 1) \bmod n + 1, \quad (155)$$

$$1 \le j \le m, \quad 1 \le \mu \le n, \quad 1 \le \rho \le mn. \quad (156)$$

We shall also use the vectors

$$\eta_j = (r_j(s_1), r_j(s_2), \ldots, r_j(s_n)) \in \mathbb{R}^n, \quad (157)$$

$$\zeta_\mu = (r_1(s_\mu), r_2(s_\mu), \ldots, r_m(s_\mu)) \in \mathbb{R}^m, \quad (158)$$

$$\xi_j = (\xi_{j1}, \xi_{j2}, \ldots, \xi_{jn}) \in \mathbb{R}^n. \quad (159)$$

The elements $(s_\mu, \zeta_\mu) \in [0, T] \times \mathbb{R}^m$ will be used as arguments for the component functions f_i and $\partial f_i / \partial r_j$.

The major steps of the procedure have the form:

Step A_ν: Set up $A \in \mathbb{R}^{mn \times mn}$,
Step B_ν: Set up $b \in \mathbb{R}^{mn}$,
Step C_ν: Solve $Ax = b$,
Step D_ν: $z := z + x$,
Step F: $y := Sz$.

The linear system in Step C_ν corresponds to the Nyström method applied on the Fredholm equation. We multiply both sides of linear system (133) in Theorem 5.1 by -1. The resulting matrix is $A = B - I$, where B is defined as in Theorem 5.1, while the constant vector b is the *negative* of b given in the theorem. Step F, which yields the Fourier coefficients of the Legendre polynomial expansion, is executed after the ν-loop is completed.

In what follows, we describe parallel algorithms for Steps A_ν, B_ν, C_ν, and F. These algorithms each partition the workload into m parts, with each part corresponding to one of the components of the m-dimensional problem. This natural partition, chosen for convenience and clarity, allows us to illustrate concepts without having to deal with cumbersome notation

Numerical Solution on Parallel Processors

and detail. We assume that the parallel system has m processors, each with access to a memory and each able to communicate with the remaining $m - 1$ processors. For our purposes, we need not consider the configuration of the dedicated or shared memories nor the manner in which the processors are interconnected.

6.1. Setup of the Matrix. The goal is to compute the coefficients of the matrix $A = [A_{\rho\sigma}] = B - I$, where B is defined in Theorem 5.1. The m processors work in unison, with the jth processor computing the $mn \times n$ submatrix

$$\begin{bmatrix} A_{1,(j-1)n+1} & \cdots & A_{1,jn} \\ & \cdots & \\ A_{mn,(j-1)n+1} & \cdots & A_{mn,jn} \end{bmatrix} = \begin{bmatrix} K_{1j} \\ \vdots \\ K_{jj} - I_n \\ \vdots \\ K_{mj} \end{bmatrix}, \tag{160}$$

where

$$K_{ij} = [a_\mu k_{ij}(s_\lambda, s_\mu)] \in \mathbb{R}^{n \times n}, \tag{161}$$

$$k_{ij}(t, s) = \gamma(t, s) \frac{\partial f_i}{\partial r_j}(s, r_1(s), \ldots, r_m(s)), \tag{162}$$

and I_n is the $n \times n$ identity matrix. We have

$$K_{ij} = CDF'_{ij}, \tag{163}$$

where

$$C = [\gamma(s_\lambda, s_\mu)], \tag{164a}$$

$$D = \text{diag}[a_\lambda], \tag{164b}$$

and

$$F'_{ij} = \text{diag}\left[\frac{\partial f_i}{\partial r_j}(s_1, \zeta_1), \ldots, \frac{\partial f_i}{\partial r_j}(s_n, \zeta_n)\right] \tag{165}$$

are all in $\mathbb{R}^{n \times n}$.

Every processor needs a copy of the matrix $E = CD$ and a copy of the last vector z. The jth processor requires routines for evaluating the m partial derivatives

$$\frac{\partial f_1}{\partial r_j}, \frac{\partial f_2}{\partial r_j}, \ldots, \frac{\partial f_m}{\partial r_j}. \tag{166}$$

It also uses a work array $[k_{\lambda\mu}] \in \mathbb{R}^{n \times n}$ and n vectors $\zeta_\lambda \in \mathbb{R}^m$.

The m processors work in parallel with the jth processor executing the following algorithm:

For $\lambda = 1, \ldots, n$ do:
$\quad \zeta_\lambda := (z_\lambda, z_{n+\lambda}, z_{2n+\lambda}, \ldots, z_{(m-1)n+\lambda})$
For $i = 1, \ldots, m$ do:
$\quad F'_{ij} := \text{diag}\left[\dfrac{\partial f_i}{\partial r_j}(s_1, \zeta_1), \ldots, \dfrac{\partial f_i}{\partial r_j}(s_n, \zeta_n)\right]$
$\quad [k_{\lambda\mu}] := EF'_{ij}$
$\quad$ For $\lambda, \mu = 1, \ldots, n$ do:
$\quad\quad A_{(i-1)n+\lambda,(j-1)n+\mu} := k_{\lambda\mu}$

For $\lambda = 1, \ldots, n$ do:
$\quad \rho := (j-1)n + \lambda$
$\quad A_{\rho\rho} := A_{\rho\rho} - 1$

The first λ-loop prepares the arguments ζ_λ needed in evaluating the partial derivatives. The last λ-loop represents the matrix subtraction $K_{jj} - I_n$.

6.2. Setup of the Constant Vector. The ith component of the function $v(t)$ in the Fredholm equation is given by

$$v_i(t) = \int_0^T \gamma(t, s) f_i(s, r_1(s), \ldots, r_m(s))\, ds - r_i(t) + w_i(t), \qquad (167)$$

$$w_i(t) = \frac{t}{T}(r_{Ti} - r_{0i}) + r_{0i}. \qquad (168)$$

Since the integral transform cannot be evaluated exactly, we use the n-point quadrature rule to approximate $v_i(t)$ by

$$\tilde{v}_i(t) = \sum_{\mu=1}^n a_\mu \gamma(t, s_\mu) f_i(s_\mu, r_1(s_\mu), \ldots, r_m(s_\mu)) - r_i(t) + w_i(t). \qquad (169)$$

The desired constant vector b is given by

$$b = -(\tilde{v}_1(s_1), \ldots, \tilde{v}_1(s_n); \ldots; \tilde{v}_m(s_1), \ldots, \tilde{v}_m(s_n)). \qquad (170)$$

We use the representation

$$b = -(\bar{v}_1; \ldots; \bar{v}_m), \quad (171)$$

where

$$\bar{v}_j = (\tilde{v}_j(s_1), \ldots, \tilde{v}_j(s_n)) \in \mathbb{R}^n, \quad 1 \leq j \leq m, \quad (172)$$

is the negative of the jth n-tuple in b.

We then have

$$\bar{v}_j = EF_j - \eta_j - \bar{w}_j, \quad (173)$$

where

$$F_j = \begin{bmatrix} f_j(s_1, \zeta_1) \\ \cdots \\ f_j(s_n, \zeta_n) \end{bmatrix} \in \mathbb{R}^n, \quad \bar{w}_j = \begin{bmatrix} w_j(s_1) \\ \cdots \\ w_j(s_n) \end{bmatrix} \in \mathbb{R}^n, \quad (174)$$

and $E \in \mathbb{R}^{n \times n}$, $\zeta_\lambda \in \mathbb{R}^m$, $\eta_j \in \mathbb{R}^n$ have been defined earlier.

Every processor needs a copy of the matrix E and a copy of the last vector z. The jth processor requires routines for evaluating the two-component functions

$$f_j(s, \zeta), \, w_j(s). \quad (175)$$

We suppose that each processor possesses the n vectors ζ_λ computed earlier in Section 6.1. The m processors work in parallel with the jth processor executing the following algorithm:

$$\begin{aligned}
F_j &:= [f_j(s_\lambda, \zeta_\lambda)], \\
\eta_j &:= (z_{(j-1)n+1}, z_{(j-1)n+2}, \ldots, z_{jn}), \\
\bar{v}_j &:= -\eta_j + \bar{w}_j, \\
\bar{v}_j &:= EF_j + \bar{v}_j.
\end{aligned} \quad (176)$$

The jth processor transmits a copy of $-\bar{v}_j$ to each of the remaining $m - 1$ processors. At the end of this broadcast operation, each processor has a complete copy of the vector b.

6.3. Solution of the Linear System.

We outline a parallel algorithm for the LU factorization of the matrix A stored n columns at a time among the m processors as indicated in Section 6.1. For ease of description, we omit certain details, including the interchange of elements in partial pivoting. The triangular matrices L and U are overwritten on A in the usual way.

For $\sigma = 1, 2, \ldots, mn$ do:
 $j := (\sigma - 1) \text{ div } n + 1$
 In the jth processor do:
  Find the pivot z of the σth column
  For $\rho = \sigma + 1, \ldots, mn$ do:
   $A_{\rho\sigma} := A_{\rho\sigma}/z$
   $M_\rho := A_{\rho\sigma}$

 The jth processor sends $M_{\sigma+1}, \ldots, M_{mn}$ to the $j+1, \ldots, m$th processors.
 In the kth processor, $j \leq k \leq m$, do in parallel:

$$q(k) := \begin{cases} \sigma + 1 & \text{if } k = j \\ (k-1)n + 1 & \text{if } k > j \end{cases}$$

 For $\mu = q(k), \ldots, kn$ do:
  $z := A_{\sigma\mu}$

$$\begin{bmatrix} A_{\sigma+1,\mu} \\ \vdots \\ A_{mn,\mu} \end{bmatrix} := \begin{bmatrix} A_{\sigma+1,\mu} \\ \vdots \\ A_{mn,\mu} \end{bmatrix} - z \begin{bmatrix} M_{\sigma+1} \\ \vdots \\ M_{mn} \end{bmatrix}$$

 End μ loop.
End σ loop.

For $\sigma = (j-1)n + 1, \ldots, jn$ the jth processor finds the pivot and $mn - \sigma$ multipliers of the σth column. During that phase, the remaining $m - 1$ processors are idle. Next comes a communication step during which the jth processor sends the multipliers to the $j+1, \ldots, m$th processors. The remaining part of the σ-loop is executed in parallel with $m - j + 1$ processors (or $m - j$ processors when $\sigma = jn$) each adding scalar multiples of the multiplier vector to each of its remaining active columns. This step, the last μ-loop in the algorithm, can be advantageously implemented in pipelined vector arithmetic.

Numerical Solution on Parallel Processors

After the LU factorization of A has been found, the machine solves the system $Ax = b$ by using forward elimination on $LW = b$ followed by backward elimination on $Ux = w$. The efficiency of parallel algorithms for these eliminations, as for the factorization itself, depends on both precedence constraints and communication needs. We refer to the literature mentioned at the end of this section for various alternatives.

6.4. Legendre Polynomial Expansion. We describe Step F of the procedure, namely, the transformation of the final approximant z to a polynomial interpolant. The desired orthogonal expansion for the jth component is given by

$$r_j^n(t) = \xi_{j1}\varphi_1(t) + \xi_{j2}\varphi_2(t) + \cdots + \xi_{jn}\varphi_n(t), \tag{177}$$

where, from Theorem 5.1,

$$\xi_j = S'\eta_j. \tag{178}$$

Each processor needs a copy of $S' = [a_\mu \varphi_\lambda(s_\mu)] \in \mathbb{R}^{n \times n}$ and a copy of the final approximant z. The m processors work in parallel, with the jth processor executing the jth vector-matrix multiplication (178) where η_j is given by (157) and ξ_j by (159). If n is very large, each processor can additionally use an algorithm based on recursive doubling (Ref. 5).

At this stage, we need the ability to evaluate the orthogonal expansion at given values of t. Our strategy is to have the jth processor evaluate the jth component (177) using a scheme based on the three-term recurrence relation described in Section 2.2. Each processor requires routines for evaluating the functions:

$$\beta(\lambda) = \frac{2\sqrt{(2\lambda + 1)(2\lambda + 3)}}{(\lambda + 1)T}, \quad \delta(\lambda) = \frac{\beta(\lambda)}{\beta(\lambda - 1)} = \frac{\lambda}{\lambda + 1}\sqrt{\frac{2\lambda + 3}{2\lambda - 1}}. \tag{179}$$

Assume that $t, n, \xi_{j1}, \ldots, \xi_{jn}$ are inputs and that $r_j^n(t)$ is the output of the jth processor. The m processors work in parallel, with the jth processor executing the following algorithm:

$\varepsilon_n := \xi_{jn}$
If $n = 1$ go to (*)
$\varepsilon_{n-1} := \xi_{j,n-1} + \beta(n-2) \cdot (t - T/2)\varepsilon_n$
If $n = 2$ go to (*)
For $\lambda = n - 2, n - 3, \ldots, 1$ do:
$\quad\quad \varepsilon_\lambda := \xi_{j\lambda} + \beta(\lambda - 1) \cdot (t - T/2)\varepsilon_{\lambda+1} - \delta(\lambda) \cdot \varepsilon_{\lambda+2}$
(*) $\quad r_j^n(t) := \varepsilon_1/\sqrt{T}$

In an implementation of the algorithm, only three memory locations are needed to keep track of the numbers $\varepsilon_n, \varepsilon_{n-1}, \ldots, \varepsilon_1$.

To demonstrate the validity of the algorithm, we proceed as follows. Verify easily that the algorithm is correct for $n = 1, 2$. For $n > 2$, solve for $\xi_{j\lambda}$, substitute in (177), and rearrange terms to get

$$r_j^n(t) = \varepsilon_1 \varphi_1(t) + \varepsilon_2[\varphi_2(t) - \beta(0) \cdot (t - T/2)\varphi_1(t)]$$

$$+ \sum_{\lambda=3}^{n-2} \varepsilon_\lambda [\varphi_j(t) - \beta(j-2)(t - T/2)\varphi_{\lambda-1}(t) + \delta(\lambda - 2) \cdot \varphi_{\lambda-2}(t)]$$

$$- \varepsilon_{n-1}\beta(n-3) \cdot (t - T/2)\varphi_{n-2}(t)$$

$$+ \sum_{\lambda=n-1}^{n} \varepsilon_\lambda \delta(\lambda - 2) \cdot \varphi_{\lambda-2}(t) + \xi_{j,n-1}\varphi_{n-1}(t) + \xi_{jn}\varphi_n(t). \qquad (180)$$

The expressions in brackets are zero because of the recursive definition of the polynomials φ_λ. Next, substitute $\xi_{jn} = \varepsilon_n$ and $\xi_{j,n-1} = \varepsilon_{n-1} - \varepsilon_n \beta(n-2) \cdot (t - T/2)$, rearrange terms, and again apply the recurrence relation to find $r_j^n(t) = \varepsilon_1 \varphi_1(t) = \varepsilon_1/\sqrt{T}$.

6.5. Odds and Ends.

Each of the m processors requires the weights $a_1, \ldots, a_n$ and nodes $s_1, \ldots, s_n$ of the n-point Gaussian quadrature rule for the interval $[0, T]$. These weights and nodes are given by

$$a_\lambda = \frac{T}{2} \alpha_\lambda, \qquad (181a)$$

$$s_\lambda = \frac{T}{2}(\sigma_\lambda + 1), \qquad (181b)$$

where α_λ and σ_λ are the weights and nodes of the rule for the interval $[-1, 1]$. These normalized parameters have been well tabulated (Refs. 10–11). A computer program stores α_λ and σ_λ, and then computes on-line the relations (181) to get a_λ and s_λ corresponding to a given input value of T. The nodes should be ordered $s_1 < s_2 < \cdots < s_n$ so that the relations

$$\alpha_{n+1-\lambda} = \alpha_\lambda, \qquad (182a)$$

$$\sigma_{n+1-\lambda} = -\sigma_\lambda. \qquad (182b)$$

can be used to save memory storage.

The matrix $C = [\gamma(s_\lambda, s_\mu)]$ is symmetric, thus allowing it to be stored in $n(n+1)/2$ memory locations rather than n^2. The vector $[a_\lambda]$ represents the diagonal matrix D used to define the matrix $E = CD$ in Sections 6.1–6.2.

The matrix S' used in Section 6.4 is given by

$$S' = [\varphi_{\lambda\mu}]D, \qquad (183a)$$

$$\varphi_{\lambda\mu} = \varphi_\lambda(s_\mu). \qquad (183b)$$

The three-term recurrence relation for the Legendre polynomials can be applied directly to compute the elements $\varphi_{\lambda\mu}$. The algorithm is:

For $\mu = 1, 2, \ldots, n$ do:
$$\varphi_{1\mu} := \sqrt{\frac{1}{T}}$$
$$\varphi_{2\mu} := \sqrt{\frac{3}{T^3}}(2s_\mu - T)$$
For $\lambda = 3, \ldots, n$ do:
$$\varphi_{\lambda\mu} := \beta(\lambda - 2) \cdot (s_\mu - T/2)\varphi_{\lambda-1,\mu} - \delta(\lambda - 2) \cdot \varphi_{\lambda-2,\mu}$$

The symmetry relations

$$\varphi_\lambda(T - s) = (-1)^{\lambda-1}\varphi_\lambda(s) \qquad (184)$$

and

$$s_{n+1-\mu} = T - s_\mu \qquad (185)$$

can be used to reduce both computation and storage.

We close this section with a few remarks. The parallel algorithms described above partition the computation into m components, in correspondence with the dimension of the problem. This natural "coarse-grain" partition provides excellent load balance for a system with m processors. There are many other ways to parallelize the computation. This may be needed if the system at hand does not conveniently allow the usage of m processors. Effective parallelization of our numerical technique depends on the architecture of the multiprocessor system and on the scheme used to store the matrix A. For example, for a distributed-memory and message-passing machine, such as the hypercube (Refs. 41-42), it may be convenient to allocate the columns of A one at a time to the processors. In this scheme, if there are p processors ($p \ll mn$), then the kth processor gets the k, $k + p$, $k + 2p, \ldots$, th columns of A. Corresponding parallel algorithms for the LU factorization and forward-backward eliminations are described in Ref. 5.

Background material on parallel processing can be found in several recent texts, for example, Refs. 43-48. In particular, see Ortega (Ref. 48) for descriptions of parallel algorithms for the solution of linear systems. Theoretical studies on parallel methods for factoring dense matrices on miscellaneous machine architecture have recently appeared in the periodical literature, for example, Refs. 49-55.

7. Numerical Examples

We illustrate in part the performance of our numerical technique on two simple prototype boundary value problems. The computations were done on a sequential computer. The numerical results for the second example were obtained by simulating the parallel algorithms described in the last section. We are unable at present to give timing and benchmarking results on actual parallel software, as such software needs yet to be tested on miscellaneous multiprocessor architectures. The numerical technique and corresponding parallel algorithms proposed in this paper should therefore be viewed as experimental. Preliminary results, however, look promising.

7.1. Earth-Mars Trajectories. The planet Mars is said to be at *opposition* when its projected position on the ecliptic is on the fictional line defined by the Sun and Earth. Oppositions of Mars are important because they define opportunities for Earth-to-Mars (or reverse) trajectories with minimum energy. A spacecraft on such a trajectory needs to leave Earth about 97 days before opposition and arrive at Mars about 162 days after opposition, though several factors will cause the actual flight time to vary by up to several weeks. Taking advantage of an opposition of Mars that occurred on 15 December, 1975, the Viking I and II probes were launched by NASA on 20 August and 9 September of that year, and arrived at Mars on 19 June and 7 August of the following year.

The geometry of such a trajectory is illustrated in Fig. 3. The path of the spacecraft is a semiellipse, with focus at the Sun, tangent to the orbit of the Earth at start of flight, and tangent to the orbit of Mars at arrival. This semiellipse, called a *Hohmann orbit*, is the trajectory with least energy for a transfer between two circular and coplanar orbits. The spacecraft will move 180° in $T = 259$ days. The Earth will have moved 259/365 of a period or 255°. Thus the Earth will be 255°–180° = 75° ahead of Mars when the spacecraft arrives. Since the mean daily motion of the Earth is 0.462°/day faster than that of Mars, opposition of Mars occurs 75°/(0.462°/day) = 162 days before the spacecraft's arrival. Consequently, a flight to Mars needs to leave Earth 97 days before opposition. Actual times may differ from these estimates by several weeks due to the noncircular and noncoplanar orbits of Earth and Mars (the eccentricities of these orbits are 0.0167 and 0.0934, respectively, and the inclination of Mars to the ecliptic is 1° 51′ 0″).

We described Hohmann transfer orbits because of their basic importance in astrodynamics and because one can use two-point boundary value problems to analyze perturbed versions of such orbits. However, we shall illustrate our numerical technique on a type of Earth-to-Mars trajectory, which exploits the orbital motions of the planets to boost the spacecraft toward its destination, and thereby considerably reduce its flight time.

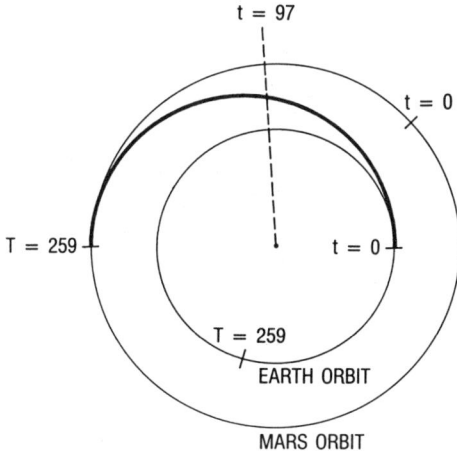

Fig. 3. Hohmann orbit to Mars. Mars is at opposition 97 days after launch of the spacecraft.

Specifically, we consider a two-point boundary value problem (1)-(2), with $m = 3$, in which

$$f(t, r) = g(r) + g^M(r) \tag{186}$$

is defined by (64), where $\sqrt{\mu} = k$ is the Gaussian gravitational constant, and where the perturbing effect of all $M = 9$ major planets is included. The initial position r_0 of the spacecraft is a point on the sphere of influence of the Earth on 11 August 1973. The end point r_T is on the sphere of influence of Mars on 27 December 1973, namely, $T = 128$ days later. The resulting trajectory of the spacecraft is a near-elliptic arc. Its semimajor axis is about 1.4 au, its eccentricity is about 0.324, and its inclination to the ecliptic is about 0° 25′. Refer to Richard and Roth (Ref. 56), Nacozy and Feagin (Ref. 57), and Feagin (Ref. 29).

Figure 4 illustrates the positions of Earth and Mars during the flight of the spacecraft. The trajectory is typical of those that occur when the planets are configured so that at the start of the trajectory the position vector of Mars is about 23° ahead of that of Earth. Such propitious configurations occur roughly every 25 months.

The positions of the $M = 9$ planets are obtained from a standard ephemeris tape. The positions $r^j(s_\lambda)$ at the Gaussian nodes are calculated by interpolation formulas and tabulated for subsequent use. The initial approximation is taken to be the rectilinear motion from the initial point r_0 to the endpoint r_T. The number of terms in the desired Legendre polynomial expansion is $n = 64$. The exit criterion in Step E_ν is taken to be

$$\sum_{\lambda=1}^{n} a_\lambda (|u_1(s_\lambda)|^2 + |u_2(s_\lambda)|^2 + |u_3(s_\lambda)|^2) \le 10^{-18} T. \tag{187}$$

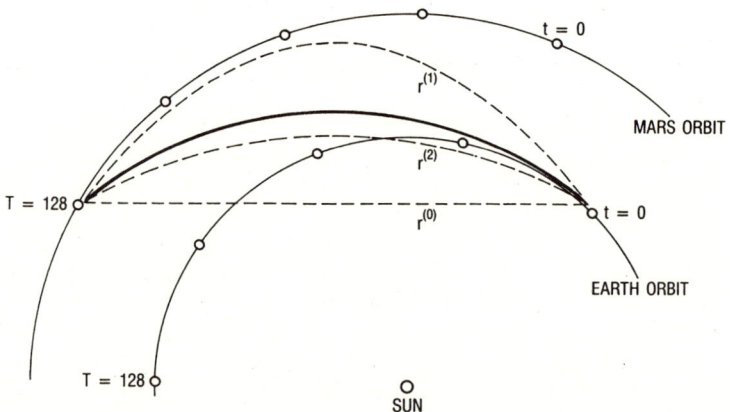

Fig. 4. An Earth-to-Mars trajectory. The positions of the planets are indicated every 32 days.

The quadrature on the left side approximates the quantity

$$T \max_{0 \le t \le T} \|u(t)\|^2. \tag{188}$$

The exit criterion thus translates to a numerical accuracy of about $\varepsilon = 10^{-9}$ au ≈ 150 m. The procedure required $N = 8$ quasi-linearizations in order to reach this accuracy. The character of the first two iterates is illustrated in Fig. 4. The numerical results compare well with those obtained by Nacozy and Feagin (Refs. 57–58).

7.2. Trajectory Optimization. The optimization of certain rendezvous maneuvers can be handled by quasi-linearization of two-point boundary value problems; see Bryson and Ho (Ref. 59), Tapley and Lewallen (Ref. 60), Radbill and McCue (Ref. 61), and Feagin (Ref. 29). We briefly describe a relatively simple case dealing with the optimization of a trajectory of a space vehicle from a given position r_0 on the sphere of influence of a planet to a given position r_T on the sphere of influence of a neighboring planet. The time of flight T, as well as the initial and terminal velocities $\dot{r}_0$ and $\dot{r}_T$, are specified to ensure that the spacecraft meets rendezvous conditions with the second planet. The motion of the vehicle is governed by the gravitational attraction of the Sun as well as a variable thrust. For simplicity, it is assumed that the trajectory is two-dimensional and that the mass of the vehicle does not vary.

The problem is to minimize the functional

$$J(\theta) = \int_0^T \|\theta(t)\|^2 \, dt, \tag{189}$$

where $\theta(t) \in \mathbb{R}^2$, assumed to be twice differentiable over $[0, T]$, is the acceleration due to the thrust. Letting $r(t) \in \mathbb{R}^2$ be the heliocentric position vector of the vehicle, a necessary condition for a minimum requires that the two vectors $r(t)$ and $\theta(t)$ satisfy the differential equations

$$\ddot{r}(t) = -\mu r(t)/R(t)^3 + \theta(t) \tag{190}$$

and

$$\ddot{\theta}(t) = -\mu \Omega(t) \theta(t)/R(t)^3, \tag{191}$$

subject to the boundary conditions

$$r(0) = r_0, \tag{192a}$$

$$r(T) = r_T, \tag{192b}$$

$$\dot{r}(0) = \dot{r}_0, \tag{192c}$$

$$\dot{r}(T) = \dot{r}_T. \tag{192d}$$

Here Ω is the matrix

$$\Omega = \begin{bmatrix} 1 - 3r_1^2/R^2 & 0 \\ 0 & 1 - 3r_2^2/R^2 \end{bmatrix}, \tag{193}$$

and R is the Euclidean norm $R(t) = \|r(t)\|$.

The system of four second-order differential equations (191) and the eight boundary conditions (192) constitute a two-point boundary value problem of dimension $m = 4$ in the unknown vector $u = (r_1, r_2, \theta_1, \theta_2)$. The equivalent Hammerstein integral equation can be solved numerically using the method of discretized quasi-linearization described earlier. The initial approximation $u^0(t)$ is taken as the rectilinear motion from r_0 to r_T with zero thrust $\theta(t) \equiv 0$. Typically, numerical accuracy to $\varepsilon \le 150$ meters requires 15-20 Newton iterations with a Legendre expansion of $n = 32$ terms. Such results compare well with those of Feagin (Ref. 29).

Figure 5 describes the character of the acceleration $\theta(t)$ (in 5×10^{-4} au/day^2 units) during a prototype Earth-to-Mars trajectory. It is interesting to observe that the thrust changes direction by nearly 180° at about the time the Euclidean norm $\|\theta(t)\|$ is at minimum.

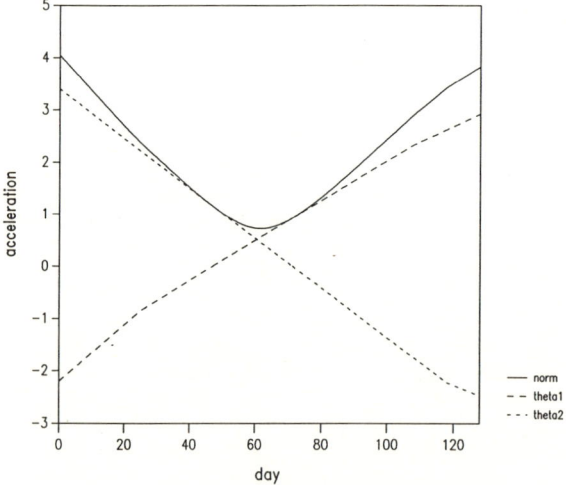

Fig. 5. Optimized acceleration due to thrust.

8. Conclusion

We presented an experimental method for numerically solving two-point boundary value problems typical of applications in astrodynamics. The numerical technique, which is applied on the equivalent Hammerstein integral equation, offers several benefits, among them the convenience of an orthogonal expansion and seemingly robust parallelization for implementation on multiprocessor systems. However, much research remains to be done. We indicate some of the areas of immediate interest:

1. Analysis and comparison of various parallelizations of the numerical method.
2. Corresponding implementations on different types of parallel architectures and related numerical experimentation.
3. Study of techniques based on parallel computing for solving ill-posed two-point boundary value problems.

As a final note, we point out that the numerical method, described in this paper within the context of astrodynamics, can be applied advantageously in diverse scientific disciplines.

References

1. GALLAGHER, L. J., AND PERLIN, I. E., *Use of Green's Function in the Numerical Solution of Two-Point Boundary Value Problems*, Proceedings of the Conference on the Numerical Solution of Ordinary Differential Equations—October 1972, Edited by D. G. Bettis, Lecture Notes in Mathematics No. 362, Springer-Verlag, New York, New York, 1974.
2. FIRNETT, P. J., AND TROESCH, B. A., *Shooting Splitting Methods for Sensitive Two-Point Boundary Value Problems*, Proceedings of the Conference on the Numerical Solution of Ordinary Differential Equations—October 1972, Edited by D.G. Bettis, Lecture Notes in Mathematics No. 362, Springer-Verlag, New York, New York, 1974.
3. RHEINBOLDT, W. C., *Methods for Solving Systems of Nonlinear Equations*, Society for Industrial and Applied Mathematics, Philadelphia, Pennsylvania, 1974.
4. KELLER, H. B., *Numerical Methods for Two-Point Boundary-Value Problems*, Blaisdell Publishing Company, Waltham, Massachusetts, 1968.
5. MIEL, G., *Parallel Solution of Fredholm Integral Equations of the Second Kind by Orthogonal Polynomial Expansions*, Applied Numerical Mathematics, Vol. 5, pp. 345-361, 1989.
6. KRASNOSEL'SKII, M. A., et al., *Approximate Solution of Operator Equations*, Translated from Russian by D. Louvish, Wolters-Noordhoff Publishing, Groningen, Netherlands, 1972.

7. NOBLE, B., *The Numerical Solution of Nonlinear Integral Equations and Related Topics*, Nonlinear Integral Equations, Edited by P. M. Anselone, University of Wisconsin Press, Madison, Wisconsin, pp. 216-318, 1964.
8. RALL, L. B., *Computational Solution of Nonlinear Operator Equations*, John Wiley and Sons, New York, New York, 1969.
9. JERRI, A. J., *Introduction to Integral Equations with Applications*, Marcel Dekker, New York, New York, 1985.
10. ABRAMOWITZ, M., AND STEGUN, I. A., *Handbook of Mathematical Functions with Formulas, Graphs, and Mathematical Tables*, Dover, New York, New York, 1968.
11. STROUD, A. H., AND SECREST, D., *Gaussian Quadrature Formulas*, Prentice-Hall, Englewood Cliffs, New Jersey, 1966.
12. KANTOROVICH, L. V., *Functional Analysis and Applied Mathematics*, Uspekhi Matematicheskikh Nauk, Vol. 3, p. 89, 1948.
13. BELLMAN, R., *Functional Equations in the Theory of Dynamic Programming, V. Positivity and Quasi-linearity*, Proceedings of the National Academy of Sciences of the United States, Vol. 41, p. 743, 1955.
14. BELLMAN, R., AND KALABA, R. E., *Quasilinearization and Nonlinear Boundary Value Problems*, American Elsevier Publishing Company, New York, New York, 1965.
15. MIEL, G., *Majorizing Sequences and Error Bounds for Iterative Methods*, Mathematics of Computation, Vol. 34, No. 149, pp. 185-202, 1980.
16. MIEL, G., *Semilocal Analysis of Equations with Smooth Operators*, International Journal of Mathematics and Mathematical Sciences, Vol. 4, No. 3, pp. 553-563, 1981.
17. YAMAMOTO, T., *A Method for Finding Sharp Error Bounds for Newton's Method Under the Kantorovich Assumptions*, Numerische Mathematik, Vol. 49, Nos. 2-3, pp. 203-220, 1986.
18. YAMAMOTO, T., *Error Bounds for Newton-like Methods Under Kantorovich Type Assumptions*, Japan Journal of Applied Mathematics, Vol. 3, No. 2, pp. 259-313, 1986.
19. GOLBERG, M. A., *A Survey of Numerical Methods for Integral Equations*, Solution Methods for Integral Equations—Theory and Applications, Edited by M. A. Golberg, Plenum Press, New York, New York, pp. 1-58, 1979.
20. BATTIN, R. H., *Astronautical Guidance*, McGraw-Hill, New York, New York, 1964.
21. BATTIN, R. H., *An Introduction to the Mathematics and Methods of Astrodynamics*, AIAA Education Series, Washington, DC, 1987.
22. ESCOBAL, P. R., *Methods of Orbit Determination*, John Wiley and Sons, New York, New York, 1965.
23. ESCOBAL, P. R., *Methods of Astrodynamics*, Robert E. Krieger Publishing Company, Huntington, New York, New York, 1979.
24. *Users' Description of Jet Propulation Laboratories Ephemeris Tapes*, JPL Technical Report No. 32-580, 1964.
25. PEABODY, P. R., SCOTT, J. F., AND OROZCO, E. G., *JPL Ephemeris Tapes E9510, E9511, and E9512*, JPL Technical Memorandum No. 33-167, 1964.

26. LUNDQUIST, C. A., AND VEIS, G., *Geodetic Parameters for a 1966 Smithsonian Institution Standard Earth*, Special Report No. 200, Volumes 1-3, 1966.
27. RIZOS, C., AND STOLZ, A., *Force Modelling for GPS Satellite Orbits*, Proceedings First International Symposium on Precise Positioning with the Global Positioning System, U.S. Department of Commerce, NOAA, Rockville, Maryland, 1985.
28. COLLATZ, L., *Functional Analysis and Numerical Mathematics*, Academic Press, New York, New York, 1966.
29. FEAGIN, T. W., *The Numerical Solution of Two Point Boundary Value Problems Using the Chebyshev Polynomial Series*, University of Texas at Austin, PhD Thesis, 1972.
30. CHATELIN, F., *Spectral Approximations of Linear Operators*, Academic Press, New York, New York, 1983.
31. SCHOCK, E., *Arbitrarily Slow Convergence, Uniform Convergence and Superconvergence of Galerkin-like Methods*, IMA Journal on Numerical Analysis (to appear).
32. SPENCE, A., AND THOMAS, K. S., *On Superconvergence Properties of Galerkin's Method for Compact Operator Equations*, IMA Journal on Numerical Analysis, Vol. 3, pp. 253-271, 1983.
33. GOLBERG, M., *Galerkin Methods for CSIES of the First Kind*, Numerical Solution of Singular Integral Equations, Edited by A. Gerasoulis and R. Vichnevetsky, IMACS, Rutgers University, 1984.
34. GOLBERG, M. A., LEA, M., AND MIEL, G., *A Superconvergence Result for the Generalized Airfoil Equation with Application to the Flap Problem*, Journal of Integral Equations, Vol. 5, pp. 175-186, 1983.
35. MIEL, G., *Rates of Convergence and Superconvergence of Galerkin's Method for the Generalized Airfoil Equation*, Numerical Solution of Singular Integral Equations, Edited by A. Gerasoulis and R. Vichnevetsky, IMACS, Rutgers University, 1984.
36. VAINIKKO, G. M., *Perturbed Galerkin Method and General Theory of Approximate Methods for Nonlinear Equations*, Zhurnal Vychislitel'noi Matematiki i Matematicheskoi Fiziki, Vol. 7, 1967.
37. MIEL, G., *Perturbed Projection Methods for Split Equations of the First Kind*, Integral Equations and Operator Theory, Vol. 8, pp. 268-275, 1985.
38. LINZ, P., *Theoretical Numerical Analysis*, Wiley-Interscience, New York, New York, 1979.
39. LINZ, P., *Stability Analysis for the Numerical Solution of Singular Integral Equations*, Numerical Solution of Singular Integral Equations, Edited by A. Gerasoulis and R. Vichnevetsky, IMACS, Rutgers University, 1984.
40. MIEL, G., *On the Galerkin and Collocation Methods for a Cauchy Singular Integral Equations*, SIAM Journal on Numerical Analysis, Vol. 23, No. 1, pp. 135-143, 1986.
41. HEATH, M. T., Editor, *Hypercube Multiprocessors 1986*, SIAM, Philadelphia, Pennsylvania, 1986.
42. HEATH, M. T., Editor, *Hypercube Multiprocessors 1987*, SIAM, Philadelphia, Pennsylvania, 1987.

43. SCHENDEL, U., *Introduction to Numerical Methods for Parallel Computers*, Translated from German by B. W. Conolly, Halstead Press, New York, New York, 1984.
44. SINGH, M. G., et al., Editors, *Parallel Processing Techniques for Simulation*, Plenum Press, New York, New York, 1986.
45. GEAR, C. W., AND VOIGT, R. G., Editors, *Parallel Processing for Scientific Computing*, SIAM Journal on Scientific and Statistical Computing, Vol. 8, No. 1 and No. 2, 1987.
46. QUINN, M. J., *Designing Efficient Algorithms for Parallel Computers*, McGraw-Hill, New York, New York, 1987.
47. GIBBONS, A., AND RYTTER, W., *Efficient Parallel Algorithms*, Cambridge University Press, New York, New York, 1988.
48. ORTEGA, J. M., *Introduction to Parallel and Vector Solution of Linear Systems*, Plenum Press, New York, New York, 1988.
49. NETA, B., AND TAI, H. M., *LU Factorization on Parallel Computers*, Computational Mathematics and Applications, Vol. 11, pp. 573-579, 1985.
50. O'LERAY, D. P., AND STEWART, G. W., *Data-Flow for Parallel Matrix Computations*, Communications of the ACM, Vol. 28, pp. 840-853, 1985.
51. IPSEN, I. C. F., YOUCEF, S., AND SCHULTZ, M. H., *Complexity of Dense-Linear-System Solution on a Multiprocessor Ring*, Linear Algebra and Applications, Vol. 77, pp. 205-239, 1986.
52. YOUCEF, S., *Communication Complexity of the Gaussian Elimination Algorithm on Multiprocessors*, Linear Algebra and Applications, Vol. 77, pp. 315-340, 1986.
53. CAPELLO, P. R., *Gaussian Elimination on a Hybercube Automaton*, Journal on Parallel and Distributed Computing, Vol. 4, pp. 288-308, 1987.
54. JOHNSSON, S. L., *Communication Efficient Basic Linear Computations on Hypercube Architectures*, Journal on Parallel and Distributed Computing, Vol. 4, pp. 133-172, 1987.
55. ORTEGA, J. M., AND ROMINE, C. H., *The ijk Forms of Factorization Methods II. Parallel Systems*, Parallel Computing, Vol. 7, pp. 149-160, 1988.
56. RICHARD, R. J., AND ROTH, R. Y., *Earth-Mars Trajectories—1973*, JPL Technical Memorandum No. 33-100, Vol. 5, Part C, 1965.
57. NACOZY, P. E., AND FEAGIN, T., *Chebyshev Series-Solutions of Swing-By Trajectories*, AIAA Paper No. 71-192, Technical Information Service, New York, New York, 1971.
58. NACOZY, P. E., AND FEAGIN, T., *Approximation of Interplanetary Trajectories by Chebyshev Series*, AIAA Journal, Vol. 10, No. 3, pp. 243-244, 1972.
59. BRYSON, A. E., AND HO, Y., *Applied Optimal Control*, Blaisdell Publishing Company, Waltham, Massachusetts, 1969.
60. TAPLEY, B. D., AND LEWALLEN, J. M., *Comparison of Several Numerical Optimization Methods*, Journal of Optimization Theory and Applications, Vol. 1, No. 1, pp. 1-32, 1967.
61. RADBILL, J. R., AND MCCUE, G. A., *Quasilinearization and Nonlinear Problems in Fluid and Orbital Mechanics*, American Elsevier, New York, New York, 1970.

5

Introduction to the Numerical Solution of Cauchy Singular Integral Equations

M. A. GOLBERG

Abstract. In this chapter we present a survey of many of the polynomial-based projection and quadrature methods that have been developed in the past 15 years for the numerical solution of Cauchy singular integral equations with constant coefficients on $(-1, 1)$. Emphasis is placed on equations of the first kind, where an elementary proof of the inversion formula for the airfoil equation due to Peters enables one to develop the theory and subsequent numerical analysis using techniques which are well known for Fredholm equations of the second kind. Convergence proofs are given for many of the algorithms, and numerical examples are drawn from the literature to illustrate the efficiency of these methods when the data are smooth. However, in many practical problems where either the kernel and/or the right-hand side may have discontinuities, this is not the case, and further work needs to be done. Some possibilities for doing this are examined as well.

1. Introduction

In the past decade there has been a substantial increase in interest in the numerical solution of Cauchy singular integral equations (CSIES) of the form

$$a(x)v(x) + \frac{b(x)}{\pi} \int_{-1}^{1} \frac{v(t)\, dt}{t-x} + \int_{-1}^{1} k(x,t)v(t)\, dt = f(x), \quad -1 < x < 1, \tag{1}$$

where $k(x, t)$ is a Fredholm kernel (Refs. 1–5). These equations occur in a wide variety of mathematical, physical, and engineering problems, and so

---

M. A. GOLBERG • Department of Mathematical Sciences, University of Nevada, Las Vegas, Nevada 89154.

there has been a growing need to find efficient methods for their solution (Refs. 1-2, 4, 6). Although much remains to be done, when $a(x)$ and $b(x)$ are constants, there is a reasonably complete theory of polynomial expansion and related quadrature methods which will be developed in some detail in this chapter. When either $a(x)$ or $b(x)$ is not constant, these methods can be generalized under certain conditions, and this topic will be taken up by Elliott in Chapter 6. Recently, a number of researchers have begun to study the use of spline approximations (Refs. 6-8), and some aspects of this approach are surveyed in Ref. 6.

In this chapter we will begin our study of numerical methods by specializing Eq. (1) to the case where $a = 0$ and $b = 1$ (equations of the first kind), where it appears that the theory and numerical implementation are most developed (Refs. 1-4). Here Eq. (1) becomes

$$\frac{1}{\pi}\int_{-1}^{1} \frac{v(t)\,dt}{t-x} + \int_{-1}^{1} k(x,t)v(t)\,dt = f(x), \qquad -1 < x < 1, \qquad (2)$$

which we refer to as the generalized airfoil equation (GAE)—a terminology coined because when $k(x,t) = 0$ in Eq. (2) it reduces to the airfoil equation

$$\frac{1}{\pi}\int_{-1}^{1} \frac{v(t)\,dt}{t-x} = f(x), \qquad -1 < x < 1, \qquad (3)$$

in aerodynamics (Refs. 4, 9).

As we proceed, it will become apparent that many of the numerical algorithms currently in use depend on knowing the analytical solution to Eq. (1) when $k(x,t) = 0$ (this is called the dominant equation). This solution is well known (at least to workers in the field), but even when $a(x)$ and $b(x)$ are constant, the standard derivation makes use of complex variable techniques which may not be familiar to all readers (Ref. 10). Since our emphasis will be on numerical methods for solving Eq. (2), and because we want this chapter to be reasonably self-contained, we shall start our analysis with a clever elementary derivation of the solution of Eq. (3) due to Peters (Ref. 11). This solution will bring out two important features that need to be accounted for when solving Eq. (1): (i) in general, no matter how smooth $k(x,t)$ and $f(x)$ are, the solution $v(x)$ will have singularities at $x = \pm 1$; (ii) the solution is usually not unique.

Having solved Eq. (3), the inversion formula can then be used in a number of ways to develop various classes of numerical algorithms for solving Eq. (2). This will be done in detail in Sections 4-16 and, as we shall see, the crucial factor is the action of the integral operator in Eq. (3) on functions of the form $v(t) = w(t)p(t)$, where $w(t)$ is a "weight" function derived from the inversion formulas and $p(t)$ is a polynomial. In Section 17 these ideas will be generalized to produce numerical algorithms for solving Eq. (1) when $a(x)$ and $b(x)$ are nonzero constants. In Section 18 some other polynomial expansion methods will be developed that do not always account for the singularities in $v(t)$. While in Section 19 we will prove the convergence of many of the algorithms discussed in Sections 4-17.

2. Analytical Solution of the Airfoil Equation

As indicated in the Introduction, we begin our study of Eq. (2) with an "elementary" derivation of the solution of Eq. (3). For this we need the classical Abel inversion formulas which we state as Theorem 2.1 (Ref. 12).

Theorem 2.1. Consider the integral equations

$$\int_0^x \frac{u(t)\, dt}{\sqrt{x-t}} = f(x), \qquad 0 \le t \le x \le 1, \tag{4}$$

and

$$\int_x^1 \frac{v(t)\, dt}{\sqrt{t-x}} = f(x), \qquad 0 \le x \le t \le 1. \tag{5}$$

Then

$$u(x) = \frac{d}{dx}\left[\frac{1}{\pi}\int_0^x \frac{f(t)\, dt}{\sqrt{x-t}}\right], \tag{6}$$

and

$$v(x) = -\frac{d}{dx}\left[\frac{1}{\pi}\int_x^1 \frac{f(t)\, dt}{\sqrt{t-x}}\right]. \tag{7}$$

Theorem 2.2. See Ref. 11. If $f(x)$ is Hölder continuous, the airfoil equation

$$\fint_{-1}^1 \frac{v(\xi)\, d\xi}{\xi - x} = f(x), \qquad -1 < x < 1, \tag{8}$$

has the solution

$$v(x) = \frac{1}{\pi^2\sqrt{1-x^2}}\fint_{-1}^1 \frac{\sqrt{1-\xi^2}f(\xi)\, d\xi}{x - \xi} + \frac{c}{\pi\sqrt{1-x^2}}, \tag{9}$$

where
$$c = \int_{-1}^{1} v(\xi)\, d\xi. \tag{10}$$

Proof. Since the proof of (9) is fairly long and algebraically tedious, we merely outline the major steps. The reader interested in all of the details should consult Ref. 11.

To begin, we make a simple scale transformation which converts Eq. (8) to

$$\fint_0^1 \frac{\phi(\xi)\, d\xi}{\xi - x} = \tilde{f}(x), \qquad 0 < x < 1, \tag{11}$$

and then Eq. (11) will be solved, and the inverse scale transformation applied to the solution yields (9).

Now multiply Eq. (11) by x, and add and subtract

$$\fint_0^1 \frac{\xi \phi(\xi)\, d\xi}{\xi - x}$$

to the left-hand side of Eq. (11) giving

$$\fint_0^1 \frac{\xi \phi(\xi)\, d\xi}{\xi - x} = x\tilde{f}(x) + c, \tag{12}$$

where $c = \int_0^1 \phi(\xi)\, d\xi$. Dividing both sides of (12) by $\sqrt{x}$, it becomes

$$\fint_0^1 \frac{\xi \phi(\xi)\, d\xi}{[\sqrt{\xi^2} - \sqrt{x^2}]\sqrt{x}} = \sqrt{x}\tilde{f}(x) + c/\sqrt{x}. \tag{13}$$

Then integrating both sides of Eq. (13) we arrive at the equation

$$-\int_0^1 \log\left|\frac{\sqrt{\xi} - \sqrt{x}}{\sqrt{\xi} + \sqrt{x}}\right| \sqrt{\xi}\, \phi(\xi)\, d\xi = \int_0^x \sqrt{\xi}\, \tilde{f}(\xi)\, d\xi + 2c\sqrt{x}. \tag{14}$$

By standard calculus manipulations it can be shown that the logarithmic part of the kernel of Eq. (14) can be written as

$$-\log\left|\frac{\sqrt{\xi} - \sqrt{x}}{\sqrt{\xi} + \sqrt{x}}\right| = \begin{cases} \displaystyle\int_0^{\xi} \frac{d\sigma}{\sqrt{\xi - \sigma}\sqrt{x - \sigma}}, & x > \xi, \\ \displaystyle\int_0^{x} \frac{d\sigma}{\sqrt{\xi - \sigma}\sqrt{x - \sigma}}, & \xi > x, \end{cases} \tag{15}$$

and substituting (15) into (14) gives

$$\int_0^x \sqrt{\xi}\phi(\xi) \left[\int_0^\xi \frac{d\sigma}{\sqrt{\xi-\sigma}\sqrt{x-\sigma}} \right] d\xi + \int_x^1 \sqrt{\xi}\phi(\xi) \left[\int_0^x \frac{d\sigma}{\sqrt{\xi-\sigma}\sqrt{x-\sigma}} \right] d\xi$$

$$= \int_0^x \sqrt{\xi}\tilde{f}(\xi)\, d\xi + 2c\sqrt{x}. \tag{16}$$

Changing the order of integration in both integrals on the left-hand side of Eq. (16) it becomes

$$\int_0^x \frac{1}{\sqrt{x-\sigma}} \left[\int_\sigma^x \frac{\sqrt{\xi}\phi(\xi)\, d\xi}{\sqrt{\xi-\sigma}} \right] d\sigma + \int_0^x \frac{1}{\sqrt{x-\sigma}} \left[\int_x^1 \frac{\sqrt{\xi}\phi(\xi)\, d\xi}{\sqrt{\xi-\sigma}} \right] d\sigma$$

$$= \int_0^x \sqrt{\xi}\tilde{f}(\xi)\, d\xi + 2c\sqrt{x}, \tag{17}$$

and combining the inner integrals shows that Eq. (17) can be written as an Abel integral equation

$$\int_0^x \frac{\psi(\sigma)\, d\sigma}{\sqrt{x-\sigma}} = \int_0^x \sqrt{\xi}\tilde{f}(\xi)\, d\xi + 2c\sqrt{x} = g(x), \tag{18}$$

where

$$\psi(\sigma) = \int_\sigma^1 \frac{\sqrt{\xi}\phi(\xi)\, d\xi}{\sqrt{\xi-\sigma}}. \tag{19}$$

Using Eq. (4)

$$\psi(\sigma) = \frac{1}{\pi} \frac{d}{d\sigma} \int_0^\sigma \frac{g(x)}{\sqrt{\sigma-x}}\, dx, \tag{20}$$

and because $g(x)$ is differentiable for $x > 0$,

$$\psi(\sigma) = \frac{1}{\pi}\left[\frac{g(0)}{\sqrt{\sigma}} + \int_0^\sigma \frac{g'(x)\, dx}{\sqrt{\sigma-x}} \right], \tag{21}$$

which may be obtained from Eq. (20) by an integration by parts. Since

$g(0) = 0$ and $g'(x) = \sqrt{x}\tilde{f}(x) + c/\sqrt{x}$,

$$\psi(\sigma) = \frac{1}{\pi}\left[\int_0^\sigma \frac{\sqrt{x}\tilde{f}(x)\,dx}{\sqrt{\sigma-x}} + c\int_0^\sigma \frac{dx}{\sqrt{x}\sqrt{\sigma-x}}\right]$$

$$= \frac{1}{\pi}\int_0^\sigma \frac{\sqrt{x}\tilde{f}(x)\,dx}{\sqrt{\sigma-x}} + c, \tag{22}$$

because $\int_0^\sigma dx/\sqrt{x}\sqrt{\sigma-x} = \pi$.

From the form of $\psi(\sigma)$, we see that Eq. (19) is an Abel equation of the form in (5) and it can be solved for $\sqrt{\xi}\phi(\xi)$ by (7) giving

$$\sqrt{\xi}\phi(\xi) = -\frac{1}{\pi}\frac{d}{d\xi}\int_\xi^1 \frac{c\,d\sigma}{\sqrt{\sigma-\xi}} - \frac{1}{\pi^2}\frac{d}{d\xi}\int_\xi^1 \frac{1}{\sqrt{\sigma-\xi}}\left[\int_0^\sigma \frac{\sqrt{x}\tilde{f}(x)\,dx}{\sqrt{\sigma-x}}\right]d\sigma$$

$$= \frac{c}{\pi\sqrt{1-\xi}} - \frac{1}{\pi^2}\frac{d\gamma(\xi)}{d\xi}, \tag{23}$$

where $\gamma(\xi)$ is the integral on the right-hand side of (23). Thus, we have solved (11), but unfortunately this is not the standard form of the solution (Ref. 9). However, a little more algebra eventually gives expression (9).

Interchanging the order of integration in $\gamma(\xi)$,

$$\gamma(\xi) = \int_\xi^1 \sqrt{x}\tilde{f}(x)\left[\int_x^1 \frac{d\sigma}{\sqrt{\sigma-\xi}\sqrt{\sigma-x}}\right]dx$$

$$+ \int_0^\xi \sqrt{x}\tilde{f}(x)\left[\int_\xi^1 \frac{d\sigma}{\sqrt{\sigma-\xi}\sqrt{\sigma-x}}\right]dx, \tag{24}$$

and then by using

$$\log\left|\frac{\sqrt{1-\xi}+\sqrt{1-x}}{\sqrt{1-\xi}-\sqrt{1-x}}\right| = \begin{cases} \int_x^1 \frac{d\sigma}{\sqrt{\sigma-\xi}\sqrt{\sigma-x}}, & x > \xi, \\ \int_\xi^1 \frac{d\sigma}{\sqrt{\sigma-\xi}\sqrt{\sigma-x}}, & \xi > x, \end{cases} \tag{25}$$

we find that

$$\sqrt{\xi}\phi(\xi) = \frac{c}{\pi\sqrt{1-\xi}} - \frac{1}{\pi^2}\frac{d}{d\xi}\int_0^1 \log\left(\left|\frac{\sqrt{1-\xi}+\sqrt{1-x}}{\sqrt{1-\xi}-\sqrt{1-x}}\right|\right)\sqrt{x}\tilde{f}(x)\,dx. \tag{26}$$

Numerical Solution of Cauchy Singular Integral Equations

Taking the differentiation under the integral sign gives

$$\phi(\xi) = \frac{c}{\pi\sqrt{\xi}\sqrt{1-\xi}} - \frac{1}{\pi^2}\int_0^1 \frac{\sqrt{x}\sqrt{1-x}\tilde{f}(x)\,dx}{\sqrt{1-\xi}\sqrt{\xi}(x-\xi)}, \qquad (27)$$

and relabeling the variables $\xi \to x$, $x \to \xi$, Eq. (27) becomes

$$\phi(x) = \frac{c}{\pi^2\sqrt{x}\sqrt{1-x}} - \frac{1}{\pi^2}\int_0^1 \frac{\sqrt{\xi}\sqrt{1-\xi}\tilde{f}(\xi)\,d\xi}{\sqrt{x(1-x)}(\xi-x)}$$

$$= \frac{1}{\pi^2\sqrt{x(1-x)}}\left[\int_0^1 \frac{\sqrt{\xi(1-\xi)}\tilde{f}(\xi)\,d\xi}{x-\xi} + \pi c\right]. \qquad (28)$$

To transform (28) back to the standard interval $[-1, 1]$, make the change of variables $t = 2x - 1$, $u = 2\xi - 1$, so that

$$\phi[(1+t)/2] = \frac{1}{\pi^2}\frac{4}{\sqrt{1-t^2}}\int_{-1}^1 \frac{\sqrt{1-u^2}\tilde{f}[(u+1)/2]\,du}{t-u} + \pi c$$

$$= \frac{1}{\pi^2}\frac{1}{\sqrt{1-t^2}}\int_{-1}^1 \frac{\sqrt{1-u^2}\tilde{f}[(u+1)/2]\,du}{t-u} + \pi c. \qquad (29)$$

Now labeling the dependent variables so that $\phi[(t+1)/2] = v(t)$ and $\tilde{f}[(u+1)/2] = f(u)$, Eq. (29) becomes

$$v(t) = \frac{1}{\pi^2\sqrt{1-t^2}}\left[\int_{-1}^1 \frac{\sqrt{1-u^2}f(u)\,du}{t-u} + \pi c\right], \qquad (30)$$

and we get the standard solution to the airfoil equation (Ref. 9). If we revert back to our original variables by changing t to x and u to t in Eq. (30), then

$$v(x) = \frac{1}{\pi^2}\frac{1}{\sqrt{1-x^2}}\left[\int_{-1}^1 \frac{\sqrt{1-t^2}f(t)\,dt}{x-t} + \pi c\right],$$

with $c = \int_{-1}^1 v(t)\,dt$. □

Since it is convenient to write the airfoil equation as in (3), replacing $f(x)$ in (8) with $\pi f(x)$ gives the solution to the airfoil equation (3) in its most common form

$$v(x) = \frac{1}{\pi\sqrt{1-x^2}}\left[\int_{-1}^{1} \frac{\sqrt{1-t^2}f(t)\,dt}{x-t} + c\right], \tag{31}$$

which is the inversion formula that we shall use in our subsequent work.

As mentioned previously, Eq. (31) clearly shows the presence of square-root singularities at $x = \pm 1$ and the fact that the solution is not unique. Again we emphasize that these two factors should be accounted for when devising numerical algorithms for solving Eq. (2), and as a first step in this process, using (31) and cross-multiplying by $\sqrt{1-x^2}$, it is convenient to take

$$u(x) = \sqrt{1-x^2}\,v(x) \tag{32}$$

as our new dependent variable. Using this, Eq. (3) can be written as

$$\frac{1}{\pi}\int_{-1}^{1} \frac{u(t)\,dt}{\sqrt{1-t^2}(t-x)} = f(x), \tag{33}$$

and in terms of $u(x)$ the inversion formula becomes

$$u(x) = \frac{1}{\pi}\int_{-1}^{1} \frac{\sqrt{1-t^2}f(t)\,dt}{x-t} + \frac{c}{\pi}. \tag{34}$$

With these representations of the solution to Eq. (3) we are now ready to consider the determination of c.

3. Determining c

Since Eq. (31) contains an unknown constant c, in applications, some additional condition(s), depending on the physical context of the problem, needs to be imposed for its determination. Mathematically, this additional condition is usually given by

$$l(v) = M, \tag{35}$$

where l is some linear functional of the solution v. The most common values of l and M seem to be:

1. $l(v) = v(1)$, $M = 0$, which is the Kutta condition, usually occurring in problems in aerodynamics and hydrodynamics (Refs. 4, 13).

2. $l(v) = \int_{-1}^{1} v(t) \, dt$, which arises in problems in solid mechanics (Refs. 7-8), and occasionally $l(v) = \int_{-1}^{1} g(t)v(t) \, dt$.
3. In some problems the solution is required to be zero at both $x = \pm 1$, and then we have two functionals $l_1(v) = v(1)$ and $l_2(v) = v(-1)$. In this case, in order for a unique solution to exist, the given function $f(x)$ must satisfy an additional condition. And, as we shall see, this eventually complicates matters numerically in comparison to the latter conditions (1) and (2).

3.1. The Kutta Condition. If $v(x)$ is required to satisfy the condition $v(1) = 0$, then the term in the brackets in Eq. (31) should equal zero at $x = 1$. Thus

$$\int_{-1}^{1} \frac{\sqrt{1-t^2}}{1-t} f(t) \, dt + c = 0, \tag{36}$$

so that

$$c = -\int_{-1}^{1} \sqrt{\frac{1+t}{1-t}} f(t) \, dt. \tag{37}$$

Substituting c back into Eq. (31), we get

$$v(x) = \frac{1}{\pi\sqrt{1-x^2}} - \int_{-1}^{1} \left[\frac{\sqrt{1-t^2}}{x-t} - \sqrt{\frac{1-t}{1+t}} \right] f(t) \, dt$$

$$= \frac{1}{\pi} \sqrt{\frac{1-x}{1+x}} \int_{-1}^{1} \sqrt{\frac{1+t}{1-t}} \frac{f(t) \, dt}{x-t}. \tag{38}$$

Equation (38) is usually referred to in the aerodynamics literature as the *Söhngen inversion formula* (Ref. 14).

From Eq. (38) we see that

$$v(x) = \sqrt{\frac{1-x}{1+x}} u(x), \tag{39}$$

and this suggests that now the airfoil equation should be written as

$$\frac{1}{\pi} \int_{-1}^{1} \sqrt{\frac{1-t}{1+t}} \frac{u(t) \, dt}{t-x} = f(x), \tag{40}$$

with $u(t)$ as the new unknown. Then Eq. (38) becomes

$$u(x) = \frac{1}{\pi} \int_{-1}^{1} \sqrt{\frac{1+t}{1-t}} \frac{f(t)\,dt}{x-t}. \tag{41}$$

One usually refers to Eq. (40) as an equation of *index zero*, the index being the negative of the sum of the powers in the weight function $(1-t)^{1/2}(1+t)^{-1/2}$. In this case if we start with the airfoil equation in the form (40), it automatically has the Kutta condition built into the solution $v(x)$, and then Eq. (3) has the *unique* solution (41).

In some problems the Kutta condition is imposed at $x = -1$. In this case Eq. (3) also has index zero with the weight being the reciprocal of that in (40). Since the mathematics is essentially the same as that when $v(1) = 0$, in the rest of the chapter we shall assume that $v(1) = 0$ when referring to the zero index case of either (2) or (3).

3.2. $l(v) = \int_{-1}^{1} v(t)\,dt = M.$ In this situation we again begin with Eq. (31). Since now $c = \int_{-1}^{1} v(x)\,dx$,

$$u(x) = \frac{1}{\pi} \int_{-1}^{1} \frac{\sqrt{1-t^2} f(t)\,dt}{x-t} + \frac{M}{\pi}, \tag{42}$$

where $u(x)$ is defined by (32). Here we say that the index is 1, which again is the negative of the sum of the exponents in the weight function $(1+t)^{-1/2}(1-t)^{-1/2}$ in (33).

In the commonly occurring case where $l(v) = \int_{-1}^{1} v(t)\,dt = 0$, the solution to Eq. (3) is

$$u(x) = \frac{1}{\pi} \int_{-1}^{1} \frac{\sqrt{1-t^2} f(t)\,dt}{x-t}, \tag{43}$$

which is analogous to the Söhngen solution, except for a difference in the weight functions in the two formulas. The astute reader will have noticed (I hope) that in both cases the weight functions in the original equation and its solution formula are reciprocals of each other—a fact that is known to be true in the general case for CSIES with constant coefficients (Ref. 10).

3.3. $v(1) = v(-1) = 0.$ If $v(t)$ satisfies $v(1) = 0$, then

$$v(x) = \frac{1}{\pi} \sqrt{\frac{1-x}{1+x}} \int_{-1}^{1} \sqrt{\frac{1+t}{1-t}} \frac{f(t)\,dt}{x-t}, \tag{44}$$

and in order for $v(-1)$ to be zero as well

$$\int_{-1}^{1} \sqrt{\frac{1+t}{1-t}} \frac{f(t)\,dt}{1+t} = \int_{-1}^{1} \frac{f(t)\,dt}{\sqrt{1-t^2}} = 0. \tag{45}$$

Thus, a solution to Eq. (3) exists only if condition (45) is satisfied. In this case equation (44) becomes

$$v(x) = \frac{1}{\pi} \sqrt{\frac{1-x}{1+x}} \int_{-1}^{1} \left[\sqrt{\frac{1+t}{1-t}} \frac{f(t)}{x-t} - \frac{f(t)}{\sqrt{1-t^2}} \right] dt$$

$$= \frac{\sqrt{1-x^2}}{\pi} \int_{-1}^{1} \frac{f(t)\,dt}{\sqrt{1-t^2}(x-t)} = \sqrt{1-x^2}\,u(x), \tag{46}$$

and using (46) in Eq. (3) gives the equation–solution pair

$$\begin{cases} \dfrac{1}{\pi} \displaystyle\int_{-1}^{1} \dfrac{\sqrt{1-t^2}\,u(t)\,dt}{t-x} = f(x), \\[2ex] \dfrac{1}{\pi} \displaystyle\int_{-1}^{1} \dfrac{f(x)\,dt}{\sqrt{1-t^2}(x-t)} = u(x). \end{cases} \tag{47}$$

Here the index is -1.

As one can see, the behavior of the solution near $x = \pm 1$ is different for each index. Denoting the index by ν: if $\nu = 0$, then the solution is bounded at $x = 1$ and unbounded at $x = -1$; if $\nu = 1$, then the solution is unbounded at $x = \pm 1$, while if $\nu = -1$, the solution is bounded at both endpoints. These properties carry over to the more general case of Eq. (2).

4. Numerical Methods for the Generalized Airfoil Equation

As we pointed out in the Introduction, because of their pervasiveness in practice, and the special properties imparted by the square-root singularities in their solutions, we feel that it is worthwhile to consider the numerical solution of the generalized airfoil equation separately, before proceeding to constant coefficient equations of the second kind and ultimately to the general case of Eq. (1) to be discussed in Chapter 6. In contrast to Fredholm integral equations, we shall see that the index (or equivalently the form of the endpoint singularities) to a large extent determines the nature of the algorithms that are used, and so throughout this part of the chapter each of these cases will be treated separately. For some work, particularly when dealing with problems of convergence and error estimates, all types may be treated in a unified fashion (Ref. 15), and if certain conditions are satisfied (Ref. 16), all may be treated as if $\nu = 0$. However, when $\nu = -1$, special problems arise numerically, and further research needs to be done (Refs. 15, 17).

Historically, many of the early algorithms developed for solving CSIES were indirect. That is, the inversion formula for the dominant equation was used to convert the complete equation to an equivalent Fredholm equation of the second kind, which could then be solved by numerous well-known methods (Refs. 18-19). The difficulty with this approach is that one is rarely able to calculate the kernel of the resulting Fredholm equation explicitly, and the kernel itself may be weakly singular leading to further numerical difficulties (Ref. 13). As a consequence, this approach has been largely abandoned, even though some recent work suggests that it may have certain theoretical and practical advantages. For instance, in Ref. 20, it was shown that weaker conditions could be imposed on the data for convergence of an indirect method than for a corresponding direct one. In addition, some recently developed iterative methods of Gerasoulis (Ref. 7) and Chatelin and Guessous (Ref. 21) require that an indirect approach be used to begin. Notwithstanding, such methods have, at present, largely been replaced by *direct ones* which do not require the conversion of Eq. (1) to an equivalent Fredholm equation, and since virtually all of the published numerical calculations in the past 15 years have been done this way, our emphasis will be on these techniques. For completeness, however, we will begin our development of numerical methods with some indirect ones because they can be useful, as will be shown in Section 10, in motivating certain direct algorithms, and in some theoretical studies as well (Refs. 15, 22).

5. Indirect Methods

For ease of exposition and to avoid repetition, we will study some indirect methods for solving Eq. (2) when the index is zero. It should then be clear how to proceed when $\nu = \pm 1$.

First, by examining the nature of the solution when $k(x, t) = 0$, it is appropriate to introduce the new independent variable $v(x) = [(1 - x)/(1 + x)]^{1/2} u(x)$ into Eq. (2), where $u(x)$ is generally the "smooth" part of the solution. In this case Eq. (2) becomes

$$\frac{1}{\pi}\int_{-1}^{1}\sqrt{\frac{1-t}{1+t}}\frac{u(t)\,dt}{t-x}$$
$$+\int_{-1}^{1}\sqrt{\frac{1-t}{1+t}}k(x,t)u(t)\,dt = f(x), \quad -1 < x < 1, \quad (48)$$

and by writing

$$w(t) = \sqrt{(1-t)/(1+t)},$$

Eq. (48) takes the form

$$\frac{1}{\pi}\int_{-1}^{1}\frac{w(t)u(t)\,dt}{t-x} + \int_{-1}^{1}w(t)k(x,t)u(t)\,dt = f(x), \quad -1 < x < 1. \quad (49)$$

As one can see from Eq. (48), the Kutta condition $v(1) = 0$ is automatically incorporated, and so we assume that it has a unique solution.

To convert (49) into a Fredholm equation of the second kind [this process is called Carleman–Vekua regularization (Ref. 10) to distinguish it from other approaches (Ref. 23)], we bring the non-Cauchy integral to the right-hand side of (49) giving

$$\frac{1}{\pi}\int_{-1}^{1}\frac{w(t)u(t)\,dt}{t-x} = f(x) - \int_{-1}^{1}w(t)k(x,t)u(t)\,dt = g(x), \quad (50)$$

and by using the Söhngen inversion formula

$$u(x) = \frac{1}{\pi}\int_{-1}^{1}\frac{g(t)\,dt}{w(t)(x-t)} = \frac{1}{\pi}\int_{-1}^{1}\frac{f(t)\,dt}{w(t)(x-t)}$$
$$-\frac{1}{\pi}\int_{-1}^{1}\frac{1}{w(s)(x-s)}\left[\int_{-1}^{1}w(t)k(s,t)u(t)\,dt\right]ds. \quad (51)$$

If $k(x, t)$ is smooth enough (Ref. 10), then the order of integration can be changed in the double integral giving

$$\mathchoice{\,\vcenter{\hbox{$\textstyle-$}}\kern-.4em\int}{\mathchoice{\,\vcenter{\hbox{$\textstyle-$}}\kern-.4em\int}}{}{}_{-1}^{1} \frac{1}{w(s)(x-s)} \left[\int_{-1}^{1} w(t)k(s,t)u(t)\,dt \right] ds$$

$$= \int_{-1}^{1} w(t)u(t) \left[\mathchoice{\,\vcenter{\hbox{$\textstyle-$}}\kern-.4em\int}{\mathchoice{\,\vcenter{\hbox{$\textstyle-$}}\kern-.4em\int}}{}{}_{-1}^{1} \frac{k(s,t)\,ds}{w(s)(x-s)} \right] dt$$

$$= \int_{-1}^{1} w(t)K(x,t)u(t)\,dt, \qquad (52)$$

where

$$K(x,t) = \mathchoice{\,\vcenter{\hbox{$\textstyle-$}}\kern-.4em\int}{\mathchoice{\,\vcenter{\hbox{$\textstyle-$}}\kern-.4em\int}}{}{}_{-1}^{1} \frac{k(s,t)\,ds}{w(s)(x-s)}. \qquad (53)$$

Thus, $u(x)$ satisfies the Fredholm equation of the second kind

$$u(x) + \int_{-1}^{1} w(t)K(x,t)u(t)\,dt = h(x), \qquad (54)$$

where

$$h(x) = \mathchoice{\,\vcenter{\hbox{$\textstyle-$}}\kern-.4em\int}{\mathchoice{\,\vcenter{\hbox{$\textstyle-$}}\kern-.4em\int}}{}{}_{-1}^{1} \frac{f(t)\,dt}{w(t)(x-t)}. \qquad (55)$$

If $k(x, t)$ is Hölder continuous, then $K(x, t)$ will generally be integrable (Ref. 10), and one may proceed to solve Eq. (54) numerically by any one of the well-known techniques for equations of the second kind (Ref. 19).

However, before this can be done, $K(x, t)$ and $h(x)$ usually have to be determined numerically, say by the Gaussian quadrature rule for Cauchy principal-value integrals given by Eq. (181), and these approximations may be poor unless $k(x, t)$ and $f(x)$ are very smooth. As one can see, this indirect process can lead to quite involved, arithmetically complicated algorithms, and it is for this reason that they have generally not been used recently for actual numerical computations. On the other hand, regularization methods often play a crucial role in the analysis of direct algorithms (Refs. 15, 22), and so have a useful place in the overall theory. Because of this, we shall give another way of converting (2) to an equation of the second kind.

In this second approach we define a new *dependent variable* $G(x)$ by

$$G(x) = \frac{1}{\pi} \mathchoice{\,\vcenter{\hbox{$\textstyle-$}}\kern-.4em\int}{\mathchoice{\,\vcenter{\hbox{$\textstyle-$}}\kern-.4em\int}}{}{}_{-1}^{1} \frac{w(t)u(t)\,dt}{t-x}, \qquad (56)$$

so that by the Söhngen inversion formula

$$u(t) = \frac{1}{\pi} \fint_{-1}^{1} \frac{G(s)\,ds}{w(s)(t-s)}, \qquad (57)$$

and substituting these values into Eq. (2) gives the following equation of the second kind for $G(x)$:

$$G(x) + \frac{1}{\pi}\fint_{-1}^{1} w(t)k(x,t)\left[\fint_{-1}^{1}\frac{G(s)\,ds}{w(s)(t-s)}\right]dt = f(x). \qquad (58)$$

Changing the order of integration in the integral in (58), it becomes

$$G(x) + \frac{1}{\pi}\int_{-1}^{1}\frac{K(x,s)G(s)\,ds}{w(s)} = f(x), \qquad (59)$$

where

$$K(x,s) = \fint_{-1}^{1}\frac{w(t)k(x,t)\,dt}{t-s}. \qquad (60)$$

Now Eq. (58) is solved numerically giving an approximation $\hat{G}(x)$ to $G(x)$, and then an approximation $\hat{u}(t)$ to $u(t)$ can be obtained from (57) by substituting $\hat{G}(s)$ for $G(s)$. For practical calculations this procedure appears to be no more attractive than the standard regularization (54). However, when applied to equations of index one, it has some interesting theoretical properties which will be taken up in Section 12.4.

6. Direct Methods

In contrast to indirect methods, direct methods work without first converting Eq. (2) to a Fredholm equation, but the methods themselves are similar in character to those used to successfully solve Fredholm equations of the second kind and can be classified in a similar way. They generally fall into the following categories (Refs. 1-2, 7):

(i) degenerate kernel methods;
(ii) expansion methods, particularly Galerkin and collocation methods;
(iii) quadrature methods;
(iv) iterative methods.

As for Fredholm equations, the procedures in (ii) and (iii) are apparently used most often, but with considerably more restrictions than in the Fredholm case, while only recently have iterative techniques begun to be employed with some frequency (Refs. 7, 21). Although degenerate kernel methods seem to be rarely used explicitly, it turns out, just as for Fredholm equations, that Galerkin and collocation methods are implicitly degenerate kernel methods, although this fact is generally not used in their implementation (Ref. 1).

Although (i)-(iv) appear to be the most common classes of direct methods, some other techniques, which are variants of these, have been used as well. For instance, in a number of papers Ioakimidis has developed a generalization of Kantorovich regularization, well known for Fredholm equations (Refs. 24-25), and an analogue of Sloan iteration has been discussed by Fromme and Golberg in (Ref. 26) and has been used by Faath in a recent PhD thesis to solve a particular equation of the first kind with index one (Ref. 27).

We shall assume that the reader is familiar with these techniques for Fredholm equations, and will proceed with the application of these ideas to equations of the first kind in Sections 9-16. Then we take up generalizations of some of them for equations of the second kind in Sections 17 and 18. At present most of the direct algorithms for CSIES that have been used in practice, and analyzed theoretically, have been based essentially on polynomial approximations (Refs. 1-3). For the GAE, this is because the integral operators in (41), (43), and (47) have nice properties when applied to polynomials, and so we proceed with a study of these next.

7. Some Mapping Properties of the Airfoil Operator

As we have stated, currently the most well-developed methods for solving CSIES are based on expansions of the approximate solution in certain orthogonal polynomials. In particular, for constant coefficient equations these are the classical Jacobi polynomials. For equations of the first kind with $\nu = \pm 1$ they are Chebyshev polynomials of the first and second kinds, while for equations of index zero they are normalized Jacobi polynomials $P_n^{(1/2,-1/2)}(x)$ or $P_n^{(-1/2,1/2)}(x)$, which we have called the *airfoil polynomials* in previous work (Ref. 13) [this terminology was chosen because equations of index zero have been used, perhaps most often, for solving problems in aerodynamics (Refs. 4, 13, 26)]. The reason for the importance of these polynomials is that, when the coefficients are constant in Eq. (1), the dominant part maps one set of such polynomials into another, and as shown by Elliott (Ref. 28), Krenk (Ref. 29), Junghanns and Silbermann (Ref. 3), and Welstead (Ref. 30), this property also holds for certain classes of equations with variable coefficients. Although the general case seems to require the use of complex variable techniques (Refs. 3, 28, 30), for equations of the first kind they can be obtained in an elementary way using the well-known recurrence relations satisfied by the Chebyshev and airfoil polynomials.

Before establishing these facts, it is convenient to express the various forms of the airfoil equation in operator notation. For this let

$$[H_\nu u](x) = \frac{1}{\pi} \int_{-1}^{1} \frac{w_\nu(t) u(t) \, dt}{t - x}, \quad -1 < x < 1, \tag{61}$$

where

$$w_\nu(x) = \begin{cases} \sqrt{1 - x^2}, & \nu = -1, \\ \sqrt{\dfrac{1 - x}{1 + x}}, & \nu = 0, \\ \dfrac{1}{\sqrt{1 - x^2}}, & \nu = 1. \end{cases} \tag{62}$$

Then Eq. (3) can be written as

$$[H_\nu u](x) = f(x), \quad \nu = -1, 0, 1, \quad -1 < x < 1. \tag{63}$$

Theorem 7.1. Let $T_n(x) = \cos[n(\cos^{-1} x)]$, $-1 \leq x \leq 1$, $n = 0, 1, 2, \ldots$, denote the Chebyshev polynomials of the first kind. Then

$$\begin{cases} H_1 T_0 = 0, & (64) \\ H_1 T_n = U_{n-1}, \quad n = 1, 2, \ldots, \quad -1 \leq x \leq 1, & (65) \end{cases}$$

where $U_n(x) = \sin[(n + 1)(\cos^{-1} x)]/\sin(\cos^{-1} x)$, $n = 0, 1, 2, \ldots$, are the Chebyshev polynomials of the second kind.

Proof. From the definition of $T_n(x)$, $T_0(x) = 1$ and $T_1(x) = x$, and for $n \geq 1$ it is well known (and easily shown) that $\{T_n(x)\}_{n=1}^{\infty}$ satisfy

$$T_{n+1}(x) = 2x T_n(x) - T_{n-1}(x). \tag{66}$$

Similarly $U_0(x) = 1$, $U_1(x) = 2x$, and for $n \geq 1$, $\{U_n(x)\}_{n=1}^{\infty}$ satisfy the same recursion (66) as $\{T_n(x)\}_{n=1}^{\infty}$. To prove the theorem, it then suffices to show that $H_1 T_0 = 0$, $H_1 T_1 = 1$ and that $V_n = H_1 T_n$ satisfies (66). Solution of this recursion with $V_0 = 0$ and $V_1 = 1$ will show that $V_n(x) = U_{n-1}(x)$, $n \geq 1$.

We begin by showing that $H_1 T_0 = 0$ implies $V_1 = 1$, and that $\{V_n\}_{n=1}^\infty$ satisfy (66). Last, we demonstrate by a clever change of variable [apparently due to Tricomi (Ref. 31)] that $H_1 T_0 = 0$. Solution of the recursion for $\{V_n\}_{n=1}^\infty$ is then elementary.

Now assuming that $H_1 T_0 = 0$, we have

$$H_1 T_1 = \frac{1}{\pi} \int_{-1}^{1} \frac{t\, dt}{\sqrt{1-t^2}(t-x)} = \frac{1}{\pi} \int_{-1}^{1} \frac{(t-x)\, dt}{\sqrt{1-t^2}(t-x)} + \frac{x}{\pi} \int_{-1}^{1} \frac{dt}{\sqrt{1-t^2}(t-x)}$$

$$= \frac{1}{\pi} \int_{-1}^{1} \frac{dt}{\sqrt{1-t^2}} + x H_1 T_0 = \pi/\pi + 0 = 1, \tag{67}$$

and applying H_1 to both sides of (66) gives

$$H_1 T_{n+1} = 2 H_1(x T_n) - H_1 T_{n-1}, \tag{68}$$

where

$$H_1(x T_n) = \frac{1}{\pi} \int_{-1}^{1} \frac{t T_n(t)\, dt}{\sqrt{1-t^2}(t-x)}$$

$$= \frac{1}{\pi} \int_{-1}^{1} \frac{(t-x) T_n(t)\, dt}{\sqrt{1-t^2}(t-x)} + \frac{x}{\pi} \int_{-1}^{1} \frac{T_n(t)\, dt}{\sqrt{1-t^2}(t-x)}$$

$$= \frac{1}{\pi} \int_{-1}^{1} \frac{T_n(t)\, dt}{\sqrt{1-t^2}} + x H_1 T_n. \tag{69}$$

Since $\{T_n(x)\}_{n=0}^\infty$ are orthogonal with respect to the weight function $(1-x^2)^{-1/2}$,

$$H_1(x T_n) = x H_1 T_n = x V_n, \quad n \geq 1. \tag{70}$$

Thus for $n \geq 1$, $\{V_n(x)\}$ satisfy

$$\begin{cases} V_{n+1}(x) = 2x V_n(x) - V_{n-1}(x), \\ V_0(x) = 0, \quad V_1(x) = 1, \end{cases} \tag{71}$$

and solving (71) by standard techniques shows that

$$V_n(x) = \sin[n(\cos^{-1} x)]/\sin(\cos^{-1} x), \quad n \geq 1.$$

Thus $H_1 T_n = U_{n-1}$, $n \geq 1$, so to complete the proof it remains to show that

$H_1 T_0 = 0$. For this we evaluate the integral explicitly using the change of variable

$$t = (1 - u^2)/(1 + u^2), \quad -1 \le u \le 1, \tag{72}$$

so that

$$u = \sqrt{\frac{1-t}{1+t}} \tag{73}$$

and

$$du = -4t\, dt/(1 + t^2)^2. \tag{74}$$

Substituting expressions (72) and (74) into

$$\frac{1}{\pi} \int_{-1}^{1} \frac{dt}{\sqrt{1-t^2}(x-t)} = H_1 T_0$$

gives

$$H_1 T_0 = \frac{1}{\pi} \int_0^\infty \frac{2\, du}{(1-x) - u^2(1+x)}, \tag{75}$$

and letting $a = [(1-x)/(1+x)]^{1/2}$, equation (75) becomes

$$\frac{2}{\pi(1+x)} \int_0^\infty \frac{du}{a^2 - u^2}, \quad -1 < x < 1. \tag{76}$$

Now an elementary integration using the definition of the Cauchy principal-value integral gives

$$\int_0^\infty \frac{du}{a^2 - u^2} = \lim_{\varepsilon \to 0} \left[\lim_{R \to \infty} \int_0^{a-\varepsilon} \frac{du}{a^2 - u^2} + \int_{a+\varepsilon}^R \frac{du}{a^2 - u^2} \right] = 0,$$

and the proof is complete. □

By similar calculations we obtain the following mapping properties of H_{-1} and H_0.

Theorem 7.2.

$$H_{-1} U_n = -T_{n+1}, \quad n \ge 0, \quad -1 \le x \le 1, \tag{77}$$

and letting

$$\alpha_n(x) = \sin[(n + 1/2) \cos^{-1} x]/\sqrt{\pi} \sin[(\cos^{-1} x)/2] \tag{78}$$

and

$$\gamma_n(x) = \cos[(n + 1/2) \cos^{-1} x]/\sqrt{\pi} \cos[(\cos^{-1} x)/2] \qquad (79)$$

denote the airfoil polynomials of the first and second kinds,

$$H_0 \alpha_n = -\gamma_n, \qquad n = 0, 1, 2, \ldots, \qquad -1 \le x \le 1. \qquad (80)$$

Proof. The proofs of (77) and (80) proceed along the same lines as for Theorem 7.1. For (77) we show by direct calculation that $H_{-1}U_0 = -T_1$, $H_{-1}U_1 = -T_2$ and that $W_n = H_{-1}U_n$ satisfies the three-term recursion (66). Solution of this recursion with the initial conditions $W_0 = -T_1$ and $W_1 = -T_2$ gives (77). The airfoil polynomials $\{\alpha_n\}_{n=0}^{\infty}$ and $\{\gamma_n\}_{n=0}^{\infty}$ also satisfy (66) (Ref. 13). Again by direct calculation one shows that $H_0\alpha_0 = -\gamma_0$, $H_0\alpha_1 = -\gamma_1$ and that $H\alpha_n$ satisfies (66). It then follows automatically that (80) holds.

Since (80) is a little less familiar than (66) or (77), let us demonstrate that $H_0\alpha_0 = -\gamma_0$ and $H_0\alpha_1 = -\gamma_1$ follow from Tricomi's integral $H_1T_0 = 0$. First, elementary trigonometry gives $\alpha_0 = \gamma_0 = 1/\sqrt{\pi}$, $\alpha_1 = (2x + 1)/\sqrt{\pi}$, and $\gamma_1 = (2x - 1)/\sqrt{\pi}$. Thus

$$H_0\alpha_0 = \frac{1}{\pi} \frac{1}{\sqrt{\pi}} \int_{-1}^{1} \sqrt{\frac{1-t}{1+t}} \frac{dt}{t-x} = \frac{1}{\sqrt{\pi}} \left[\frac{1}{\pi} \int_{-1}^{1} \frac{(1-t)\, dt}{\sqrt{1-t^2}(t-x)} \right]$$

$$= (H_1T_0 - H_1T_1)/\sqrt{\pi} = -U_0/\sqrt{\pi} = -1/\sqrt{\pi} = -\gamma_0 \qquad (81)$$

and

$$H_0\alpha_1 = \frac{1}{\sqrt{\pi}} \left[\frac{1}{\pi} \int_{-1}^{1} \frac{(1+2t)(1-t)}{\sqrt{1-t^2}(t-x)}\, dt \right] = \frac{1}{\sqrt{\pi}} \left[\frac{1}{\pi} \int_{-1}^{1} \frac{1+t-2t^2}{\sqrt{1-t^2}(t-x)}\, dt \right]$$

$$= [H_1T_0 + H_1T_0 - H_1(2t^2)]/\sqrt{\pi} = [1 - H_1(2t^2)]/\sqrt{\pi}. \qquad (82)$$

But $T_2 = 2t^2 - 1$, so that $2t^2 = T_2 + 1$ and using this in (82) gives

$$H_0\alpha_1 = [1 - H_1(T_2 + 1)]/\sqrt{\pi} = [1 - H_1T_2 - H_1T_0]/\sqrt{\pi}$$

$$= [1 - U_1]/\sqrt{\pi} = (1 - 2x)/\sqrt{\pi} = -\gamma_1.$$

The fact that $\{H\alpha_n\}_{n=1}^{\infty}$ satisfy (66) follows in exactly the same fashion as the proof of (71), and we leave the details to the reader. □

Before proceeding to use (64)-(65), (77), and (80), in developing numerical algorithms for the GAE, let us point out several other useful properties of the polynomials $\{T_n\}_{n=0}^{\infty}$, $\{U_n\}_{n=0}^{\infty}$, $\{\alpha_n\}_{n=0}^{\infty}$, and $\{\gamma_n\}_{n=0}^{\infty}$.

If we introduce inner products $\langle f, g \rangle_{w_\nu} = \int_{-1}^{1} w_\nu(t) f(t) g(t)\, dt$, $\nu = 0$, ± 1, on the spaces of real functions square-integrable with respect to w_ν, then $\{T_n\}_{n=0}^{\infty}$ are orthogonal with respect to w_1, $\{U_n\}_{n=0}^{\infty}$ are orthogonal with respect to w_{-1}, $\{\alpha_n\}_{n=0}^{\infty}$ are orthonormal with respect to w_0, and $\{\gamma_n\}_{n=0}^{\infty}$ are orthonormal with respect to $1/w_0$ (Refs. 13, 32). Also, as it is sometimes more convenient to work with orthonormal functions, let

$$\tau_n(x) = \begin{cases} T_0(x)/\sqrt{\pi}, & \quad (83) \\ \sqrt{2}\,T_n(x)/\sqrt{\pi}, & n \geq 0, \quad (84) \end{cases}$$

and

$$\mu_n(x) = \sqrt{2}\,U_n(x)/\sqrt{\pi}, \qquad n \geq 0, \qquad (85)$$

then it is easily verified that $\{\tau_n\}_{n=0}^{\infty}$ are orthonormal with respect to w_1, and $\{\mu_n\}_{n=0}^{\infty}$ are orthonormal with respect to w_{-1}. Similarly (64)-(65) and (77) hold with $\{T_n\}_{n=0}^{\infty}$ and $\{U_n\}_{n=0}^{\infty}$ replaced by $\{\tau_n\}_{n=0}^{\infty}$ and $\{\mu_n\}_{n=0}^{\infty}$ respectively.

To give an indication of how the mapping relations (64)-(65), (77), and (80) can be useful in solving the airfoil equation numerically, suppose we consider the equation of index zero,

$$\frac{1}{\pi} \int_{-1}^{1} \sqrt{\frac{1-t}{1+t}} \frac{u(t)\, dt}{t-x} = f(x).$$

If f is continuous, it can be expanded in the polynomials $\{\gamma_n\}_{n=0}^{\infty}$ as

$$f(x) = \sum_{n=0}^{\infty} \langle f, \gamma_n \rangle_{1/w_0} \gamma_n(x) = \sum_{n=0}^{\infty} f_n \gamma_n(x), \qquad (86)$$

and we assume that $u(t)$ can be expanded as

$$u(t) = \sum_{n=0}^{\infty} a_n \alpha_n(t). \qquad (87)$$

Applying H_0 to $u(t)$ gives (at least formally)

$$H_0 u = \sum_{n=0}^{\infty} a_n H \alpha_n = \sum_{n=0}^{\infty} -a_n \gamma_n = \sum_{n=0}^{\infty} f_n \gamma_n, \qquad (88)$$

so that $a_n = -f_n$, $n \geq 0$. Thus

$$u(t) = -\sum_{n=0}^{\infty} f_n \gamma_n(t), \qquad (89)$$

and if f is approximated by the first $N + 1$ terms of (86), then

$$u_N = -\sum_{n=0}^{N} f_n \gamma_n \qquad (90)$$

may be taken as an approximation to u, with similar relations holding for $\nu = \pm 1$. From this, it appears quite reasonable to try to approximate the solution of Eq. (2) in terms of $\{U_n\}_{n=0}^{\infty}$, $\{T_m\}_{n=0}^{\infty}$, or $\{\alpha_n\}_{n=0}^{\infty}$ depending on the index. After some preliminary analysis which follows from the results of this section, we shall take up this type of approximation in great detail.

8. Operator Formulation of the Generalized Airfoil Equation

In order to make our derivations more concise, and ultimately for the discussion of the convergence of various algorithms, it is convenient to reformulate the generalized airfoil equation as an operator equation between appropriate pairs of Hilbert spaces. We shall assume that our Hilbert spaces consist of real functions, although in solving unsteady problems in fluid dynamics it is necessary to use complex Hilbert spaces (Ref. 13). This is obviously a minor modification, and the reader can make the necessary changes as the need arises.

If w is one of the weight functions w_ν in (62), then L_w will denote the Hilbert space of functions square-integrable respect to w. That is (the appropriate value of ν will generally be known from the problem context),

$$L_w = \left\{ f:[-1, 1] \to R, \int_{-1}^{1} w(t) f^2(t) \, dt < \infty \right\}, \qquad (91)$$

with the inner product of two functions f and g in L_w denoted by

$$\langle f, g \rangle_w = \int_{-1}^{1} w(t) f(t) g(t) \, dt, \qquad (92)$$

and the corresponding norm

$$\|f\|_w = \left[\int_{-1}^{1} w(t) f^2(t) \, dt \right]^{1/2}. \qquad (93)$$

It then follows from (64)–(65) and (77) and (80) that H_ν, $\nu = \pm 1, 0$, can be extended as bounded linear operators from L_w to $L_{1/w}$. For instance, if $\nu = 0$, then the extended operator, also called H_0, is given by (Ref. 13)

$$H_0 u = - \sum_{n=0}^{\infty} a_n \gamma_n, \qquad (94)$$

where $u = \sum_{n=0}^{\infty} \langle u, \alpha_n \rangle_{w_0} \alpha_n = \sum_{n=0}^{\infty} a_n \alpha_n$, with analogous expressions for $H_{\pm 1} u$.

Depending on the index of the problem, which in turn depends on the endpoint singularities of v, we make the change of variable

$$u = w_\nu v \qquad (95)$$

in Eq. (2) so that it becomes

$$\frac{1}{\pi} \int_{-1}^{1} \frac{w_\nu(t) u(t)\, dt}{t - x} + \int_{-1}^{1} w_\nu(t) k(x, t) u(t)\, dt = f(x). \qquad (96)$$

If we define

$$Ku = \int_{-1}^{1} w_\nu(t) k(x, t) u(t)\, dt, \qquad (97)$$

then (96) can be written in operator notation

$$H_\nu u + Ku = f, \qquad (98)$$

where we shall assume that K defines a *compact operator* from $L_w \to L_{1/w}$. A sufficient condition for this is that $k(x, t)$ be Hilbert–Schmidt so that

$$\int_{-1}^{1} \int_{-1}^{1} \frac{w_\nu(t)}{w_\nu(x)} k^2(x, t)\, dx\, dt < \infty \qquad (99)$$

(Ref. 13). In particular (99) holds if $k(x, t)$ is continuous.

Further properties of H_ν follow from (64)–(65), (77), and (80), they are:

(i) If $\nu = 0$, then H_0 is invertible, and moreover H_0 is unitary, so that $\langle H_0 f, H_0 g \rangle_{1/w_0} = \langle f, g \rangle_{w_0}$, for all $(f, g) \in L_{w_0}$. In particular $\|H_0 u\|_{1/w_0} = \|u\|_{w_0}$, and the operator norm $\|H_0\| = 1$.

(ii) If $\nu = 1$, then H_1 is bounded, and its Hilbert space adjoint H_1^* is a right inverse of H_1. That is

$$H_1 H_1^* u = u, \qquad (100)$$

and additionally

$$H_1^* H_1 u = u + cT_0, \tag{101}$$

where c is a constant.

(iii) If $\nu = -1$, then H_{-1}^* is a left inverse of H_{-1}, so that

$$H_{-1}^* H_{-1} u = u. \tag{102}$$

In addition to the Hilbert spaces L_w, we will also need various spaces of differentiable functions. If $r \geq 0$ is a real number, let $[r]$ be the largest integer less than or equal to r. We say that a function belongs to C^r on $[-1, 1]$ if it has $[r]$ continuous derivatives and its rth derivative is Hölder continuous of order $r - [r]$, $0 \leq \alpha < 1$. If $\alpha = 0$; then the rth derivative is continuous. In particular, C^0 denotes the space of continuous functions, and if r is an integer, then C^r is just the usual space of r times continuously differentiable functions on $[-1, 1]$.

Now multiplying (98) on the left by H_ν^* we find that u satisfies

$$u + H_\nu^* K u = H_\nu^* f + b_\nu, \tag{103}$$

where $b_\nu = 0$ if $\nu = 0$ or $\nu = -1$. Thus (98) is equivalent to an equation of the second kind and it follows from standard Fredholm theory (Ref. 19) that it has a unique solution iff the null space $N(I + H_\nu^* K) = 0$, a condition assumed to hold from now on. When $\nu = 1$, the constant b_1 needs to be determined and, as for the case $K = 0$, this is usually done by imposing an additional condition of the form $l(u) = M$ on the solution, where l is a bounded linear functional on L_{w_1}. In this case we must solve the simultaneous equations

$$\begin{cases} H_1 u + K u = f, & (104) \\ l(u) = M. & (105) \end{cases}$$

When $\nu = -1$, it follows from (45) that u must satisfy the consistency condition (Ref. 10)

$$\langle f - Ku, T_0 \rangle_{w_1} = 0. \tag{106}$$

Since this condition depends on the unknown u, it generally cannot be verified *a priori*, and this makes solving equations of index -1 somewhat more difficult numerically than when $\nu = 0$ or $\nu = 1$. If the range of K, $R(K)$, is orthogonal to T_0, then (106) reduces to $\langle f, T_0 \rangle_{w_1} = 0$, which is the same consistency condition as when $K = 0$. In this case, some of the theory becomes no more difficult than in the others (Refs. 16, 30).

With these preliminaries out of the way, we now turn our attention to the development of direct numerical methods for solving Eq. (2).

9. Degenerate Kernel Methods

Although degenerate kernel methods play a fundamental role in the theoretical analysis of CSIES (Ref. 13), little explicit practical use has been made of them numerically. However, since they are virtually identical to the corresponding ones for Fredholm equations of the second kind, we will give a short exposition of this approach and indicate later that other more popular algorithms are nothing more than degenerate kernel methods in "disguise." Although we could work in other function space settings (Ref. 15), the Hilbert space approach, as indicated in the previous section, is most convenient for the algebraic and analytic development of these and subsequent algorithms; so we shall assume that we are working within this framework. For simplicity, in this section we consider only the case where $\nu = 0$. Since $K: L_w \to L_{1/w}$ is compact, it is well known that it can be approximated in operator norm by a sequence of finite rank operators $\{K_N\}_{N=0}^\infty$. That is, $\|K - K_N\| \to 0$, $N \to \infty$, where $\|T\|$ denotes the operator norm of a bounded operator from $L_w \to L_{1/w}$ (we will generally leave operator norms unsubscripted. When it becomes necessary to indicate which spaces are involved, then $\|T\|_{X \to Y}$ will denote the operator norm of a bounded operator from $X \to Y$). In this case we say that K_N is a *degenerate kernel operator* if

$$K_N u = \sum_{k=1}^{N} \int_{-1}^{1} w_0(t) a_k(x) b_k(t) u(t) \, dt. \tag{107}$$

Replacing K by K_N in (99) we obtain an approximation u_N to u by solving (f could be replaced an approximation f_N as well)

$$H_0 u_N + K_N u_N = f. \tag{108}$$

Since H_0 is bounded and has a bounded inverse, it is straightforward to show for all N sufficiently large that Eq. (108) has a unique solution in L_w and that $\|u - u_N\|_w \to 0$, $N \to \infty$.

Assuming that N has been chosen to satisfy this condition, we are now faced with the task of solving Eq. (108), and for K_N of the form in (107) this can be done in a fashion analogous to that for Fredholm equations of the second kind.

Writing out (108) explicitly, we find that u_N satisfies

$$H_0 u_N(x) + \sum_{k=1}^{N} a_k(x) \int_{-1}^{1} w_0(t) b_k(t) u_N(t) \, dt = f(x), \qquad (109)$$

or equivalently

$$H_0 u_N + \sum_{k=0}^{N} a_k c_k = f, \qquad (110)$$

where

$$c_k = \int_{-1}^{1} w_0(t) b_k(t) u_N(t) \, dt, \qquad k = 1, 2, \ldots, N. \qquad (111)$$

If we can determine $\{c_k\}_{k=1}^{N}$, then u_N can be obtained from

$$u_N = -\sum_{k=1}^{N} c_k H_0^{-1} a_k + H_0^{-1} f. \qquad (112)$$

Thus it remains to calculate $\{c_k\}_{k=1}^{N}$.

To do this, operate on Eq. (110) with H_0^{-1} giving

$$u_N + \sum_{k=1}^{N} c_k H_0^{-1} a_k = H_0^{-1} f, \qquad (113)$$

and observing from (111) that $c_k = \langle b_k, u_N \rangle_w$, take the inner product of (113) with b_j, $j = 1, 2, \ldots, N$, which yields

$$\langle b_j, u_N \rangle_w + \sum_{k=0}^{N} c_k \langle b_j, H_0^{-1} a_k \rangle_w = \langle b_j, H_0^{-1} f \rangle_w, \qquad j = 1, 2, \ldots, N. \quad (114)$$

Thus $\{c_k\}_{k=1}^{N}$ satisfy the system of linear equations

$$c_j + \sum_{k=0}^{N} h_{jk} c_k = f_j, \qquad j = 1, 2, \ldots, N, \qquad (115)$$

where

$$h_{jk} = \langle b_j, H_0^{-1} a_k \rangle_w, \qquad (k, j) = 1, \ldots, N, \qquad (116)$$

and

$$f_j = \langle b_j, H_0^{-1} f \rangle_w, \qquad j = 1, 2, \ldots, N. \qquad (117)$$

Note that this method, while conceptually simple, requires rather extensive arithmetic to set up because we must evaluate $H_0^{-1}a_k$, $k = 1, 2, \ldots, N$, $H_0^{-1}f$, and compute the inner products (integrals) $\{h_{jk}\}_{(j,k)=1}^N$ and $\{f_j\}_{j=1}^N$. Since these calculations can rarely be done analytically, numerical methods must be used. Although this approach has been applied in practice (Ref. 33), it has the same order of magnitude of arithmetic as an indirect method, and since it has been essentially abandoned for constant coefficient equations we will not pursue it any further.

As a last comment, we note that other finite rank approximations to K have been considered for solving Eq. (2), particularly in the Soviet literature (Ref. 20). For instance, one might use

$$K_N u = \sum_{k=1}^N w_k k(x, t_k) u(t_k), \tag{118}$$

where $(\{w_k\}_{k=1}^N, \{t_k\}_{k=1}^N)$ are the weights and nodes of a suitable quadrature rule for approximating Ku. In this case, if $\nu = 0$, u_N satisfies

$$u_N(x) + \sum_{k=1}^N w_k (H_0^{-1}k)(x, t_k) u_N(t_k) = (H_0^{-1}f)(x), \tag{119}$$

and evaluating (119) at $x = t_j$, $j = 1, 2, \ldots, N$, gives the N equations for $\{u_N(t_j)\}_{j=1}^N$:

$$u_N(t_j) + \sum_{k=1}^N w_k (H_0^{-1}k)(t_j, t_k) u_N(t_k) = (H_0^{-1}f)(t_j), \quad j = 1, 2, \ldots, N. \tag{120}$$

Then $u_N(x)$ may be obtained from (119), which is a Nyström form of interpolation.

10. Galerkin's Method

Galerkin's method using polynomial bases and weights was one of the first direct methods for solving CSIES, going at least as far back as Karpenko's paper in 1966 (Ref. 34), and was later used in a series of papers for single and systems of equations of the first and second kinds by Erdogan, Gupta, and Cook in the late 1960s and early 1970s (Refs. 35–36).

In 1977 Linz proved the uniform convergence of this method for equations of the first kind and index one under certain differentiability conditions on $k(x, t)$ and $f(x)$ (Ref. 37), while mean-square convergence was considered under weaker conditions on the data for equations of the first kind in Ref. 13. In 1978 Fromme and Golberg considered Galerkin's method for the GAE with $\nu = 0$ and its relation to a collocation method for solving this equation (Ref. 13). In Refs. 22 and 38, Ioakimidis extended Linz's result on uniform convergence to equations of the second kind, and this analysis was expanded upon by Venturino in Ref. 39. In Refs. 40, 3, and 30, Elliott, Junghanns and Silbermann, and Welstead generalized previous mean-square convergence theorems to some equations with nonconstant coefficients, and uniform convergence of these polynomial algorithms was established by Junghanns in Ref. 41. In all of these papers it was assumed that all the inner products involved were calculated without error, and in Ref. 42 Miel reconsidered the convergence problem for the GAE with $\nu = 0$ when the inner products were evaluated by numerical integration (however, we feel that one of his assumptions is too weak, and believe that not all of the results of that paper are correct). Miel's analysis was generalized by us in Refs. 16 and 43 for equations of the first and second kinds with constant coefficients, and these results will be improved upon in this chapter, bringing the error estimates in line with Miel's by changing one of his assumptions.

In Ref. 44 Thomas developed spline-based Galerkin methods for some equations of the second kind with zero index, and his results have been extended to CSIES with strongly elliptic dominant parts by Prössdorf and Elschner in Ref. 6.

Despite the large number of papers on this method in the literature, it has rarely been used for actual calculations, no doubt because of its relative inefficiency in comparison to collocation and quadrature methods. On the other hand, it does have a number of interesting theoretical properties; for example, it is generally stable (Ref. 45) and can be used as a basis for "relatively" simple proofs of the convergence of the standard polynomial-based collocation and quadrature methods as will be shown in Section 19. In addition, its superconvergence properties make it attractive for doing some calculations in aerodynamics, where other techniques have had little success (Refs. 46-47) (this will be elaborated on in Section 10.2).

10.1. Galerkin's Method: $\nu = 0$. To motivate the derivation of this method, we will begin with $\nu = 0$ and use the classical Galerkin method on the regularized Fredholm equation. After a little algebra, using the unitarity of H_0, this indirect method becomes equivalent to a direct one, which strictly speaking is a Galerkin-Petrov method (Ref. 40), because it uses different basis and weight functions. When the basis functions are the airfoil polynomials, the algorithm simplifies considerably, and this is the only version that we study in any detail.

Recalling that $H_0^* = H_0^{-1}$, Eq. (98) can be written as

$$u + H_0^* K u = H_0^* f. \tag{121}$$

Now assume that X_N, $N \geq 0$, $\dim(X_N) = N + 1$, is a sequence of finite-dimensional subspaces of L_w with $\bigcup_{N \geq 0} X_N$ dense in L_w, and let $\{\beta_k\}_{k=0}^N$ be a basis for X_N. Proceeding in the usual way for equations of the second kind, we approximate u by

$$u_N = \sum_{k=0}^N a_k \beta_k, \tag{122}$$

and determine $\{\beta_k\}_{k=0}^N$ by setting the residual

$$r_N = u_N + H_0^* K u_N - H_0^* f$$

orthogonal to $\{\beta_j\}_{j=0}^N$. Thus

$$\langle r_N, \beta_j \rangle_w = 0, \quad j = 0, 1, 2, \ldots, N, \tag{123}$$

and writing (123) out in full gives the following $N + 1$ equations for the $N + 1$ unknowns $\{a_k\}_{k=0}^N$:

$$\sum_{k=0}^N \langle \beta_k, \beta_j \rangle_w a_k + \sum_{k=0}^N \langle H_0^* K \beta_k, \beta_j \rangle_w a_k = \langle H_0^* f, \beta_j \rangle_w, \quad j = 0, 1, 2, \ldots, N. \tag{124}$$

Now $\langle H_0^* K \beta_k, \beta_j \rangle_w = \langle K \beta_k, H_0 \beta_j \rangle_{1/w}$, $\langle H_0^* f, \beta_j \rangle_w = \langle f, H_0 \beta_j \rangle_{1/w}$, and $\langle \beta_k, \beta_j \rangle_w = \langle H_0 \beta_k, H_0 \beta_j \rangle_{1/w}$, because H_0 is unitary. Thus Eqs. (124) become

$$\sum_{k=0}^N \langle H_0 \beta_k, H_0 \beta_j \rangle_{1/w} a_k + \sum_{k=0}^N \langle K \beta_k, H_0 \beta_j \rangle_{1/w} a_k$$

$$= \langle f, H_0 \beta_j \rangle_{1/w}, \quad j = 0, 1, 2, \ldots, N. \tag{125}$$

We now note that Eqs. (125) can be obtained directly, without regularization, by expanding u_N as in (122), and then setting the residual r_N orthogonal to the functions $\{H_0 \beta_j\}_{j=0}^N$ in $L_{1/w}$. In this formulation, (122) and (125) constitute a Galerkin–Petrov method because the weight functions $\{H_0 \beta_j\}_{j=0}^N$ differ in general from the basis functions $\{\beta_k\}_{k=0}^N$. However, in keeping with standard usage, we shall refer to (122) and (125) as just Galerkin's method.

From the standard theory for Galerkin's method (Ref. 19) (see also Section 19.1.1), u_N exists for all N sufficiently large; Eqs. (125) then have a unique solution, and $\|u - u_N\|_w \to 0$, $N \to \infty$. In addition, the convergence in L_w is quasi-optimal in the sense that

$$\|u - Q_N u\|_w \leq \|u - u_N\|_w \leq [(1 + \gamma(N)]\|u - Q_N u\|_w, \quad (126)$$

where Q_N is the operator of orthogonal projection onto X_N, and $\gamma(N) \to 0$ as $N \to \infty$. Thus for large N, u_N converges in L_w at essentially the same rate as the best approximation $Q_N u$ to u.

Despite its nice convergence properties, implementing Galerkin's method numerically is still a nontrivial task. First we must compute $\{H_0 \beta_j\}_{j=0}^N$, and then the inner products in (125) must be calculated. Generally all of this has to be done numerically, and can lead to a rather time-consuming algorithm. However, when $X_N = \text{span}\{\alpha_k\}_{k=0}^N$, $H_0 \alpha_k = -\gamma_k$, where $\{\alpha_k\}_{k=0}^\infty$ and $\{\gamma_k\}_{k=0}^\infty$ are the airfoil polynomials. In this case $\{H_0 \beta_j\}_{j=0}^N$ are known explicitly, and because $\{\gamma_j\}_{j=0}^\infty$ are orthogonal, (125) simplifies to

$$a_j - \sum_{k=0}^N \langle K\alpha_k, \gamma_j \rangle_{1/w} a_k = -\langle f, \gamma_j \rangle_{1/w}, \quad j = 0, 1, 2, \ldots, N. \quad (127)$$

Here we can derive more precise rates of convergence than given by (126). For example, it is shown in Ref. 40 that if $k(x, t) \in C^r$ and $f(x) \in C^r$, $r > 0$, then $\|u - u_N\|_w = O(N^{-r})$ and if $\|\ \|_\infty$ denotes the uniform norm, then $\|u - u_N\|_\infty = O(N^{-r+2})$, $r > 2$, which will be established in Section 19. Moreover, if all the data are analytic, then u_N converges exponentially fast, both in L_w and in the uniform norm.

Further numerical work requires that we calculate $\{\langle K\alpha_k, \gamma_j \rangle_{1/w}\}$ and $\{\langle f, \gamma_j \rangle_{1/w}\}$, which are double and single integrals, respectively. If the kernel $k(x, t)$ and $f(x)$ are smooth, then we may proceed numerically in the following way. Let $(\{\sigma_l\}_{l=0}^{L(N)}, \{x_l\}_{l=0}^{L(N)})$ be the weights and nodes of the $L(N) + 1$ point Gaussian quadrature rule for $[(1 + x)/(1 - x)]^{1/2}$ and let $(\{w_m\}_{m=0}^{M(N)}, \{t_m\}_{m=0}^{M(N)})$ be the weights and nodes of the $M(N) + 1$ point Gaussian quadrature rule for $[(1 - t)/(1 + t)]^{1/2}$ (these are known analytically), then

$$\langle f, \gamma_j \rangle_{1/w} = \int_{-1}^1 \sqrt{\frac{1+x}{1-x}} f(x) \gamma_j(x) \, dx \simeq \sum_{l=0}^{L(N)} \sigma_l f(x_l) \gamma_j(x_l), \quad (128)$$

and

$$\langle K\alpha_k, \gamma_j \rangle_{1/w} = \int_{-1}^{1} \int_{-1}^{1} \sqrt{\frac{1+x}{1-x}} \sqrt{\frac{1-t}{1+t}} k(x,t)\alpha_k(t)\gamma_j(x)\,dx\,dt$$

$$\simeq \sum_{m=0}^{M(N)} \sum_{l=0}^{L(N)} w_m \sigma_l k(x_l, t_m)\alpha_k(t_m)\gamma_j(x_l). \tag{129}$$

Substituting (128) and (129) into (127) and letting $\{b_k\}_{k=0}^{N}$ denote the approximations to $\{a_k\}_{k=0}^{N}$ obtained by solving

$$b_j - \sum_{k=0}^{N} \left[\sum_{m=0}^{M(N)} \sum_{l=0}^{L(N)} w_m \sigma_l k(x_l, t_m)\alpha_k(t_m)\gamma_j(x_l) \right] b_k$$

$$= -\sum_{l=0}^{L(N)} \sigma_l f(x_l)\gamma_j(x_l), \qquad j = 0, 1, 2, \ldots, N, \tag{130}$$

we obtain the approximation

$$v_N = \sum_{k=0}^{N} b_k \alpha_k(x) \tag{131}$$

to $u(x)$. In the terminology used for Fredholm equations, we refer to (130)-(131) as a *discrete Galerkin method* (Ref. 16). In Section 19 it will be shown that if $k(x,t) \in C^r$, $f(x) \in C^r$, $M(N) \geq N$, and $L(N) \geq N$, then v_N converges uniformly to u, and $\|u - v_N\|_\infty = O(N^{-r+3})$, $r > 3$. Thus, for smooth kernels and right-hand sides, evaluating the inner products by Gaussian quadrature (or any other quadrature rule with positive weights and polynomial precision at least $2N$) almost preserves the rate of convergence of the full Galerkin method. If the kernel and/or right-hand side is not smooth, then as far as we know there are no published convergence results concerning the discrete Galerkin method. This question is not just an academic one, as such situations occur in practice (Refs. 46-48).

In aerodynamics, for instance, many of the common kernels are of the form

$$k(x, t) = a(x, t) \log(|x - t|) + b(x, t), \tag{132}$$

where $a(x, t)$ and $b(x, t)$ are analytic, while at the same time $f(x)$ may be discontinuous, most often a step-function (Refs. 46-47). These problems occur, for example, in studying unsteady flows over airfoils with flaps, and are quite difficult to solve (Ref. 46). In this case, the solution has a logarithmic singularity at the location of the discontinuity of $f(x)$, and neither u_N nor v_N can converge uniformly to $u(x)$. Essentially a Gibbs-type phenomenon occurs, and we have no proof of even pointwise convergence to $u(x)$. In Example 10.1 we exhibit the results of some numerical calculations for this type of problem, and Table 2 clearly illustrates the erratic "convergence" behavior of the computed solutions.

Example 10.1. To illustrate the difference in the convergence behavior of Galerkin's method when $f(x)$ is a step-function as opposed to the case where $f(x)$ is smooth, we solved $H_0 u_i + K u_i = f_i$, $i = 1, 2$, with the kernel

$$k(x, t) = 1/\sinh(x - t) - 1/(x - t) \tag{133}$$

[which arises in an aerodynamics problem (Ref. 26)], and

$$f_1(x) = 1, \quad -1 \leq x \leq 1, \tag{134}$$

and

$$f_2(x) = \begin{cases} 0, & -1 \leq x < 0.5, \\ 1, & 0.5 \leq x \leq 1. \end{cases} \tag{135}$$

The numerical calculations were carried out for $N = 5$, $N = 10$, and $N = 15$, with the inner products $\{\langle f, \gamma_j \rangle_{1/w}\}_{j=0}^N$ evaluated analytically, and $\{\langle K\alpha_k, \gamma_j \rangle_{1/w}\}_{(k,j)=0}^N$ approximated by using Gaussian quadrature with 15 nodes in both x and t.

The results are displayed in Tables 1 and 2, and while rapid convergence is evident for $f_1(x)$, those for $f_2(x)$ are behaving as indicated above. In the latter case it is not even clear if convergence is occurring at all.

However, in many aerodynamics problems linear functionals of the solution, such as the lift, pitching moment, etc., may be of primary engineering interest (Ref. 13), and for approximating these one can often establish rapid convergence, regardless of the smoothness of $f(x)$. This is a generalization of a superconvergence result for Fredholm equations of the second kind discussed by Sloan in Chapter 2. We take up some aspects of this topic next.

10.2. A Superconvergence Result for $\nu = 0$. Suppose that $l(u)$ is a bounded linear functional of u, and that u_N is the Galerkin approximation determined by (122) and (127). If we approximate $l(u)$ by $l(u_N)$, then

$$|l(u) - l(u_N)| \leq \|H_0 + K\| \|u^* - Q_N u^*\|_{1/w} \|u - u_N\|_w, \tag{136}$$

Table 1. Galerkin Solution of the GAE with $f_1(x)$

x	$u_5(x)$	$u_{10}(x)$	$u_{15}(x)$
−0.99	1.275937	1.275912	1.275912
−0.89	1.347865	1.347964	1.347964
−0.79	1.420485	1.420581	1.420581
−0.69	1.493323	1.493341	1.493341
−0.59	1.565881	1.565816	1.565816
−0.49	1.637692	1.637578	1.637578
−0.39	1.708319	1.708202	1.708202
−0.29	1.777355	1.777277	1.777277
−0.19	1.844421	1.844409	1.844409
−0.09	1.909172	1.909232	1.909232
0.01	1.971289	1.971406	1.971406
0.11	2.030486	2.030627	2.030627
0.21	2.086504	2.086628	2.086628
0.31	2.139116	2.139181	2.139181
0.41	2.188125	2.188100	2.188100
0.51	2.233363	2.233242	2.233242
0.61	2.274692	2.274504	2.247504
0.71	2.312005	2.311823	2.311824
0.81	2.345224	2.345179	2.345179
0.91	2.347301	2.345179	2.374585
0.99	2.394567	2.395298	2.395298

where u^* is the unique solution to the adjoint equation $H_0^* u^* + K^* u^* = g$, $l(u) = \langle g, u \rangle_w$, $g \in L_w$, and Q_N is the operator of orthogonal projection onto $\text{span}\{\gamma_k\}_{k=0}^N$.

Proof of (136). See Ref. 47. Since l is bounded linear, there exists a function $g \in L_w$ such that $l(u) = \langle g, u \rangle_w$. Thus

$$l(u) - l(u_N) = \langle g, u - u_N \rangle_w. \tag{137}$$

Now let u^* be the unique solution to the adjoint equation

$$(H_0^* + K^*) u^* = g = A^* u^*$$

[that u^* exists follows from the Fredholm alternative, and our assumptions on $H_0 + K$ (Ref. 19)]. Thus

$$\langle g, u - u_N \rangle_w = \langle A^* u^*, u - u_N \rangle_w = \langle u^*, A(u - u_N) \rangle_{1/w}$$

$$= \langle u^*, Au - Au_N \rangle_{1/w} = \langle u^*, f - Au_N \rangle_{1/w}. \tag{138}$$

Table 2. Galerkin Solution of the GAE with $f_2(x)$

x	$u_5(x)$	$u_{10}(x)$	$u_{15}(x)$
−0.99	0.349170	0.39421	0.389371
−0.89	0.492665	0.444727	0.443582
−0.79	0.565741	0.536527	0.507430
−0.69	0.598909	0.563231	0.570293
−0.59	0.617921	0.596547	0.626112
−0.49	0.643775	0.683899	0.708639
−0.39	0.692711	0.808333	0.796222
−0.29	0.776214	0.921694	0.861533
−0.19	0.901012	0.992262	0.941211
−0.09	1.069074	1.033868	1.076587
0.01	1.277616	1.104194	1.227187
0.11	1.519097	1.275502	1.330589
0.21	1.781217	1.591513	1.456562
0.31	2.046922	2.029445	1.851287
0.41	2.294400	2.486454	2.530251
0.51	2.497084	2.804763	3.064040
0.61	2.623649	2.838741	2.876690
0.71	2.638014	2.560028	2.237855
0.81	2.499342	2.148512	2.232836
0.91	2.162042	2.047539	2.222236
0.99	1.715509	2.598282	2.588894

Writing $u^* = Q_N u^* + (I - Q_N)u^*$, (138) becomes

$$\langle Q_N u^*, f - Au_N \rangle_{1/w} + \langle (I - Q_N)u^*, f - Au_N \rangle_{1/w}. \tag{139}$$

Now $f - Au_N$ is the residual r_N, and by definition of Galerkin's method $\langle Q_N u^*, r_N \rangle_{1/w} = 0$, so that

$$l(u) - l(u_N) = \langle (I - Q_N)u^*, A(u - u_N) \rangle_{1/w}. \tag{140}$$

Finally, using the Cauchy–Schwarz inequality gives

$$|l(u) - l(u_N)| \le \|A\| \|u^* - Q_N u^*\|_{1/w} \|u - u_N\|_w.$$

Suppose now that f is a step-function, but that $u^* \in C^r$, then it follows from (136) and the expansion of f in terms of $\{\gamma_n\}_{n=0}^\infty$ that (Ref. 47)

$$|l(u) - l(u_N)| = O(N^{-(r+1/2)}), \tag{141}$$

which shows that $l(u_N)$ can converge rapidly, even if u_N does not. Example 10.2 exhibits this phenomenon.

Example 10.2. Since the coefficients $u_k = \langle u, \alpha_k \rangle_w$ are obvious continuous linear functionals of the solution, $a_k = \langle u_N, \alpha_k \rangle_w$ can be used as an approximating linear functional for u_k. From (141) we expect rapid convergence of these if the kernel of Eq. (2) is smooth and f is a step-function, because the representor of $l(u) = \langle u, \alpha_k \rangle_w$ is just α_k, which is C^∞.

To illustrate this we solved Eq. (2) with $k(x, t) = 1/\sinh(x - t) - 1/(x - t)$ and

$$f(x) = \begin{cases} 0, & -1 \le x < 0.5, \\ 1, & 0.5 \le x \le 1, \end{cases}$$

again.

As in Example 10.2 the matrix elements $\{\langle K\alpha_k, \gamma_j \rangle_{1/w}\}$ were approximated using Gaussian quadrature with 15 nodes in both the x and t variables (on the basis of the results of Theorem 19.11 we expect the computed values of $\{a_k\}_{k=0}^N$ to differ little from their true values). The rapid numerical convergence of $\{a_k\}_{k=0}^N$ is illustrated in Table 3.

However, if $k(x, t)$ is not smooth, then proper evaluation of the inner products $\{\langle K\alpha_k, \gamma_j \rangle_{1/w}\}$ is essential if superconvergence is to be maintained (Ref. 43). As we have pointed out previously, in unsteady fluid-flow problems kernels of the form (132) are commonplace, and then we need an accurate method for calculating integrals of the form

$$I = \int_{-1}^{1} \int_{-1}^{1} \sqrt{\frac{1+x}{1-x}} \sqrt{\frac{1-t}{1+t}} \log(|x - t|) f(x, t) \, dx \, dt, \tag{142}$$

Table 3. Convergence of $\{a_k^N\}$ for $f(x)$ a Step-function

k	a_k^2	a_k^3	a_k^4	a_k^5	a_k^6	a_k^7	a_k^8
1	1.50209	1.495810	1.494970	1.494800	1.4948100	1.4948000	1.4948000
2	1.05552	1.050830	1.050880	1.050740	1.0507400	1.0507400	1.0507400
3	—	0.230363	0.230520	0.230551	0.2305460	0.2305470	0.2305470
4	—	—	−0.129218	−0.129199	−0.1291980	−0.1291980	−0.1291980
5	—	—	—	−0.219546	−0.2195480	−0.2195480	−0.2195480
6	—	—	—	—	−0.0975991	−0.0975991	−0.0975992
7	—	—	—	—	—	−0.0697948	−0.0697947
8	—	—	—	—	—	—	0.1308740

where $f(x, t)$ is "smooth." In Section 11.2 on collocation, wheren only single integrals are present, we investigate a procedure which uses Chebyshev expansions of $f(x, t)$ and a possible extension to evaluate (142) is outlined below.

First write

$$\int_{-1}^{1}\int_{-1}^{1} \sqrt{\frac{1+x}{1-x}} \sqrt{\frac{1-t}{1+t}} \log(|x-t|) f(x,t)\, dx\, dt$$

as

$$\int_{-1}^{1}\int_{-1}^{1} \frac{(1+x)(1-t)\log(|x-t|) f(x,t)\, dx\, dt}{\sqrt{1-x^2}\sqrt{1-t^2}}, \tag{143}$$

and then approximate $g(x, t) = (1 + x)(1 - t)f(x, t)$ by a finite Chebyshev series

$$g(x, t) \simeq \sum_{k=0}^{P} \sum_{l=0}^{P} a_{kl} T_l(x) T_k(t).$$

Then

$$I \simeq \int_{-1}^{1}\int_{-1}^{1} \sum_{k=0}^{P} \sum_{l=0}^{P} \frac{a_{kl} T_l(x) T_k(t) \log(|x-t|)\, dx\, dt}{\sqrt{1-x^2}\sqrt{1-t^2}}$$

$$= \sum_{k=0}^{P} \sum_{l=0}^{P} a_{kl} \int_{-1}^{1}\int_{-1}^{1} \frac{T_l(x) T_k(t) \log(|x-t|)\, dx\, dt}{\sqrt{1-x^2}\sqrt{1-t^2}}, \tag{144}$$

and using

$$\frac{1}{\pi} \int_{-1}^{1} \frac{\log(|x-t|) T_k(t)\, dt}{\sqrt{1-t^2}} = \begin{cases} -\log 2, & k = 0, \\ -T_k(x)/k, & k \geq 1, \end{cases} \tag{145}$$

and the orthogonality of the Chebyshev polynomials with respect to the weight function $(1 - x^2)^{-1/2}$ (Ref. 49), the integrals in (144) can be determined in closed form. On theoretical grounds we would expect good results, but for realistic kernels the algorithm would no doubt be time consuming.

10.3. Galerkin's Method: $\nu = 1$. For the rest of this section we will confine our attention to Galerkin's method using polynomial bases and weights. When $\nu = 1$, we look for approximations of the form

$$u_N = \sum_{k=0}^{N} a_k \tau_k, \qquad (146)$$

where $\{\tau_k\}_{k=0}^{N}$ are the normalized Chebyshev polynomials of the first kind. Here, $\{a_k\}_{k=0}^{N}$ are obtained by setting the residual $r_N = H_1 u_N + K u_N - f$ orthogonal to $\{\mu_j\}_{j=0}^{N-1}$, the normalized Chebyshev polynomials of the second kind. Using (64)-(65) and the orthonormality of $\{\mu_j\}_{j=0}^{N}$ gives the N equations

$$a_{j+1} + \sum_{k=0}^{N} \langle K\tau_k, \mu_j \rangle_{1/w} a_k = \langle f, \mu_j \rangle_{1/w}, \qquad j = 0, 1, 2, \ldots, N-1, \qquad (147)$$

and appending the constraint equation $l(u_N) = M$, we get a system of $N + 1$ equations in $N + 1$ unknowns. Usually

$$l(u_N) = \int_{-1}^{1} (u_N/\sqrt{1-t^2})\, dt = M,$$

so that using (146), $a_0 = M/\sqrt{\pi}$, and so can be explicitly eliminated from (147). In this case we need only solve N equations to obtain the remaining N unknowns, a_k, $k = 1, 2, \ldots, N$.

For practical implementation, as for $\nu = 0$, the inner products in (147) usually have to be obtained numerically, leading to the corresponding discrete Galerkin method (Ref. 16).

As to convergence, quasi-optimal rates can be obtained in the L_w norm (Refs. 3, 11, 16), while uniform convergence can be shown to hold under slightly weaker conditions than for $\nu = 0$. That is, if $k(x, t) \in C^r$, and $f(x) \in C^r$, then

$$\|u - u_N\|_w = O(N^{-r}), \qquad r > 0,$$

and

$$\|u - u_N\|_\infty = O(N^{-r+1}), \qquad r > 1.$$

Slightly slower convergence occurs if quadrature errors are present (Ref. 16). Proofs of these results will be given in Section 19.

10.4. Galerkin's Method: $\nu = -1$. When $\nu = -1$ and polynomial approximations are sought, the standard approach is to use the approximation

$$u_N = \sum_{k=0}^{N} a_k \mu_k, \qquad (149)$$

setting the residual orthogonal to $\{\tau_j\}_{j=0}^{N+1}$ to give the $N+2$ equations in $N+1$ unknowns

$$-a_{j-1} + \sum_{k=0}^{N} \langle K\mu_k, \tau_j \rangle_{1/w} a_k = \langle f, \tau_j \rangle_{1/w}, \qquad j = 0, 1, 2, \ldots, N+1, \qquad (150)$$

where $a_{-1} = 0$ (Ref. 1). From (150) it can be shown that the zeroth equation is just the discrete analogue of the consistency condition (106).

In this situation a new problem arises, in that (150) is an overdetermined system and it is not clear how to go about solving it if the equations are consistent. Some authors suggest dropping one equation (Ref. 35), however, it is not obvious which one should go, but the zeroth is the most natural (Ref. 30). Perhaps in this case an indirect method is more appropriate, as the resulting Galerkin system is square, and is equivalent to the equations in (150) for $j = 1, 2, \ldots, N+1$.

As the corresponding indirect Galerkin method converges, this again suggests that the zeroth equation is superfluous, and that for N sufficiently large, the last $N+1$ equations in (150) have a unique solution.

11. Collocation

Owing to the large amount of arithmetic often needed to set up the Galerkin equations, collocation methods, where applicable, have generally been preferred for solving integral equations, and the situation for CSIES is no different (Refs. 1, 19). Historically, particularly when $\nu = 0$, these methods have been standard (Ref. 14), and in aerodynamics and hydrodynamics it seems to be the accepted approach (Refs. 4, 14). However, when $\nu = 1$, the case most frequently arising in solid mechanics (Refs. 2, 7), quadrature techniques appear to be the current method of choice. For the above-mentioned reasons, our emphasis in this section will be on polynomial-based collocation methods for $\nu = 0$, while in Section 12 we will concentrate on equations with index one.

11.1. $\nu = 0$: Continuous Data.
When $\nu = 0$, the standard polynomial collocation method for solving the GAE is to approximate u by

$$u_N = \sum_{k=0}^{N} a_k \alpha_k, \tag{151}$$

form the residual $r_N = H_0 u_N + K u_N - f$, and then set it equal to zero at the points $\{x_j\}_{j=0}^{N}$ which are the zeros of γ_{N+1}. Doing this leads to the system of $N+1$ equations in $N+1$ unknowns,

$$\sum_{k=0}^{N} a_k H \alpha_k(x_j) + \sum_{k=0}^{N} a_k K \alpha_k(x_j) = f(x_j), \qquad j = 0, 1, 2, \ldots, N, \tag{152}$$

and using (80), relation (152) becomes

$$-\sum_{k=0}^{N} a_k \gamma_k(x_j) + \sum_{k=0}^{N} a_k K \alpha_k(x_j) = f(x_j), \qquad j = 0, 1, 2, \ldots, N. \tag{153}$$

One should observe that $\{x_j\}_{j=0}^{N}$ are the nodes of the $N+1$ point Gaussian quadrature rule for the weight $[(1+x)/(1-x)]^{1/2}$ (Ref. 1), and this choice of collocation points can be motivated by the observation made in Refs. 11, 13 that mathematically (153) are equivalent to the discrete Galerkin equations (129) with $M(N) = l(N) = N$.

If $k(x, t)$ and $f(x)$ are continuous, $\|u - u_N\|_w \to 0$, $N \to \infty$, and it has been shown in Refs. 1, 3 that if these functions are C^r, $r > 0$, then

$$\|u - u_N\|_w = 0(N^{-r}). \tag{154}$$

Uniform convergence has also recently been established by Miel in Ref. 42, and

$$\|u - u_N\|_\infty = O(N^{-r+2}), \qquad r > 2, \tag{155}$$

as for Galerkin's method. More generally, if K maps L_{w_0} compactly into C^0, the space of continuous functions with the uniform norm, and $f(x)$ is Riemann integrable, then the mean-square convergence of u_N to u also holds. If it happens that $k(x, t)$ is discontinuous, but $u \in C^r$, then error estimates of the form (154) and (155) can also be obtained. Interestingly, this is true for the case of the logarithmically singular kernels (132) (Ref. 13).

As for Galerkin's method, usually the most difficult part of an actual numerical implementation is the ability to compute

$$(K\alpha_k)(x_j) = \int_{-1}^{1} \sqrt{\frac{1-t}{1+t}} k(x_j, t)\alpha_k(t) \, dt \tag{156}$$

accurately. If $k(x_j, t)$ is smooth in t, then the Gaussian quadrature rule for $w_0(t) = [(1-t)/(1+t)]^{1/2}$ is a good choice, and is usually used with $N+1$ nodes (Ref. 13). The resulting numerical method is called a *discrete collocation method* but, in contrast to the discrete Galerkin method, no general error analysis seems to have been published. However, using Gaussian quadrature with $N+1$ nodes is mathematically equivalent to using the Gaussian quadrature method (see Section 12.2), and so convergence follows from known convergence theorems for that method (Refs. 3, 5, 16).

11.2. $\nu = 0$: $k(x, t) = a(x, t) \log(|x - t|) + b(x, t)$. When $k(x, t)$ is of the form (132), then

$$(K\alpha_k)(x_j) = \int_{-1}^{1} \sqrt{\frac{1-t}{1+t}} a(x_j, t)\alpha_k(t) \log(|x_j - t|) \, dt$$

$$+ \int_{-1}^{1} \sqrt{\frac{1-t}{1+t}} b(x_j, t)\alpha_k(t) \, dt. \tag{157}$$

If we assume that $a(x, t)$ and $b(x, t)$ are analytic, as occurs in the Küssner-Schwarz kernel ($i^2 = -1$)

$$k(x, t) = \exp[-i\mu(x-t)]\{\text{Ci}(\mu|x-t|) + i \, \text{Si}[\mu(x-t)] - i\pi/2\},$$

where

$$\begin{cases} \text{Ci}(z) = \gamma + \log z + \sum_{n=1}^{\infty} (-1)^n z^{2n}/[(2n)(2n!)], \\ \text{Si}(z) = \sum_{n=0}^{\infty} (-1)^n z^{2n+1}/\{(2n+1)[(2n+1)!]\}, \\ \gamma = 0.5772\ldots \quad \text{(Euler's constant)}, \end{cases}$$

in aerodynamics (Ref. 13), then the second integral in (157) can be evaluated by Gaussian quadrature with respect to $[(1-t)/(1+t)]^{1/2}$. Letting $a(t) =$

$a(x_j, t)$, we are now faced with the task of computing

$$I_k = \int_{-1}^{1} w_0(t) \log(|x - t|) a(t) \alpha_k(t) \, dt, \qquad k = 0, 1, 2, \ldots, N, \qquad (158)$$

where $a(t)$ is analytic.

In early work in aerodynamics it was customary to calculate I_k by writing $a(t) = [a(t) - a(x)] + a(x)$, and then it becomes (Ref. 13)

$$I_k = \int_{-1}^{1} w_0(t) \log(|x - t|) [a(t) - a(x)] \alpha_k(t) \, dt$$

$$+ a(x) \int_{-1}^{1} w_0(t) \log(|x - t|) \alpha_k(t) \, dt. \qquad (159)$$

Since the integrand in the first integral in (159) is continuous, it could be approximated by Gaussian quadrature, while the second was obtained analytically (Ref. 13). However, the slow convergence of the numerical integration tended to quickly dominate the truncation error in the numerical solution, and it was difficult to obtain more than 3-4 figure accuracy using an acceptable number of basis elements (≤ 10) (Ref. 13). This was in contrast to the six or more figure accuracy typically attainable in steady-flow problems without the logarithmic term in k (Ref. 13). These rather poor results suggested that one look for a more accurate way of computing the first integral in (156).

In Refs. 26, 50 Fromme and Golberg made a series of elementary transformations to write $[f(t) = \alpha_k(t) a(t)]$

$$\int_{-1}^{1} w_0(t) \log(|x - t|) f(t) \, dt$$

$$= (1 + a)^{1/2} \int_{0}^{1} u^{-1/2} [(1 - t)^{1/2} \log(x - t) f(t)] \, du \qquad [t = -1 + (1 + a)u]$$

$$+ (x - a) \int_{0}^{1} -\log u [w_0(t) f(t)] \, du \qquad [t = x - (x - a)a]$$

$$+ (x - a) \log[(x - a)] \int_{0}^{1} w_0(t) f(t) \, du \qquad [t = x - (x - a)a]$$

$$+ (b - x) \log(b - x) \int_{0}^{1} w_0(t) f(t) \, du \qquad [t = x + (b - x)u]$$

$$-(b-x)\int_0^1 -\log u[w_0(t)f(t)]\,du \qquad [t = x + (b-x)u]$$

$$+ (1-b)^{1/2}\int_0^1 u^{-1/2}\{w_0(t)(1-t)\log[(t-x)]f(t)\}\,du$$

$$[t = 1 - (1-b)u]$$

$$= I_1 + I_2 + I_3 + I_4 + I_5 + I_6, \qquad (160)$$

where $-1 < a < x < b < 1$. Integrals I_1 and I_6 were then approximated by ordinary Gaussian quadrature after the additional substitution $u = v^2$ was made, I_2 and I_5 were approximated using Gaussian quadrature with respect to the weight function $-\log u$, while I_3 and I_4 were obtained by using ordinary Gaussian quadrature as well. With this formula they were often able to achieve 12 figure accuracy for aerodynamic moments with as few as eight basis elements (Ref. 26).

Unfortunately, because the weights and nodes of ordinary Gaussian quadrature on $[0, 1]$ and Gaussian quadrature with respect to the weight $-\log u$ are not known analytically, tables of these were written into their program, which became very long and unwieldy (Refs. 26, 50). In addition, this numerical integration method does not converge uniformly in x (this can be seen in Tables 4-6 below), and so there is some concern as to what will happen to the accuracy of the algorithm given in Refs. 26, 50 as the number of basis elements increases and the collocation points move closer to ± 1.

Example 11.1. To illustrate the behavior of the Fromme-Golberg quadrature rule (160) it was used to calculate the integral (Ref. 49)

$$I = \frac{1}{\pi}\int_{-1}^1 \sqrt{\frac{1-t}{1+t}}\log(|x-t|)T_5(t)\,dt$$

$$= -T_5(x)/5 + [T_6(x)/6 + T_4(x)/4]. \qquad (161)$$

In all cases the integrals in (160) were approximated using the appropriate Gaussian rule with 1-8 nodes, and numerical results are given in Tables 4-6 for $x = -0.99$, $x = 0.0$, and $x = 0.99$. The resulting approximations are denoted by $\hat{I}_N$, $N = 1, 2, \ldots, 8$, and one can clearly see the difference in accuracy obtained near ± 1 as distinct from that when $x = 0$.

To overcome the convergence problem of the Fromme-Golberg rule, McKenna in Ref. 49 developed another method for approximating (158)

Table 4. Convergence of the Fromme–Golberg Rule for (161): $x = -0.99$

Nodes N	$\hat{I}_N$	Error	I
2	0.1151601	0.4276748	0.3125147
3	0.1069479	0.2055668	0.3125147
4	0.2076872	0.1048274	0.3125147
5	0.2516463	$6.086838E - 02$	0.3125147
6	0.2754945	$3.702015E - 02$	0.3125147
7	0.2896029	$2.291176E - 02$	0.3125147
8	0.2981505	$1.436418E - 02$	0.3125147

Table 5. Convergence of the Fromme–Golberg Rule for (161): $x = 0.00$

Nodes N	$\hat{I}_N$	Error	I
2	$2.926924E - 02$	$1.239742E - 02$	$4.166667E - 02$
3	$4.500535E - 02$	$-3.338687E - 02$	$4.166667E - 02$
4	$4.146993E - 02$	$1.967326E - 04$	$4.166667E - 02$
5	$4.166902E - 02$	$-2.354384E - 06$	$4.166667E - 02$
6	$4.166668E - 02$	$-1.490116E - 08$	$4.166667E - 02$
7	$4.166667E - 02$	$-7.450581E - 09$	$4.166667E - 02$
8	$4.166665E - 02$	$1.490116E - 08$	$4.166667E - 02$

Table 6. Convergence of the Fromme–Golberg Rule for (161): $x = 0.99$

Nodes N	$\hat{I}_N$	Error	I
2	-0.1397303	0.1483005	$8.570209E - 03$
3	$3.580811E - 02$	-0.0272379	$8.570209E - 03$
4	$3.900199E - 03$	$4.670010E - 03$	$8.570209E - 03$
5	$8.395526E - 03$	$1.746826E - 04$	$8.570209E - 03$
6	$8.429481E - 03$	$1.407284E - 04$	$8.570209E - 03$
7	$8.501762E - 03$	$6.844662E - 05$	$8.570209E - 03$
8	$8.535216E - 03$	$3.499258E - 05$	$8.570209E - 03$

similar to one considered by Elliott and Paget in Ref. 51. To derive it, let

$$\int_{-1}^{1} w_0(t) \log(|x-t|) f(t) \, dt = \int_{-1}^{1} \frac{(1-t) f(t) \log(|x-t|) \, dt}{\sqrt{1-t^2}}, \quad (162)$$

and approximate $(1-t)f(t)$ by a finite Chebyshev series

$$(1-t) f(t) \simeq \sum_{j=0}^{M'} c_{M,j} T_j(t), \quad (163)$$

where the prime in (163) indicates that the first term is halved. Substituting (163) into (162) gives

$$\int_{-1}^{1} w_0(t) \log(|x-t|) f(t) \, dt \simeq \sum_{j=0}^{M'} c_{M,j} \int_{-1}^{1} \frac{\log(|x-t|) T_j(t) \, dt}{\sqrt{1-t^2}}, \quad (164)$$

and then using (145)

$$\int_{-1}^{1} w_0(t) \log(|x-t|) f(t) \, dt \simeq -c_{M,0} \log 2/2 - \sum_{j=1}^{M} c_{M,j} T_j(x)/j. \quad (165)$$

The remaining problem is to choose $\{c_{M,j}\}_{j=0}^{M}$.

For this, McKenna used an "almost best" approximation (in the sense of uniform approximation) given by Ref. 52,

$$c_{M,j} = \frac{2}{M+1} \sum_{i=0}^{M+1''} g(\theta_i) \cos[ij\pi/(M+1)], \quad j = 0, 1, 2, \ldots, M, \quad (166)$$

where $g(t) = (1-t)f(t)$ and

$$\theta_i = \cos[i\pi/(M+1)], \quad i = 0, 1, 2, \ldots, M+1, \quad (167)$$

and the double prime in (166) indicates that the first and last terms are halved [Elliott and Paget in Ref. 51 used interpolation of $g(t)$ at the zeros of $T_{M+1}(t)$, but no numerical results were given].

In Ref. 49, it was shown that $\hat{I}_M(x)$, the quadrature rule given by (165)-(167), converges uniformly to $I(x) = \int_{-1}^{1} w_0(t) \log(|x-t|) g(t) \, dt$, and if $f(t) \in C^r$, $r \geq 2$, then

$$|I(x) - \hat{I}_M(x)| \leq c/M^r.$$

This result improves on the pointwise convergence of the Fromme-Golberg rule (160), for $-1 < x < 1$, and it was also shown in Ref. 49 that $\hat{I}_M(x)$ always used fewer function evaluations than theirs for a given degree of polynomial precision for all $M \geq 1$. In this sense $\hat{I}_M(x)$ is more efficient than using Gaussian quadrature in (160).

Example 11.2. To illustrate the behavior of $\hat{I}_M(x)$ it was used in Ref. 49 to calculate a variety of integrals, and here we present a number of results which show that excellent accuracy can be obtained using relatively few nodes.

In Table 7 values of

$$s_8 = \frac{1}{\pi} \int_{-1}^{1} \sqrt{\frac{1-t}{1+t}} \log(|x-t|) T_8(t) \, dt$$

$$= -T_8(x)/8 + [T_9(x)/9 + T_7(x)/7] \tag{168}$$

are given, where s_8 was computed using $M = 9$ nodes in (165). Here we expect to get numerical results accurate to machine precision. The calculations were done in double precision on an IBM-PC compatible computer where the maximum precision obtainable is about 17 significant figures. As one can see from Table 7, this accuracy was essentially achieved uniformly for $-1 \leq x \leq 1$.

For examples whose analytic expressions were not available in a simple closed form, the integrals

$$u_k(x) = \frac{1}{\pi} \int_{-1}^{1} \sqrt{\frac{1-t}{1+t}} \log(|x-t|) e^{kt} \, dt \tag{169}$$

Table 7. Numerical Evaluation of s_8 by McKenna's Rule

x	s_8 exact*	Error
−1.0	−2.519840	−2.08169E − 17
−0.9	0.216870	−4.85723E − 17
−0.5	0.082341	4.68375E − 17
0.0	0.125000	1.38778E − 17
0.5	0.042659	−1.56125E − 17
0.9	0.006456	1.12757E − 17
1.0	0.001984	1.73472E − 17

* For convenience we have tabulated values to only six figures.

Table 8. Numerical Evaluation of $u_k(0.5)$

k	$\hat{u}_{30}(0.5)$*	$\hat{u}_{60}(0.5)$†	$\hat{u}(0.5)$‡
1	−0.4622842326270537	−0.4622842326270538	−0.4622843179723732
5	−4.515227967460854	−4.515227967460854	−4.515228294819275
10	−166.2514215338906	−166.2514215338905	−166.2514363342769

* $\hat{u}_{30}(0.5)$ McKenna rule-30 nodes.
† $\hat{u}_{60}(0.5)$ McKenna rule-60 nodes.
‡ $\hat{u}(0.5)$ Fromme–Golberg rule-8 nodes in I_1-I_6.

were approximated. In Table 8 we show the results obtained for $u_k(0.5)$, $k = 1, 5, 10$, using 30 and 60 nodes respectively in (165)–(166). Almost full machine precision was obtained for 30 nodes.

11.3. Polynomial Collocation: $\nu = \pm 1$. As we have already observed, polynomial collocation algorithms do not seem to be used extensively in practice when $\nu = \pm 1$, so our treatment here will be brief, as the details are almost identical to those for $\nu = 0$.

For $\nu = 1$, the standard collocation method is to approximate u by

$$u_N = \sum_{k=0}^{N} a_k T_k, \qquad (170)$$

and then set the residual $r_N = H_1 u_N + K u_N - f$ equal to zero at $\{x_j\}_{j=0}^{N-1}$, the zeros of $U_N(x)$. Writing these conditions out explicitly gives the N equations

$$\sum_{k=0}^{N} a_k H_1 T_k(x_j) + \sum_{k=0}^{N} a_k K T_k(x_j) = f(x_j), \qquad j = 0, 1, 2, \ldots, N-1, \qquad (171)$$

and using (64)–(65) these become (recall $H_1 T_0 = 0$)

$$\sum_{k=1}^{N} a_k U_{k-1}(x_j) + \sum_{k=0}^{N} a_k K T_k(x_j) = f(x_j), \qquad j = 0, 1, 2, \ldots, N-1. \qquad (172)$$

As for Galerkin's method, we obtain a square system by adding the constraint equation $l(u_N) = M$. If $l(u)$ is of the form $\int_{-1}^{1} [u(t)/\sqrt{1-t^2}] \, dt$, then $l(u_N) = a_0 \pi$, and a_0 can be eliminated from (172) leaving a system of N equations in N unknowns for $\{a_k\}_{k=1}^{N}$.

Numerical Solution of Cauchy Singular Integral Equations

As usual, actual implementation generally requires the numerical approximation of $KT_k(x_j)$, and this can be done using Gaussian integration with respect to $w_1(t)$ when $k(x, t)$ is smooth. If $k(x, t)$ is of the form in (132), then the method proposed in the previous subsection is applicable.

As for $\nu = 0$, mean-square convergence can be established for continuous kernels or for kernels mapping L_w compactly into C^0, and the convergence rates are of the form

$$\|u - u_N\|_w = O(N^{-r}), \qquad r > 0$$

and

$$\|u - u_N\|_\infty = O(N^{-r+1}), \qquad r > 1$$

if the data are C^r.

When $\nu = -1$ use

$$u_N = \sum_{k=0}^{N} a_k U_k, \tag{173}$$

and collocate at the zeros $\{x_j\}_{j=0}^{N+1}$ of T_{N+2} to get the $N + 2$ equations in $N + 1$ unknowns,

$$-\sum_{k=0}^{N} a_k T_{k+1}(x_j) + \sum_{k=0}^{N} a_k K T_k(x_j) = f(x_j), \qquad j = 0, 1, 2, \ldots, N + 1. \tag{174}$$

Before leaving this section on collocation, it is interesting to note that Cuminato (Ref. 53) has recently proved that one can always use the zeros of either $T_{N+1}(x)$ ($\nu = 0$) or $T_N(x)$ ($\nu = 1$) as collocation points (rather than the standard ones given in this section) to obtain a uniformly convergent collocation algorithm under Hölder continuity conditions on $k(x, t)$ and $f(x)$. This is probably not very important for equations of the first kind, since explicit formulas are known for the zeros of $\gamma_{N+1}(x)$ and $U_N(x)$. However, in the same paper he has shown that these same nodes can be used for equations of the second kind as well. Since we will show in Section 17 that the standard collocation methods for those equations use the zeros of Jacobi polynomials as collocation points, which usually must be obtained numerically, this result simplifies those algorithms considerably. To the author's knowledge no such result has been proven for the "collocation" nodes of the quadrature method to be discussed next, and so perhaps collocation methods should be given more consideration than they have for solving general CSIES.

12. Quadrature Methods

12.1. Quadrature Rules for Cauchy Principal-Value Integrals. For Fredholm equations of the second kind with smooth kernels, perhaps the most natural and well-known approximation technique is the Nyström method. That is, if $u(x)$ satisfies

$$u(x) + \int_a^b k(x, t)u(t)\, dt = f(x), \tag{175}$$

then one can obtain an approximation to $u(x)$ by replacing $\int_a^b k(x, t)u(t)\, dt$ with a numerical integration

$$\int_a^b k(x, t)u(t)\, dt \simeq \sum_{j=1}^{N} w_j k(x, t_j)u(t_j), \tag{176}$$

where $(\{w_j\}_{j=1}^N, \{t_j\}_{j=1}^N)$ are the weights and nodes of the integration rule. Substituting the right-hand side of (176) for the integral in (175) generates the functional equation

$$u_N(x) + \sum_{j=1}^{N} w_j k(x, t_j)u_N(t_j) = f(x), \tag{177}$$

where $u_N(x)$ is an approximation to $u(x)$. From Eq. (177) it is seen that once $\{u_N(t_j)\}_{j=1}^N$ are known, then it can be used as an interpolation formula to obtain $u_N(x)$ for all $x \in [a, b]$. This may be accomplished if Eq. (177) is evaluated at $\{t_i\}_{i=1}^N$, showing that $\{u_N(t_i)\}_{i=1}^N$ can be obtained by solving

$$u_N(t_i) + \sum_{j=1}^{N} w_j k(t_i, t_j)u_N(t_j) = f(t_i), \quad i = 1, 2, \ldots, N. \tag{178}$$

As one can clearly see, the Nyström method generally requires considerably less arithmetic to set up than either Galerkin or collocation methods, so it is reasonable to ask if such algorithm exist for CSIES.

They do, but in contrast to Fredholm equations of the second kind, these techniques are of relatively recent origin, stemming from the paper on Gaussian quadrature methods for equations of the first kind by Erdogan and Gupta in 1972 (Ref. 54), although a non-Gaussian procedure had been given as early as 1959 by Kalandiya (Ref. 55) (see Section 15).

The difficulty in developing such algorithms comes from the problem of trying to find appropriate numerical integration rules for Cauchy

Numerical Solution of Cauchy Singular Integral Equations

principal-value integrals of the form

$$\fint_a^b \frac{w(t)u(t)\,dt}{t-x}, \qquad a < x < b, \tag{179}$$

where $w(t)$ is a nonnegative weight function and $\int_a^b w(t)\,dt < \infty$ [this problem itself has been an active area of research for many years, as the recent review paper of Monegato shows (Ref. 56)].

The first "Gaussian rule" for (179) was apparently given by Hunter in Ref. 57 for $w(t) = 1$, and was then generalized by Chawla and Ramakrishnan in Ref. 58 and Krenk in Ref. 59, who considered Jacobi weights, $w(t) = (1-t)^\alpha (1+t)^\beta$, $\alpha > -1$, $\beta > -1$. Various extensions were then given by Ioakimidis and Theocaris (Refs. 60-61) who developed Lobatto and Radau rules for (179) with Jacobi weights, while Gaussian rules for arbitrary weights were given by Elliott and Paget in Ref. 62. Before proceeding with a derivation of these rules, it is interesting to point out that in the original papers of Erdogan and Gupta (Ref. 54) and Krenk (Ref. 59), on the development of quadrature methods for CSIES, these quadrature rules were not used explicitly, and it appears that Theocaris and Ioakimidis in Refs. 63-64 first recognized the relation of the Erdogan-Gupta-Krenk approach to that based on rules of Hunter's type.

Theorem 12.1. Let $w(t)$ be a nonnegative weight function such that $\int_a^b t^n w(t)\,dt < \infty$, $n \geq 0$, assume that $u(t)$ is such that the integral in (179) exists and let $Q_N = (\{w_k\}_{k=1}^N, \{t_k\}_{k=1}^N)$ be an N point polynomial interpolatory quadrature rule for $\int_a^b w(t)u(t)\,dt$. That is,

$$\int_a^b w(t)u(t)\,dt \simeq \sum_{k=1}^N w_k u(t_k), \tag{180}$$

where $w_k = \int_a^b h_k(t)\,dt$ and $\{h_k(t)\}_{k=1}^N$ are the fundamental polynomials of Lagrange interpolation for the point set $\{t_1 < t_2 \cdots < t_N\}$. Then

$$S(u) = \fint_{-a}^b \frac{w(t)u(t)\,dt}{t-x}, \qquad a < x < b,$$

can be approximated by $S_N(u)$, where

$$S_N(u) = \sum_{k=1}^N \frac{w_k u(t_k)}{t_k - x} + \frac{q_N(x)u(x)}{\sigma_N(x)}, \qquad t_k \neq x, \quad k = 1, 2, \ldots, N, \tag{181}$$

and $S_N(u) = S(u_N)$, where $u_N(t)$ is the unique polynomial which interpolates to $u(t)$ at $\{t_1, t_2, \ldots, t_N, x\}$, $\sigma_N(t) = \prod_{k=1}^{N}(t - t_k)$, and $q_N = S(\sigma_N)$. Moreover, if Q_N is the Gaussian rule for $\int_a^b w(t)u(t)\,dt$, then S_N is Gaussian in the sense that it integrates exactly all polynomials of degree $\leq 2N$.

Proof. Let $\rho_{N+1}(t) = \sigma_N(t)(t - x)$, then

$$u_N(t) = \sum_{k=1}^{N} l_k(t)u(t_k) + l_{N+1}(t)u(x), \qquad (182)$$

where

$$l_k(t) = \rho_{N+1}(t)/\rho'_{N+1}(t_k)(t - t_k)$$
$$= \sigma_N(t)(t - x)/\sigma'_N(t_k)(t - t_k)(t_k - x), \qquad k = 1, 2, \ldots, N,$$

and

$$l_{N+1}(t) = \sigma_N(t)(t - x)/\sigma_N(x)(t - x) = \sigma_N(t)/\sigma_N(x).$$

Thus,

$$S_N(u) = S(u_N) = \sum_{k=1}^{N} u(t_k) \int_a^b \frac{w(t)l_k(t)}{t - x}\,dt + u(x)\int_a^b \frac{w(t)l_{N+1}(t)}{t - x}\,dt$$

$$= \sum_{k=1}^{N} u(t_k) \int_a^b [w(t)\sigma_N(t)/\sigma'_N(t_k)(t - t_k)(t_k - x)]\,dt$$

$$+ [S(\sigma_N)/\sigma_N(x)]u(x)$$

$$= \sum_{k=1}^{N} u(t_k) \int_a^b [w(t)\sigma_N(t)/\sigma'_N(t_k)(t - t_k)(t_k - x)]\,dt$$

$$+ q_N(x)u(x)/\sigma_N(x),$$

$$x \neq t_k, \quad k = 1, 2, \ldots, N. \qquad (183)$$

Now $h_k(t) = \sigma_N(t)/\sigma'_N(t_k)(t - t_k)$, so that

$$S_N(u) = \sum_{k=1}^{N} \frac{w_k u(t_k)}{t_k - x} + \frac{q_N(x)u(x)}{\sigma_N(x)}, \qquad t_k \neq x, \quad k = 1, 2, \ldots, N. \qquad (184)$$

To prove that S_N is Gaussian if Q_N is, we first note by definition that S_N has precision at least N, so that letting $u(t) = 1$ in (184) yields

$$S_N(1) = \fint_a^b \frac{w(t)\,dt}{t-x} = \sum_{k=1}^N \frac{w_k}{t_k - x} + \frac{q_N(x)}{\sigma_N(x)}. \tag{185}$$

Now let $p(t)$ be a polynomial of degree $\leq 2N$; then $[p(t) - p(x)]/(t-x)$ is a polynomial of degree at most $2N - 1$. Since Q_N is Gaussian,

$$\fint_a^b w(t)\left[\frac{p(t) - p(x)}{t-x}\right]dt = \sum_{k=1}^N \frac{w_k[p(t_k) - p(x)]}{t_k - x}$$

$$= \sum_{k=1}^N \frac{w_k p(t_k)}{t_k - x} - p(x)\left(\sum_{k=1}^N \frac{w_k}{t_k - x}\right). \tag{186}$$

But

$$\sum_{k=1}^N \frac{w_k}{t_k - x} = \fint_a^b \frac{w(t)\,dt}{t-x} - \frac{q_N(x)}{\sigma_N(x)}, \tag{187}$$

and substituting (187) into (186) gives

$$\fint_a^b \frac{w(t)p(t)\,dt}{t-x} - p(x)\fint_a^b \frac{w(t)}{t-x}dt$$

$$= \sum_{k=1}^N \frac{w_k p(t_k)}{t_k - x} - p(x)\left[\fint_a^b \frac{w(t)\,dt}{t-x} - \frac{q_N(x)}{\sigma_N(x)}\right],$$

so that

$$\fint_a^b \frac{w(t)p(t)\,dt}{t-x} = \sum_{k=1}^N \frac{w_k p(t_k)}{t_k - x} + \frac{q_N(x)p(x)}{\sigma_N(x)},$$

as required. $\square$

12.2. The Gaussian Quadrature Method.
Now consider the GAE

$$\frac{1}{\pi}\fint_{-1}^1 \frac{w_\nu(t)u(t)\,dt}{t-x} + \int_{-1}^1 w_\nu(t)k(x,t)u(t)\,dt = f(x), \quad -1 < x < 1, \tag{188}$$

where $k(x, t)$ is continuous. In the Gaussian quadrature method for solving (188), we begin by approximating $\int_{-1}^{1} w_\nu(t)k(x, t)u(t)\, dt$ by the $N + 1$ point Gaussian rule Q_{N+1} for $w_\nu(t)$, so that

$$\int_{-1}^{1} w_\nu k(x, t)u(t)\, dt \simeq \sum_{k=0}^{N} w_k k(x, t_k) u(t_k),$$

and the Cauchy integral by (181). Substituting these into (188) and denoting the resulting approximation to $u(x)$ by $u_N(x)$, we find that it satisfies (for $x \neq t_k$, $k = 0, 1, 2, \ldots, N$)

$$\frac{1}{\pi} \sum_{k=0}^{N} \frac{w_k u_N(t_k)}{t_k - x} + \sum_{k=0}^{N} w_k k(x, t_k) u_N(t_k)$$

$$+ q_{N+1}(x) u_N(x) / \sigma_{N+1}(x) = f(x), \qquad -1 < x < 1. \qquad (189)$$

Note that we now have a problem somewhat different than in the Fredholm case in that (189) is defined only for $x \neq t_k$, $k = 0, 1, 2, \ldots, N$, and if we attempt to determine $\{u_N(t_k)\}_{k=0}^{N}$ by setting $x = t_k$, $k = 0, 1, 2, \ldots, N$ in (189), then the first term is undefined. Since we would like equations involving only $\{u_N(t_k)\}_{k=0}^{N}$, then a reasonable approach is to choose x so that $q_{N+1}(x) = 0$. Our next problem then is to determine these zeros.

Now observe that since we are considering Gaussian integration, if $\nu = 0$, then $\sigma_{N+1}(x) = a\alpha_{N+1}(x)$. If $\nu = 1$, then $\sigma_{N+1}(x) = bT_{N+1}(x)$, while if $\nu = -1$, then $\sigma_{N+1}(x) = cU_{N+1}(x)$, where a, b, and c are constants. By definition, $q_{N+1}(x) = S(\sigma_{N+1}) = \pi H_\nu(\sigma_{N+1})$, so that

$$q_{N+1}(x) = \begin{cases} -\pi c T_{N+2}(x), & \nu = -1, \\ -\pi a \gamma_{N+1}(x), & \nu = 0, \\ \pi b U_N(x), & \nu = 1. \end{cases}$$

Thus $q_{N+1}(x)$ has $N + 2$ distinct roots in $(-1, 1)$ if $\nu = -1$, $N + 1$ distinct roots if $\nu = 0$, and N distinct roots if $\nu = 1$. Denoting these roots generically by $\{x_j\}_{j=0}^{N-\nu}$, $\{u_N(t_k)\}_{k=0}^{N}$ are then obtained by solving the linear equations

$$\frac{1}{\pi} \sum_{k=0}^{N} \frac{w_k u_N(t_k)}{t_k - x_j} + \sum_{k=0}^{N} w_k k(x_j, t_k) u_N(t_k) = f(x_j),$$

$$j = 0, 1, 2, \ldots, N - \nu, \qquad (190)$$

where in all cases $t_k \neq x_j$ for all values of k and j. From (190) we see that if $\nu = 0$, there are $N + 1$ equations in $N + 1$ unknowns, for $\nu = 1$ there are only N equations so that (190) has to be supplemented by the auxiliary relation $l(u_N) = M$, and for $\nu = -1$, the system, as we have found before, is overdetermined, and if it is consistent, one equation may be dropped. In Erdogan and Gupta's 1972 paper (Ref. 54) they recommended taking $N + 1$ even, and eliminating the equation for $j = (N + 1)/2 + 1$. If the equations (190) for $\nu = -1$ are inconsistent, then some authors have suggested using a least-squares solution (Refs. 15, 17) but, as with the other methods we have discussed, this problem appears not to be completely settled (Ref. 18). In some problems, when $\nu = -1$ there is an unknown constant c in the solution to Eq. (2) that must be determined along with $u(x)$ (Ref. 66). Then (190) gives $N + 2$ equations (usually nonlinear in c) to determine $(\{u_N(t_k)\}_{k=0}^{N+1}, c)$.

Having solved (190) for $\{u_N(t_k)\}_{k=0}^{N}$, one can then use (189) as a *natural* or Nyström-type interpolation formula for $u_N(x)$ (Ref. 67). That is,

$$u_N(x) = [\sigma_{N+1}(x)/q_{N+1}(x)]\left[f(x) - \frac{1}{\pi}\sum_{k=0}^{N}\frac{w_k u_N(t_k)}{t_k - x} - \sum_{k=0}^{N} w_k k(x, t_k) u_N(t_k)\right],$$

(191)

$$x \neq x_j, \quad j = 0, 1, 2, \ldots, N - \nu, \quad x \neq t_k, \quad k = 0, 1, 2, \ldots, N.$$

When this method is used for equations with index one, it is necessary to consider the appropriate discretization of the auxiliary condition $l(u_N) = M$ and, as we have noted previously, this condition must be of the form (if l is bounded linear)

$$\int_{-1}^{1} w_1(t) g(t) u(t) \, dt = M,$$

(192)

where $g(t) \in L_{w_1}$. If g is at least continuous, then (192) can be discretized by using $N+1$ point Gaussian quadrature and doing this gives

$$\sum_{k=0}^{N} w_k g(t_k) u_N(t_k) = M.$$

(193)

The convergence of the Gaussian quadrature method has been discussed by a number of authors under a variety of conditions on the data. For instance, in Ref. 68 Gerasoulis proved that the sequence of natural interpolants converges uniformly if $\nu = 1$, $g(t) = 1$, and $k(x, t)$ and $f(x)$ are C^1. On the other hand, if $u_N(x)$ is the Lagrange polynomial interpolant of

$\{u_N(t_k)\}_{k=0}^{N}$, then Elliott in Ref. 5 and Junghanns in Ref. 41 have shown that it converges uniformly if $\nu = 1$ and the data are Hölder continuous. Some aspects of this problem will be taken up in Section 19 and a more detailed discussion will be given by Elliott in Chapter 6.

12.3. Lobatto Quadrature: $\nu = 1$. Another popular quadrature method for solving Eq. (2) when $\nu = 1$ is one based on using Lobatto rather than Gaussian quadrature in (181). For problems in solid mechanics where only the values $u(\pm 1)$ may be of importance, this method is useful because it enables one to obtain approximations to $u(\pm 1)$ without interpolation (Ref. 61).

Recall that in N point Lobatto quadrature one fixes two nodes of the N available at $t = \pm 1$, while the remaining ones are allowed to vary to give maximal polynomial precision. If there are N nodes, Lobatto quadrature has polynomial precision $2N - 3$, and the corresponding quadrature rule (181) then has precision $2N - 2$ (Ref. 61).

For the weight function $w_1(t) = 1/\sqrt{1 - t^2}$ the nodes of the Lobatto rules are found as the zeros of $(1 - t^2)U_{N-2}(t) = d\sigma_N(t)$, $N \geq 2$ (Ref. 69), and so the $N - 1$ collocation points for setting up the analog of (190) are obtained by solving

$$\fint_{-1}^{1} \frac{(1 - t^2)U_{N-2}(t) \, dt}{\sqrt{1 - t^2}(t - x)} = 0. \tag{194}$$

Using (77), the integral in (194) equals

$$\fint_{-1}^{1} \frac{U_{N-2}(t)\sqrt{1 - t^2} \, dt}{t - x} = -\pi T_{N-1}(x), \tag{195}$$

which gives the collocation points as the zeros of $T_{N-1}(x)$.

If we continue our indexing convention of using $N + 1$, rather than N quadrature points, then $\{\cos[(2j + 1)\pi/2N]\}_{j=0}^{N-1}$ are the N distinct roots of $T_N(x)$ in $(-1, 1)$, and collocating at these (which we again call $\{x_j\}_{j=0}^{N-1}$) gives the N equations in $N + 1$ unknowns for $\{u_N(t_k)\}_{k=0}^{N}$:

$$\frac{1}{\pi} \sum_{k=0}^{N} \frac{\lambda_k u_N(t_k)}{t_k - x_j} + \sum_{k=0}^{N} \lambda_k k(x_j, t_k) u_N(t_k) = f(x_j), \quad j = 0, 1, 2, \ldots, N - 1, \tag{196}$$

where ($\{\lambda_k\}_{k=0}^{N}$, $\{t_k\}_{k=0}^{N}$) are the weights and nodes of the $N + 1$ point ordinary Lobatto quadrature rule. Here the auxiliary condition $l(u_N) = M$ is discretized by

$$\int_{-1}^{1} \frac{g(t)u_N(t)\,dt}{\sqrt{1-t^2}} \simeq \sum_{k=0}^{N} \lambda_k g(t_k) u_N(t_k) = M, \tag{197}$$

and if approximations to $u(\pm 1)$ are wanted, then they are obtained automatically by solving (196) and (197) since ± 1 are in the nodal set.

Example 12.1. In this example we present some numerical results obtained by Gerasoulis in Ref. 68 comparing three methods for approximating $u(\pm 1)$ using quadrature techniques for solving Eq. (2) when $\nu = 1$. They are Gaussian quadrature with Lagrange and natural interpolation, and Lobatto quadrature. Of the three equations solved in Ref. 68 we consider

$$\begin{cases} \dfrac{1}{\pi} \int_{-1}^{1} \dfrac{w_1(t)}{t-x} u(t)\,dt + \dfrac{1}{\pi} \int_{-1}^{1} \dfrac{w_1(t)t(t^2-s^2)u(t)\,dt}{(t^2+s^2)^2} = 1, \\ \int_{-1}^{1} w_1(t) u(t)\,dt = 0, \end{cases} \tag{198}$$

which arises in an elasticity problem (Ref. 68).

The values of $u(1)$ obtained by solving (198) using the Gaussian quadrature method with the Lagrange and natural interpolants are given in the first two columns of Table 9, while those calculated by Lobatto

Table 9. Comparison of Three Quadrature Methods for Solving (198)—Values of $u_N(1)$

N	Gauss quadrature-Lagrange interpolation	Lobatto quadrature	Gauss quadrature-natural interpolation
6	0.83363	0.85970	0.86261
8	0.87264	0.86387	0.86435
10	0.86289	0.86449	0.86448
12	0.86381	0.86441	0.86433
14	0.86257	0.86424	0.86415
16	0.86281	0.86396	0.86401
18	0.86503	0.86387	0.86391
20	0.86283	0.86380	0.86383
22	0.86463	—	0.86372
40	0.86335	—	0.86358
60	0.86348	—	0.86355

quadrature are shown in the third. From this it appears that the natural interpolant is converging the fastest. [We note that the kernel in (198) is not even continuous, so that known uniform convergence proofs do not cover this case.]

12.4. Endpoint Convergence of the Lobatto Quadrature Method.

Although we shall defer our general discussion of convergence until Section 19, it is of interest to examine the convergence of the Lobatto quadrature method at $x = \pm 1$ for equations of index one. In Example 12.2 it can be seen that the values of $u_N(\pm 1)$ appear to converge rather fast even if $f(x)$ in (210) is discontinuous. This is certainly surprising in light of our discussion for equations with index zero in Section 10.1.

This strange convergence property was apparently first observed by Ioakimidis in Ref. 23, and here we give a heuristic argument as to why this can occur. The crux of the matter seems to be that when $\nu = 1$, $u(\pm 1)$ can be represented as an integral of a given function, and the Lobatto quadrature method produces numerical integration approximations of these.

To be more precise, let u satisfy Eq. (2) with $\nu = 1$ and $l(u) = \int_{-1}^{1} w_1(t) u(t) \, dt = 0$. Then

$$G(x) = \frac{1}{\pi} \int_{-1}^{1} \frac{w_1(t) u(t) \, dt}{t - x} \tag{199}$$

satisfies

$$G(x) + \frac{1}{\pi} \int_{-1}^{1} \frac{K(x, s) G(s) \, ds}{w(s)} = f(x), \tag{200}$$

where

$$K(x, s) = \int_{-1}^{1} \frac{w_1(t) k(x, t) \, dt}{t - s}. \tag{201}$$

From (34)

$$u(x) = \frac{1}{\pi} \int_{-1}^{1} \frac{\sqrt{1 - t^2} G(t) \, dt}{x - t}, \tag{202}$$

which gives (the case $x = -1$ is similar)

$$u(1) = \frac{1}{\pi} \int_{-1}^{1} \sqrt{\frac{1 + x}{1 - x}} G(x) \, dx. \tag{203}$$

Now if u is approximated by u_N using the N point Lobatto quadrature method, then (Ref. 23)

$$u_N(1) = \frac{1}{N-1} \sum_{k=1}^{N-1} (1+x_k) G_N(x_k), \qquad (204)$$

where $x_k = \cos[(2k-1)\pi/2(N-1)]$, $k = 1, 2, \ldots, N-1$, and $G_N(x)$ satisfies

$$\begin{cases} G_N(x_j) + \sum_{k=1}^{N-1} b_k K_N(x_j, x_k) G_N(x_k) = f(x_j), & j = 1, 2, \ldots, N-1, \\ b_k = (1-x_k^2)\pi/(N-1), & k = 1, 2, \ldots, N-1, \end{cases} \qquad (205)$$

with $K_N(x, s)$ determined by applying the N point Lobatto rule to (201). Also

$$u(1) = \frac{1}{\pi} \int_{-1}^{1} \frac{(1+x)G(x)\,dx}{\sqrt{1-x^2}} \simeq \frac{1}{N-1} \sum_{k=1}^{N-1} (1+x_k) G(x_k), \qquad (206)$$

where the last term in (206) is obtained by using $N-1$ point Gaussian integration with respect to $1/\sqrt{1-x^2}$. Thus

$$u(1) - u_N(1) = \left[\frac{1}{\pi} \int_{-1}^{1} \frac{(1+x)G(x)\,dx}{\sqrt{1-x^2}} - \frac{1}{N-1} \sum_{k=1}^{N-1} (1+x_k) G(x_k) \right]$$

$$- \frac{1}{N-1} \sum_{k=1}^{N-1} (1+x_k)[G(x_k) - G_N(x_k)] = I_N + J_N. \qquad (207)$$

If $G(x)$ is Riemann integrable, then $I_N \to 0$ as $N \to \infty$ (Ref. 65). However, we have no guarantee that J_N converges to zero because in general it is not known if $G_N(x)$ converges to $G(x)$. On the other hand, J_N is the average of the errors $(1+x_k)[G(x_k) - G_N(x_k)]$, and intuitively one might expect the average error to converge zero. If this is the case, then $u_N(1) \to u(1)$, even if $u_N(x) \not\to u(x)$, $-1 < x < 1$.

A similar argument is given in Ref. 70 for the Gaussian quadrature method, which again suggests that it is the integral representation of $u(\pm 1)$ that is responsible for this "superconvergent" behavior.

Example 12.2. Here we consider solving the equations

$$\begin{cases} \dfrac{1}{\pi} \displaystyle\int_{-1}^{1} \dfrac{w_1(t)u(t)\,dt}{t-x} + \displaystyle\int_{-1}^{1} w_1(t)k(x,t)u(t)\,dt = f(x), \\ \displaystyle\int_{-1}^{1} w_1(t)u(t)\,dt = 0, \end{cases} \quad (208)$$

where

$$k(x,t) = \frac{a}{b}\cot\left[\frac{\pi a(t-x)}{b}\right] - \frac{1}{\pi(t-x)}, \quad a/b = 0.4, \quad (209)$$

and

$$f(x) = \begin{cases} |x|, \\ |x|^{1/2}, \\ (1+\operatorname{sign} x)/2. \end{cases} \quad (210)$$

The Lobatto quadrature method is used to solve these equations, and the results for $u_N(\pm 1)$ for various values of N are given in Table 10.

12.5. Lobatto Quadrature: $\nu = 0$. When $\nu = 0$, using Lobatto quadrature to solve Eq. (2) runs into difficulty. In this case, as shown by Ioakimidis and Theocaris in Ref. 61, $q_N(x)$ in (181) is a polynomial of degree N, but one of its roots lies outside of $[-1, 1]$ and so is not a valid collocation point. Consequently we have only $N - 1$ equations in N unknowns if N quadrature points are used.

Table 10. Value of $u_N(1)$ Obtained by Solving Eqs. (208)–(210) Using Lobatto Quadrature

$f(x)$ N	$\lvert x\rvert$ $u_N(1)$	$\lvert x\rvert^{1/2}$ $u_N(1)$	$(1+\operatorname{sign} x)/2$ $u_N(1)$
2	0.0000	0.0000	1.2246
8	0.8625	1.0316	1.0762
14	0.8759	1.0817	1.0794
20	0.8819	1.0925	1.0801
26	0.8832	1.1020	1.0804
30	0.8836	1.1045	1.0805
Exact values			1.080744

To see this in the simplest case, let $N = 2$ so that $\sigma_2(x) = 1 - x^2$ in (181). Then

$$q_2(x) = \int_{-1}^{1} \sqrt{\frac{1-t}{1+t}} \frac{(1-t^2)\, dt}{t-x} = \int_{-1}^{1} \frac{\sqrt{1-t^2}(1-t)\, dt}{t-x}$$

$$= \pi[H_{-1}(1) - H_{-1}(t)]$$

$$= [\pi H_{-1}(U_0) - H_{-1}(U_1/2)] = \pi[-T_1 + T_2/2] = \pi[-x + x^2 - 1/2].$$

Thus the roots of $q_2(x)$ are given by solving $2x^2 - 2x - 1 = 0$, and are $(1 \pm \sqrt{3})/2$. But $(1 + \sqrt{3})/2 \simeq 1.37$ and is not a valid collocation point.

Since there is no natural auxiliary formula for $\nu = 0$ as for $\nu = 1$, it is not clear if this procedure can be applied in this circumstance. However, in Ref. 71 Ioakimidis showed that when $\nu = 0$, $u(x)$ satisfies

$$\int_{-1}^{1} w_0(t) \left[\int_{-1}^{1} \frac{k(t, x)\, dx}{w_0(t)} - 1 \right] u(t)\, dt = \int_{-1}^{1} \frac{f(t)\, dt}{w_0(t)}, \tag{211}$$

which gives an additional equation to obtain a square system [of course one should discretize (211) by the same Lobatto rule as for the GAE]. Although no formal convergence proof is given in Ref. 71, numerical results are given indicating rapid convergence for smooth data.

13. Kantorovich Regularization

We have stated a number of times in this chapter that the polynomial algorithms currently used for solving the GAE give approximations which converge rapidly when the data are highly differentiable. However, when this is not the case, particularly when $f(x)$ is not smooth, these methods may perform poorly, and as yet (as far as we know) no simple remedy has been found for all such situations. For instance, the superconvergence results of Sections 10.2 and 12.4 show that good results can be achieved if one's interest lies just in certain linear functionals of the solution, but generally the polynomial approximations themselves behave rather erratically.

For Fredholm equations of the second kind a classical method for improving convergence in such circumstances is the use of Kantorovich regularization (Ref. 18). Briefly, assume that u satisfies the integral equation

$$u = Ku + f, \tag{212}$$

where K is smooth but f is not. Then if we define the new dependent variable

$$v = u - f, \tag{213}$$

v satisfies

$$v + f = K(v + f) + f, \tag{214}$$

or equivalently

$$v = Kv + Kf, \tag{215}$$

where we assume that u and v belong to some Banach space. In general Kf will be as smooth as the kernel of K, and if $\{v_N\}_{N=1}^{\infty}$ is a sequence of convergent approximations to v, then $u_N = v_N + f$ may converge rapidly to u even if the same method applied to solving (212) does not.

To see this, suppose we approximate v by a projection method (Galerkin's method or collocation, for instance), where $\{P_N\}_{N=1}^{\infty}$ is the defining sequence of projections (Ref. 19). Then v_N, the approximation to v, satisfies

$$v_N = P_N K v_N + P_N Kf, \tag{216}$$

and if $\|K - P_N K\| \to 0$, then for all N sufficiently large $(I - P_N K)^{-1}$ exists, and

$$v - v_N = (I - P_N K)^{-1}[v - P_N v] \tag{217}$$

and

$$u - u_N = (v + f) - (v_N + f) = v - v_N, \tag{218}$$

so that

$$\|u - u_N\|_X = \|v - v_N\|_X \le c\|v - P_N v\|_X. \tag{219}$$

Now suppose that K is defined by a kernel in C^r, $r > 0$, and for simplicity assume that $P_N g \to g$ in X, for all $g \in X$. Then $u_N \to u$ and its rate of convergence is dependent on that of $\|v - P_N v\|_X$.

To specialize further, assume that $X = L_2[a, b]$ with the usual norm (Ref. 19). If Range$(P_N) = X_N$ is the set of polynomials of degree $\le N$, let e_N be the best uniform approximation to v in X_N. Then, since $P_N e_N = e_N$,

$$\|v - P_N v\|_X = \|v - e_N - (P_N v - P_N e_N)\|_X, \tag{220}$$

which gives

$$\|u - u_N\|_X = \|v - v_N\|_X \le (1 + \|P_N\|)\|v - e_N\|_X. \tag{221}$$

However, from Jackson's theorem (Ref. 72), $\|v - e_N\|_X = O(N^{-r})$ [because it follows from Eq. (215) that $v \in C^r$], so that

$$\|u - u_N\|_X = \|v - v_N\|_X = O(N^{-r}). \tag{222}$$

Since u itself need not belong to C^r, applying the same projection method to (212) will generally give a much slower convergence rate than in (222). Thus, Kantorovich regularization can dramatically improve the convergence properties of a given algorithm. Unfortunately, real life is not so simple, because this analysis assumes that $(Kf)(x) = \int_a^b k(x,y)f(y)\,dy$ is calculated exactly. If this is not the case, the $O(N^{-r})$ convergence in Eq. (222) may be seriously degraded.

On the basis of these observations it is interesting to speculate if such a procedure can be devised for solving CSIES. In Ref. 26 we proposed a generalization of (215) for the case of the GAE with index zero, and in Ref. 21 Ioakimidis developed the same method for equations with $\nu = 1$, and gave numerical results supporting the possibility of achieving improved convergence rates if f is a step-function.

To illustrate this technique, we first consider the GAE with index zero. Suppose that one can solve the dominant equation

$$H_0 v = f \tag{223}$$

analytically. Then define the new unknown u_1 by

$$u_1 = u - v, \tag{224}$$

and observe that

$$H_0 u_1 = H_0 u - H_0 v = -Ku + f - f = -Ku = -Ku_1 - Kv, \tag{225}$$

so that u_1 satisfies

$$H_0 u_1 + K u_1 = -Kv, \tag{226}$$

and if K is smoothing, then Kv will be smooth.

When $\nu = 1$, we begin by solving

$$\begin{cases} H_1 v = f, & (227) \\ 1(v) = M, & (228) \end{cases}$$

explicitly, and then an argument similar to that for $v = 0$ shows that

$$u_1 = u - v \qquad (229)$$

satisfies

$$\begin{cases} Hu_1 + Ku_1 = -Kv, & (230) \\ l(u_1) = 0, & (231) \end{cases}$$

since $l(u_1) = l(u) - l(v) = M - M = 0$.

Again if K is smoothing, then Kv will also be smooth, and if (230)–(231) are solved numerically by any of the methods discussed so far, then we can expect more rapid convergence than by solving $H_1 u + Ku = f$, $l(u) = M$, directly.

Example 13.1. To illustrate these ideas numerically, we use an example taken from Ref. 24. The integral equation to be solved is

$$\begin{cases} \dfrac{a}{b} \displaystyle\int_{-1}^{1} \cot\left[\dfrac{\pi a(t-x)}{b}\right] u(t)\, dt = f(x), & (232) \\ \displaystyle\int_{-1}^{1} u(t)\, dt = 0, & (233) \end{cases}$$

which has index one, and $f(x)$ is the step-function

$$f(x) = \begin{cases} 0, & -1 \le x < 0, \\ 1, & 0 \le x \le 1. \end{cases} \qquad (234)$$

To solve (232)–(233) the kernel of (232) is decomposed as

$$\dfrac{1}{\pi(t-x)} + \left\{ \dfrac{a}{b} \cot\left[\dfrac{\pi a(t-x)}{b}\right] - \dfrac{1}{\pi(t-x)} \right\} = \dfrac{1}{\pi(t-x)} + k(t, x), \qquad (235)$$

where $k(t, x)$ is analytic, and v is obtained by solving

$$\begin{cases} \dfrac{1}{\pi} \displaystyle\int_{-1}^{1} \dfrac{v(t)\, dt}{t - x} = f(x), & -1 < x < 1, \quad (236) \\ \displaystyle\int_{-1}^{1} v(t)\, dt = 0. & (237) \end{cases}$$

[It is noteworthy that in Ref. 24 the weight function $w_1(t)$ is not introduced until the regularized equation for u_1 is solved.]

The solution to (236) and (237) can be computed from the inversion formula (34), and is given by (Ref. 24)

$$v(t) = \left[\left(\frac{t}{2} + \frac{1}{\pi}\right)\right] / \sqrt{1-t^2} - \frac{1}{\pi}\tanh^{-1}[(1-t^2)^{1/2}]. \tag{238}$$

Since

$$\tanh^{-1}[(1-x^2)^{1/2}] = \tfrac{1}{2}\log\{[1+(1-x^2)^{1/2}]/[1-(1-x^2)^{1/2}]\}, \tag{239}$$

$v(x)$ has a complicated logarithmic singularity at $x = 0$ so that, using $v(x)$ in the equation for u_1, one is required to evaluate the integral

$$\int_{-1}^{1} k(t, x) v(t) \, dt.$$

To solve the equation for u_1 the usual change of variable $u_1(x) = u_2(x)/\sqrt{1-x^2}$ was made in Ref. 24, and since the values of $u_2(\pm 1)$ were of primary importance, the equations for u_2 were solved using the Lobatto quadrature method of the previous subsection. The function $(Kv)(x)$ was calculated by writing it as

$$(Kv)(x) = \int_{-1}^{1} (1-t^2)^{-1/2} k(t,x) \left(\frac{t}{2} + \frac{1}{\pi}\right) dt$$

$$- \frac{1}{\pi} \int_{-1}^{1} k(t,x) \tanh^{-1}[(1-t^2)^{1/2}] \, dt, \tag{240}$$

with the first integral in (240) approximated using Lobatto quadrature with 32 nodes, while the second was obtained using Gaussian quadrature with respect to the weight function $-\log|t|$ with 16 nodes (Ref. 24).

In Table 11 the computed values of

$$k_0(-1) = \frac{1}{2} - \frac{1}{\pi} u_2(-1), \quad k_0(1) = \frac{1}{2} + \frac{1}{\pi} u_2(1), \tag{241}$$

are displayed for various numbers of nodes used in solving for $u_2(x)$ [$k_0(\pm 1)$ are the stress intensity factors and were the quantities of physical interest in this problem]. As can be seen, the convergence is remarkably rapid, with

Table 11. Stress Intensity Factors for (232)-(234) Computed Using Regularization and Lobatto Quadrature*

N	$k_0(-1)$	$k_0(1)$	$k_0(-1) + k_0(1)$†
2	0.533136	1.697555	1.702891
3	0.495876	1.102262	1.598138
4	0.486298	1.084632	1.570930
5	0.484563	1.081378	1.565941
6	0.484281	1.080843	1.565124
7	0.484238	1.080759	1.564997
8	0.484231	1.080746	1.564977
9	0.484230	1.087444	1.564974
10	0.484230	1.087444	1.564974

* $a/b = 0.4$.
† The exact value of $k_0(-1) + k_0(1)$ is 1.564974.

about 6-7 significant figures being obtained for $N = 10$. For this example the regularization method appears to be quite successful and certainly warrants further investigation.

14. Product Quadrature

So far in our discussion of quadrature methods we have assumed that the kernel is at least continuous, and have observed that existing proofs of uniform convergence require even stronger smoothness conditions on $k(x, t)$. On the other hand, we have noted in our development of the Galerkin and collocation methods that the kernels which arise in practice are often weakly singular, yet this complication seems to have been paid little attention in the literature when quadrature methods are used (Refs. 7, 29). For Fredholm equations, a well-known technique in this situation is the use of *product quadrature* but, with the exception of a brief discussion by Krenk in Ref. 29, no other work seems to have been published on this method for CSIES.

Returning to Eq. (2), suppose that $k(x, t)$ is weakly singular, and following the treatment for Fredholm equations assume that it can be factored as $k(x, t) = M(x, t)L(x, t)$, where $L(x, t)$ is smooth and $M(x, t)$ carries the singularities of $k(x, t)$ (Refs. 19, 72). Then the integral $\int_{-1}^{1} w_\nu(t)k(x, t)u(t)\, dt$ can be approximated as follows. Let $Q_N = (\{w_k\}_{k=0}^{N}, \{t_k\}_{k=0}^{N})$ be an interpolatory quadrature rule which will be used in (181) to approximate the Cauchy singular part of Eq. (2). For the sake of argument assume that Q_N is the $N + 1$ point Gaussian rule for w_ν. Then approximate $L(x, t)u(t)$ by interpolation as

$$L(x, t)u(t) \simeq \sum_{k=0}^{N} s_k(t) L(x, t_k) u(t_k), \qquad (242)$$

where $\{s_k(t)\}_{k=0}^{N}$ is a cardinal basis for a suitable set of interpolating functions on $\{t_k\}_{k=0}^{N}$ (Ref. 19). Substituting (242) into $\int_{-1}^{1} w_\nu(t) k(x, t) u(t) \, dt$ gives

$$\int_{-1}^{1} w_\nu(t) k(x, t) u(t) \, dt \simeq \sum_{k=0}^{N} \left[\int_{-1}^{1} w_\nu(t) M(x, t) s_k(t) \, dt \right] L(x, t_k) u(t_k) \, dt$$

$$= \sum_{k=0}^{N} W_k(x) L(x, t_k) u(t_k), \qquad (243)$$

where

$$W_k(x) = \int_{-1}^{1} w_\nu(t) M(x, t) s_k(t) \, dt. \qquad (244)$$

Using (243) in Eq. (2) with the Cauchy integral replaced by (181), then $\{u_N(t_k)\}_{k=0}^{N}$, the approximations to $\{u(t_k)\}_{k=0}^{N}$, can be determined by collocating at the zeros of $q_{N+1}(x)$. This leads, for instance, when $\nu = 1$, to the equations

$$\frac{1}{\pi} \sum_{k=0}^{N} \frac{w_k u_N(t_k)}{t_k - x_j} + \sum_{k=0}^{N} W_k(x_j) L(x_j, t_k) u_N(t_k) = f(x_j),$$

$$j = 0, 1, 2, \ldots, N-1, \qquad (245)$$

where $q_{N+1}(x_j) = 0$, $j = 0, 1, 2, \ldots, N-1$.

15. Kalandiya's Method

The quadrature methods developed in Section 12 are of recent origin, currently widely used, and are characterized, as we have seen, by the property that the collocation and integration nodes are distinct, which makes these procedures somewhat different than their Fredholm counterparts. However, one of the earliest direct methods for solving Eq. (2) was a quadrature method given by Kalandiya in 1959 (Ref. 55) that used the same quadrature and collocation points.

Although this method has generally not been used in the West, a similar one was given by Cohen in Ref. 73 for solving equations of the second kind. Since this technique is of some historical interest, we present a derivation here. However, no published comparison of this procedure with current quadrature methods seems to have appeared.

Because the details are essentially the same for all three indices, we will consider only $\nu = 1$. In this case $u(t)$ satisfies (for symmetry we include a factor of $1/\pi$ in the regular integral)

$$\begin{cases} \dfrac{1}{\pi}\!\!\!\int_{-1}^{1} \dfrac{u(t)\,dt}{\sqrt{1-t^2}(t-x)} + \dfrac{1}{\pi}\int_{-1}^{1} \dfrac{k(x,t)u(t)\,dt}{\sqrt{1-t^2}} = f(x), & (246) \\ l(u) = M, & (247) \end{cases}$$

and we begin by approximating $u(t)$ in (246) by its Lagrange interpolation polynomial on $\{t_k\}_{k=1}^{N}$, where $\{t_k\}_{k=1}^{N}$ are the zeros of $T_N(t)$, $N > 1$. Thus

$$u(t) \simeq u_N(t) = \sum_{k=1}^{N} \frac{T_N(t)u(t_k)}{(t-t_k)T'_N(t_k)} = \frac{1}{N}\sum_{k=1}^{N} \frac{T_N(t)u(t)}{(t-t_k)U_{N-1}(t_k)}, \quad (248)$$

where we have used $T'_N(t) = NU_{N-1}(t)$ in (248), and $U_{N-1}(t)$ is the Chebyshev polynomial of the second kind of degree $N-1$. Substituting (248) into the Cauchy integral in (246) gives

$$\frac{1}{\pi}\!\!\!\int_{-1}^{1}\frac{u(t)\,dt}{\sqrt{1-t^2}(t-x)} \simeq \frac{1}{\pi N}\sum_{k=1}^{N}\frac{u(t_k)}{U_{N-1}(t_k)}\int_{-1}^{1}\frac{T_N(t)\,dt}{\sqrt{1-t^2}(t-t_k)(t-x)}. \quad (249)$$

By partial fractions,

$$\frac{1}{(t-t_k)(t-x)} = \frac{1}{(t_k-x)}\left(\frac{1}{t-t_k} - \frac{1}{t-x}\right), \quad (250)$$

and using (250) and (65) in (249)

$$\frac{1}{\pi}\!\!\!\int_{-1}^{1}\frac{T_N(t)\,dt}{\sqrt{1-t^2}(t-t_k)(t-x)} = \frac{1}{t-x}[U_{N-1}(t_k) - U_{N-1}(x)], \quad (251)$$

Numerical Solution of Cauchy Singular Integral Equations

so that

$$\frac{1}{\pi}\int_{-1}^{1}\frac{u(t)\,dt}{\sqrt{1-t^2}(t-x)} \simeq \frac{1}{N}\sum_{k=1}^{N}\frac{u(t_k)}{U_{N-1}(t_k)}\left[\frac{U_{N-1}(t_k)-U_{N-1}(x)}{t_k-x}\right]. \quad (252)$$

Using (252) in (246) with the regular integral replaced by the N point Gaussian quadrature rule for $1/\sqrt{1-t^2}$, and collocating at $\{t_j\}_{j=1}^{N}$ gives the following equations to determine $u_k \simeq u(t_k)$, $k = 1, 2, \ldots, N$:

$$\frac{1}{N}\left\{\sum_{k=1}^{N}\frac{u_k}{U_{N-1}(x_k)}\left[\frac{U_{N-1}(t_k)-U_{N-1}(t_j)}{t_k-t_j}\right] + \sum_{k=1}^{N}k(t_j,t_k)u_k\right\} = f(t_j),$$

$$j = 1, 2, \ldots, N. \quad (253)$$

[When $k = j$, the term in square brackets is given by L'Hospital's rule as $U'_{N-1}(t_j)$.]

It should be noted that in contrast to (190) we have N equations in N unknowns, and so it is not clear how the uniqueness condition is to be accounted for. The usual suggestion is to drop one equation and replace it with $l(u) = M$ (Ref. 74) but, as Ioakimidis points out in Ref. 74, this procedure is likely to produce large errors near the node whose equation is removed. For equations with $\nu = -1, 0$ this problem should not occur.

16. Some Other Numerical Methods for the GAE

Although most of the methods that we have discussed so far for the numerical solution of Eq. (2) have been justified theoretically—that is, convergence proofs have been given and some numerical stability results are known—a number of others have been published, are in common use, but have apparently not been analyzed theoretically. In this section we will discuss some of these, but generally in lesser detail than for those given previously.

These techniques seem to fall into three classes:

1. Methods based on converting Eq. (2) to an equivalent Fredholm equation of the first kind with a logarithmically singular kernel.
2. Collocation and Galerkin methods using piecewise polynomials (Refs. 7, 8, 44).
3. Polynomial expansion methods where the endpoint singularities are ignored (Refs. 75-77).

We will confine ourselves to categories (1) and (3).

16.1. Conversion to a Logarithmic Equation. If we integrate Eq. (2) with respect to x, and interchange the order of integration, it becomes

$$\frac{1}{\pi} \int_{-1}^{1} \log(|x-t|) v(t) \, dt + \int_{-1}^{1} k^{(x)}(x,t) v(t) \, dt = f^{(x)}(x) + c, \quad (254)$$

where $k^{(x)}(x,t)$ and $f^{(x)}(x)$ are any antiderivatives of $k(x,t)$ and $f(x)$ with respect to x, and c is an arbitrary constant. Thus $v(t)$ satisfies a Fredholm equation of the first kind with a weakly singular kernel.

Technically (254) is an ill-posed problem (Ref. 72), so one might expect numerical difficulties in its numerical solution. However, this does not seem to have caused any problems with actual computations so far (Refs. 4, 78).

To simplify notation for further work, we denote $k^{(x)}(x,t)$ by $l(x,t)$ and $f^{(x)}(x)$ by $g(x)$. Using these definitions (254) becomes

$$\frac{1}{\pi} \int_{-1}^{1} \log(|x-t|) v(t) \, dt + \int_{-1}^{1} l(x,t) v(t) = g(x) + c. \quad (255)$$

16.1.1. Polynomial Algorithms for (255). From the theory of the airfoil equation, $v(t)$ can be expected to have singularities of the form $1/\sqrt{1-t^2}$, so that the change of variable

$$v(t) = u(t)/\sqrt{1-t^2} \quad (256)$$

appears reasonable, where $u(t)$ is the "smooth" part of the solution. In terms of $u(t)$, (255) becomes

$$\frac{1}{\pi} \int_{-1}^{1} \frac{\log(|x-t|) u(t) \, dt}{\sqrt{1-t^2}} + \int_{-1}^{1} \frac{l(x,t) u(t) \, dt}{\sqrt{1-t^2}} = g(x) + c. \quad (257)$$

Now the solution of an equation with a logarithmically singular kernel is generally unique (Ref. 72), so that if this is true,

$$u(t) = u_1(t) + c u_2(t), \quad (258)$$

where $u_1(t)$ satisfies (255) with $c = 0$, and $u_2(t)$ satisfies (255) with $g(x) = 0$ and $c = 1$. Since the left-hand sides of the equations for u_1 and u_2 will then be the same in both cases, any algorithm which reduces (257) to solving a finite set of linear equations can generally be implemented by solving for both u_1 and u_2 simultaneously. Consequently, we will just consider the numerical solution of (257) with an arbitrary right-hand side to be called

$h(x)$. As for the generalized airfoil equation itself, the square-root singularities in (257) naturally suggest an expansion in Chebyshev polynomials. That is, we look for approximations

$$u_N(x) = \sum_{k=0}^{N} a_k T_k(x), \tag{259}$$

with $\{a_k\}_{k=0}^{N}$ to be determined by a Galerkin or a collocation method.

In Ref. 78 Moss and Chistiansen considered the Galerkin method obtained by setting the residual

$$\frac{1}{\pi} \sum_{k=0}^{N} a_k \int_{-1}^{1} \frac{\log(|x-t|) T_k(t) \, dt}{\sqrt{1-t^2}} + \sum_{k=0}^{N} a_k \int_{-1}^{1} \frac{l(x,t) T_k(t) \, dt}{\sqrt{1-t^2}} - h(x) \tag{260}$$

orthogonal to $\{T_j\}_{j=0}^{N}$ in the L_{w_1} inner product. This leads to the equations for $\{a_k\}_{k=0}^{N}$:

$$\frac{1}{\pi} \sum_{k=0}^{N} a_k \int_{-1}^{1} \int_{-1}^{1} \frac{\log(|x-t|) T_k(t) T_j(x) \, dx \, dt}{\sqrt{1-x^2}\sqrt{1-t^2}}$$

$$+ \sum_{k=0}^{N} a_k \int_{-1}^{1} \int_{-1}^{1} \frac{l(x,t) T_k(t) T_j(x) \, dx \, dt}{\sqrt{1-x^2}\sqrt{1-t^2}}$$

$$= \int_{-1}^{1} \frac{h(x) T_j(x) \, dx}{\sqrt{1-x^2}}, \quad j = 0, 1, \ldots, N. \tag{261}$$

Using (145) the first set of integrals in (261) can be evaluated exactly as

$$\frac{1}{\pi} \int_{-1}^{1} \int_{-1}^{1} \frac{\log(|x-t|) T_k(t) T_j(x) \, dx \, dt}{\sqrt{1-x^2}\sqrt{1-t^2}} = \begin{cases} -\pi \log 2, & j = k = 0, \\ -\pi/2j, & j = k \neq 0, \\ 0, & \text{otherwise,} \end{cases} \tag{262}$$

while the others can be approximated by Gaussian integration if $l(x,t)$ and $h(x)$ are smooth. In Ref. 78 it was shown that if the kernel $l(x,t)$ was of a special convolution form, then the double integrals could be determined more efficiently by solving a complicated set of difference equations. A convergence analysis was given in Ref. 78 and the reader is referred there for further details. As in Ref. 77 the use of fast Fourier transform techniques are also possible.

A less complicated polynomial algorithm can be obtained by collocating the residual at the zeros $\{x_j\}_{j=0}^{N}$ of $T_{N+1}(x)$, leading to the set of linear equations

$$-a_0 \log 2 - \sum_{k=1}^{N} a_k T_k(x_j)/k + \sum_{k=0}^{N} a_k \int_{-1}^{1} \frac{l(x_j, t) T_k(t) \, dt}{\sqrt{1-t^2}} = h(x_j),$$

$$j = 0, 1, 2, \ldots, N. \qquad (263)$$

In Ref. 79 Guseinov established convergence of this method under certain integrability conditions on the data, provided that the integrals in (263) were evaluated exactly; but no numerical results were given.

For purposes of solving the GAE, neither of these algorithms seems to have any advantage over the analogous ones given in Sections 10-11. However, to complete our discussion, if such an algorithm were used, then the remaining task is to determine c, and this problem only seems to have a simple solution if either the Kutta condition needs to be satisfied, or if we have an auxiliary condition of the form $l(u) = \int_{-1}^{1} [u(t)/\sqrt{1-t^2}] \, dt = M$, which is the usual situation if $\nu = 1$.

In the first case, the Kutta condition is imposed on u_N (either the Galerkin or collocation approximation), and then $u_N(1) = 0$, so that $c = -u_{1,N}(1)/u_{2,N}(1)$. In the second, $\int_{-1}^{1} [u_{i,N}(t)/\sqrt{1-t^2}] \, dt = \pi a_{i,0}$, $i = 1, 2$, which gives $\pi(a_{1,0} + ca_{2,0}) = M$, so that $c = (M/\pi - a_{1,0})/a_{2,0}$.

If $\nu = -1$, then a reasonable way to obtain c would be from the consistency condition with u replaced by $u_{1,N} + cu_{2,N}$.

One nice thing about this approach to solving the GAE is that all three indices are treated by the same algorithm, the only difference being in the determination of c. Again it would be interesting to compare this method to those in Sections 10-11; but to the best of our knowledge this has not been done.

16.1.2. Tuck's Method. In a number of papers (Refs. 4, 80) Tuck and his co-workers have used the logarithmic conversion of the GAE to solve a variety of problems arising in aerodynamics, hydrodynamics, and heat transfer. Basically, his method of solution of (255) may be viewed as approximating $v(t)$ [not $u(t)$] by a piecewise constant function using a nonuniform mesh and collocation points to account for the square-root singularities in the solution.

For convenience in describing this method, let $K(x, t) = \log(|x - t|)/\pi + l(x, t)$ in (255), so that it becomes [recall $h(x) = g(x) + c$]

$$\int_{-1}^{1} K(x, t) v(t) \, dt = h(x). \qquad (264)$$

To solve (264) numerically Tuck makes a nonuniform partition of the interval $[-1, 1]$, $-1 = t_0 < t < \cdots < t_{N-1} < t_N = 1$, and writes the integral in (264) as

$$\int_{-1}^{1} K(x, t)v(t)\, dt = \sum_{k=1}^{N} \int_{t_{k-1}}^{t_k} K(x, t)v(t)\, dt. \tag{265}$$

Now if $t_k - t_{k-1}$ is sufficiently small,

$$\int_{t_{k-1}}^{t_k} K(x, t)v(t)\, dt \simeq v_k \int_{t_{k-1}}^{t_k} K(x, t)\, dt, \tag{266}$$

where in (266) it is assumed that $v(t) \simeq v_k$, $t \in (t_{k-1}, t_k)$. Thus

$$\int_{-1}^{1} K(x, t)v(t)\, dt \simeq \sum_{k=1}^{N} v_k \int_{t_{k-1}}^{t_k} K(x, t)\, dt, \tag{267}$$

and we obtain approximations $\{\hat{v}_k\}_{k=1}^{N}$ to $\{v_k\}_{k=1}^{N}$ by choosing N collocation points $\{x_j\}_{j=1}^{N}$, $x_j \in (t_{j-1}, t_j)$, and then solving the equations

$$\sum_{k=1}^{N} \hat{v}_k \int_{t_{k-1}}^{t_k} K(x_j, t)\, dt = h(x_j), \qquad j = 1, 2, \ldots, N. \tag{268}$$

The mesh and collocation points now need to be chosen, and this should be done in such a way as to account for the square-root singularities in $v(t)$. To motivate Tuck's choice we let $u(t) = v(t)/\sqrt{1 - t^2}$ in (264) again, and then make the change of variables $t = -\cos\theta$ and $x = -\cos\psi$ so that

$$\int_{-1}^{1} K(x, t)v(t)\, dt = \int_{0}^{\pi} K(-\cos\psi, -\cos\theta) u(-\cos\theta)\, d\theta. \tag{269}$$

Since the integral on the right in (269) is free of square-root singularities, using a uniform mesh in θ seems to be a reasonable way to approximate it. Doing this gives

$$\int_{0}^{\pi} K(-\cos\psi, -\cos\theta)u(-\cos\theta)\, d\theta = \int_{0}^{\pi} K(-\cos\psi, -\cos\theta) w(\phi)\, d\theta$$

$$= \sum_{k=1}^{N} \int_{\theta_{k-1}}^{\theta_k} K(-\cos\psi, -\cos\theta) w(\theta)\, d\theta, \tag{270}$$

where $\theta_k = k\pi/N$, $k = 0, 1, 2, \ldots, N$. For collocation points Tuck chooses the midpoints $\psi_j = \pi(j - \frac{1}{2})/N$, $j = 1, 2, \ldots, N$, of (θ_{j-1}, θ_j), so that $x_j = -\cos \psi_j = -\cos[\pi(j - \frac{1}{2})/N]$, $j = 1, 2, \ldots, N$. Thus

$$\sum_{k=1}^{N} \int_{\theta_{k-1}}^{\theta_k} K(-\cos \psi_j, -\cos \theta) w(\theta) \, d\theta = h(-\cos \psi_j), \qquad j = 1, 2, \ldots, N, \tag{271}$$

and making the inverse transformations back to x and t in (271) gives

$$\sum_{k=1}^{N} \int_{t_{k-1}}^{t_k} K(x_j, t) v(t) \, dt = h(x_j), \qquad j = 1, 2, \ldots, N, \tag{272}$$

where

$$t_k = -\cos(k\pi/N), \qquad k = 0, 1, 2, \ldots, N, \tag{273}$$

and

$$x_j = -\cos[(j - \tfrac{1}{2})\pi/N], \qquad j = 1, 2, \ldots, N. \tag{274}$$

Approximating the integrals in (272) as before leads to the system

$$\sum_{k=1}^{N} \hat{v}_k \int_{t_{k-1}}^{t_k} K(x_j, t) \, dt = h(x_j), \qquad j = 1, 2, \ldots, N, \tag{275}$$

for the determination of $\{\hat{v}_k\}_{k=1}^{N}$.

Although no formal convergence proof of this method has apparently been given (Ref. 4), MaCaskill in Ref. 80 found experimentally that $O(N^{-3})$ convergence could be obtained for the GAE and, as Tuck remarks in Ref. 4, "this is probably as high a convergence rate that can be achieved with a step-function approximation."

Last we must consider the determination of c in (255). Since this method has apparently been used most often when $\nu = 0$, we shall deal only with that case. If $\{\hat{v}_{i,k}\}_{k=1}^{N}$, $i = 1, 2$, denote the approximations to $v_i(t)$, $i = 1, 2$, where $v_1(t)$ satisfies (255) with $c = 0$ and $v_2(t)$ satisfies (255) with $g(x) = 0$, $c = 1$, respectively, then c can be obtained from

$$c = -\hat{v}_{1,N}/\hat{v}_{2,N}. \tag{276}$$

A more sophisticated approach for obtaining c is also discussed in Ref. 4, which makes use of the known asymptotic behavior of $v(t)$ near $t = 1$.

16.2. Some Other Polynomial Methods for Solving the GAE.

So far in our discussion of algorithms for solving the GAE we have emphasized the importance of explicitly accounting for the singularities of $v(x)$ in Eq. (2) at $x = \pm 1$. However, a number of methods have been proposed for solving Eq. (1) which ignore the behavior of $v(x)$ at these points, yet seem to give reasonable numerical approximations, except near ± 1 (Refs. 75–77). Algorithms of this type have been developed using both polynomial and piecewise polynomial approximations and here we shall briefly discuss some polynomial algorithms of this type due to Majumdar (Ref. 75), Bose and Kundu (Ref. 76), and Hashmi and Delves (Ref. 77).

In the first two of these methods the kernel $k(x, t)$ is assumed to be of convolution type; that is, $k(x, t) = k(x - t)$, and smooth enough so that it may be expanded in a finite Taylor series

$$k(x - t) \simeq \sum_{m=0}^{M} k_m (x - t)^m. \tag{277}$$

Quantity $k(x - t)$ is then replaced in Eq. (2) by the expansion (277) giving the equation

$$-\frac{1}{\pi} \int_{-1}^{1} \frac{v(t)\, dt}{x - t} + \sum_{m=0}^{M} k_m \int_{-1}^{1} v(t)(x - t)^m\, dt = f(x) \tag{278}$$

to solve. Majumdar then approximates $v(t)$ by a finite Chebyshev series

$$v(t) \simeq v_N(t) = \sum_{k=0}^{N} b_k T_k(t), \tag{279}$$

with the coefficients $\{b_k\}_{k=0}^{N}$ determined by collocation (although he fails to mention in Ref. 75 which collocation points were used, the zeros of $T_{N+1}(x)$ seem to be a natural choice). Doing this yields the $N + 1$ equations for $\{b_k\}_{k=0}^{N}$:

$$\sum_{k=0}^{N} \left[-\frac{b_k}{\pi} \int_{-1}^{1} \frac{T_k(t)}{x_j - t}\, dt + \sum_{k=0}^{N} \left[\sum_{m=0}^{M} k_m \int_{-1}^{1} T_k(t)(x_j - t)^m\, dt \right] b_k = f(x_j), \right.$$

$$j = 0, 1, 2, \ldots, N, \tag{280}$$

where $\{x_j\}_{j=0}^{N}$ are the collocation points.

A novel feature of this method is that the matrix elements

$$\lambda_k(x) = \int_{-1}^{1} \frac{T_k(t)\, dt}{x - t} \tag{281}$$

and

$$\beta_k^m(x) = \int_{-1}^{1} T_k(t)(x-t)^m \, dt \qquad (282)$$

are computed recursively, thus potentially saving considerable time over methods using numerical integration. These formulas may be derived by using the three-term recursion relation (66) for $\{T_k\}_{k=0}^{\infty}$. They are (Ref. 75):

$$\lambda_{k+1}(x) - 2x\lambda_k(x) + \lambda_{k-1}(x) = 2[1 + \cos(k\pi)]/(k^2 - 1), \qquad k \geq 2, \qquad (283)$$

$$\lambda_0(x) = \log[|(1+x)/(1-x)|], \qquad (284)$$

$$\lambda_1(x) = -2 + x\lambda_0(x), \qquad (285)$$

and

$$\left(\frac{m+k+2}{k+1}\right)\beta_{k+1}^m(\theta) - 2\cos\theta \beta_k^m(\theta) - \left(\frac{m-k+2}{k-1}\right)\beta_{k-1}^m(\theta)$$

$$= 4[(\cos\theta + 1)^{m+1}\cos(k\pi) + (\cos\theta - 1)^{m+1}]/(k^2 - 1), \qquad k \geq 2, \qquad (286)$$

where $\theta = \cos^{-1}x$. Initial conditions β_k^m, $k = 0, 1$, can be obtained by direct integration, and a table for various values of m can be found in Ref. 75.

For simplicity we will derive (283), while (286) can be proved in similar, but more complicated fashion. From (66)

$$\lambda_{k+1} = \int_{-1}^{1} \frac{T_{k+1}(t) \, dt}{x-t} = \int_{-1}^{1} \frac{2tT_k(t) \, dt}{x-t} - \int_{-1}^{1} \frac{T_{k-1}(t) \, dt}{x-t}. \qquad (287)$$

But

$$\int_{-1}^{1} \frac{tT_k(t) \, dt}{x-t} = \int_{-1}^{1} \frac{t-x}{x-t} T_k(t) \, dt + x\lambda_k(x), \qquad (288)$$

so that

$$\lambda_{k+1}(x) = 2x\lambda_k(x) - \lambda_{k-1}(x) - 2\int_{-1}^{1} T_k(t) \, dt, \qquad (289)$$

and using the fact that $-2\int_{-1}^{1} T_k(t) \, dt = 2[1 + \cos(k\pi)]/(k^2 - 1)$, $k \geq 2$ (which can be shown by elementary integration), we arrive at (283). Equations (284)–(285) are found by explicit integration.

Numerical Solution of Cauchy Singular Integral Equations

In Ref. 75 Majumdar solved a number of equations with singular right-hand sides (but having nonsingular solutions), and using a ten-term expansion he obtained errors of order 10^{-2} except near $x = \pm 1$ (in Figure 5.1 of Ref. 75 it appears that the approximate solution is diverging from the true one, but we cannot be sure, since the numerical values are not quoted).

In Ref. 76 Bose and Kundu gave a similar algorithm using expansions in terms of Chebyshev polynomials of the second kind. They claim better accuracy than Majumdar, but observe that this may be due in part to an error in one of Majumdar's exact solutions.

No convergence proofs are given in either of these papers, and it appears that such algorithms should be used only when it is known that the solution has no endpoint singularities and the kernel $k(x, t)$ is analytic. In most cases the collocation algorithms of Section 11 should be superior.

Last we point out that a related procedure using Chebyshev expansions has been given by Hashmi and Delves in Ref. 77, where a Galerkin rather than a collocation method is used. In Ref. 77 the coefficients in (279) were obtained by solving

$$-\frac{1}{\pi} \sum_{k=0}^{N} b_k \int_{-1}^{1} \frac{T_j(x)}{\sqrt{1-x^2}} \left[\int_{-1}^{1} \frac{T_k(t)\, dt}{x-t} \right] dx$$

$$+ \sum_{k=0}^{N} b_k \int_{-1}^{1} \int_{-1}^{1} \frac{k(x,t) T_k(t) T_j(x)\, dx\, dt}{\sqrt{1-x^2}}$$

$$= \int_{-1}^{1} \frac{f(x) T_j(x)\, dx}{\sqrt{1-x^2}}, \qquad j = 0, 1, 2, \ldots, N. \qquad (290)$$

The singular integrals were calculated exactly by interchanging the order of integration and using (64)-(65), while Delves's fast Fourier transform techniques were used to evaluate the remaining ones (Ref. 77). They also extended this procedure to solve equations of the second kind with constant coefficients and obtained good numerical results (see Table 1 of Ref. 77) if the solutions were analytic. Again no formal convergence analysis was given.

17. Numerical Solution of CSIES of the Second Kind with Constant Coefficients

In this section we will show how to extend a number of the algorithms developed in Sections 10-12 to solve CSIES of the second kind with constant coefficients. Our emphasis will be on generalizing the Galerkin, collocation and quadrature methods discussed in those sections since they have been extensively studied, both practically and theoretically (Refs. 1-2, 7). At present the quadrature methods introduced by Krenk (Ref. 59) and Theocaris and Ioakimidis (Refs. 61, 64) seem to be the most popular techniques for solving such equations (Ref. 2). However, recent work by Thomas (Ref. 44), Gerasoulis (Ref. 7), and others (Ref. 8), suggests that piecewise-polynomial methods may be preferable for many problems.

17.1. The Standard Polynomial Algorithms. As we shall show, the Galerkin, collocation, and quadrature methods developed for equations of the first kind in Sections 10-12 can be generalized, at least theoretically, in a straightforward fashion to equations of the second kind

$$av(x) + \frac{b}{\pi} \oint_{-1}^{1} \frac{v(t)\,dt}{t-x} + \int_{-1}^{1} k(x,t)v(t)\,dt$$

$$= f(x), \quad -1 < x < 1, \quad a \neq 0, \quad b \neq 0, \quad (291)$$

where for simplicity we assume that $a^2 + b^2 = 1$ [this normalization can always be achieved by dividing both sides of (291) by $(a^2 + b^2)^{1/2}$ and so entails no loss of generality]. In contrast to equations of the first kind, the appropriate expansion functions are Jacobi polynomials $\{P_k^{(\alpha,\beta)}(x)\}$ where (α, β) depend on a and b. Thus each equation presents a new problem, since the basis functions, quadrature and collocation nodes are different in each case, and are generally not known explicitly (Ref. 69). Consequently, they must be computed numerically, by methods such as those in Ref. 69 or Ref. 81. This makes these techniques considerably more complex computationally than those for equations of the first kind, and may be difficult to implement if high-degree polynomials are needed (Ref. 81).

As for equations of the first kind, our emphasis in this section will be on the development of the methods—details of convergence will be left to Section 19.

Because the basic ideas are so similar to those for equations of the first kind, we shall treat the indices $\nu = -1, 0, 1$ simultaneously for each technique and, with the previous background, the reader (we hope) should have no difficulty in handling this simplification.

The nature of these algorithms, as for equations of the first kind, depends crucially on the singularity structure of solutions of the dominant equation $[k(x, t) = 0]$ and then on polynomial mapping properties generalizing (64)-(65), (77), and (80). Since these facts are well known, we will merely state the necessary results along with sources where proofs may be found. In this regard, we point out that the singularity structure of the solution of (291) at $x = \pm 1$ is usually determined by solving the dominant equation by reducing it to the Riemann–Hilbert problem of complex variable theory (Ref. 10). However, Peters in Ref. 82 has solved this equation, as for the airfoil equation, using only the Abel inversion formulas (6)–(7). Both of Peters's papers in this area should be more widely known.

Now returning to the dominant equation

$$av(x) + \frac{b}{\pi}\int_{-1}^{1}\frac{v(t)\,dt}{t-x} = f(x), \qquad -1 < x < 1, \qquad (292)$$

it is known that if f is smooth the solution to (292) has the form

$$v(x) = w(x)u(x), \qquad (293)$$

where ($i^2 = -1$)

$$w(x) = (1-x)^\alpha (1+x)^\beta, \qquad (294)$$

$$\alpha = \frac{1}{2\pi i}\log\left(\frac{a-ib}{a+ib}\right) + N, \qquad \beta = \frac{-1}{2\pi i}\log\left(\frac{a-ib}{a+ib}\right) + M, \qquad (295)$$

and N and M are integers determined so that the index $\nu = -(\alpha + \beta) = -(M+N)$ is restricted to the values $-1, 0, 1$, $u(x)$ is "smooth," and $-1 < (\alpha, \beta) < 1$ (Ref. 10). This guarantees that the solution $v(x)$ is integrable on $[-1, 1]$.

As for equations of the first kind, (294) suggests that we make the change of variable $v(x) = w(x)u(x)$ in (291), giving $u(x)$ as the solution to

$$aw(x)u(x) + \frac{b}{\pi}\int_{-1}^{1}\frac{w(t)u(t)\,dt}{t-x} + \int_{-1}^{1} w(t)k(x,t)u(t)\,dt = f(x),$$

$$-1 < x < 1. \qquad (296)$$

If the index of (296) is ν, then when necessary we shall denote $w(x)$ by $w_\nu(x)$ and

$$aw_\nu(x)u(x) + \frac{b}{\pi}\int_{-1}^{1}\frac{w_\nu(t)u(t)\,dt}{t-x}, \qquad (297)$$

by $H_\nu u$, so that (296) can be written in operator form as

$$H_\nu u + Ku = f. \tag{298}$$

Now let $P_n^{(\alpha,\beta)}(x)$, $n = 0, 1, 2, \ldots$, be the nth Jacobi polynomial for the weight function $(1-x)^\alpha(1+x)^\beta$ and let $\psi_n(x)$ denote $P_n^{(\alpha,\beta)}(x)$ normalized so that

$$\int_{-1}^{1} w_\nu(t)\psi_n^2(t)\,dt = 1. \tag{299}$$

[For convenience in writing we will assume that the values of (α, β) associated with $\psi_n(x)$ are understood.] If we operate on $P_n^{(\alpha,\beta)}(x)$ by H_ν (Ref. 9),

$$H_\nu P_n^{(\alpha,\beta)}(x) = \frac{-2^{-\nu}b}{\sin \pi\alpha} P_{n-\nu}^{(-\beta,-\alpha)}(x), \tag{300}$$

which the reader can check generalizes the mapping properties of H_ν when $a = 0$ and $b = 1$.

Now it follows from the normalization factors

$$\int_{-1}^{1} w(t)[P_n^{(\alpha,\beta)}(t)]^2 \,dt = \theta_n \tag{301}$$

that (Ref. 65)

$$\theta^*_{n-\nu} = 2^{-2\nu}\theta_n, \tag{302}$$

where θ_n^* is the normalization factor for $P_n^{(-\alpha,-\beta)}(x)$. Thus dividing both sides of (300) by $(\theta_n)^{1/2}$ gives

$$H_\nu[P_n^{(\alpha,\beta)}(x)/(\theta_n)^{1/2}] = -bP_{n-\nu}^{(-\alpha,-\beta)}(x)/(\theta_n^*)^{1/2} \sin \pi\alpha, \tag{303}$$

where $\sin \pi\alpha = \pm b$. Letting $\Phi_n(x)$ denote the Jacobi polynomial $P_n^{(-\alpha,-\beta)}(x)$ normalized so that

$$\int_{-1}^{1} \{[\Phi_n(t)]^2/w(t)\}\,dt = 1, \tag{304}$$

(303) gives

$$H_\nu \psi_n = \pm \Phi_{n-\nu}, \tag{305}$$

Numerical Solution of Cauchy Singular Integral Equations

with the sign of $\Phi_{n-\nu}$ determined by correctly choosing that of $\sin \pi\alpha$. If $\chi_n = \pm\Phi_n$, $n = 0, 1, \ldots$, with the appropriate sign, then (305) becomes

$$H\psi_n = \chi_{n-\nu}, \quad n = 0, 1, 2, \ldots, \tag{306}$$

where by definition $\chi_{-1} = 0$.

If we define the real Hilbert spaces $L_{w_\nu} = \{f:[-1,1] \to R, \int_{-1}^{1} w_\nu(t)f^2(t)\,dt < \infty\}$, where $w_\nu(t)$ is of the form in (294), then $\{\psi_n\}_{n=0}^{\infty}$ and $\{\chi_n\}_{n=0}^{\infty}$ are orthonormal bases for L_{w_ν} and L_{1/w_ν} with respect to the inner products

$$\langle f, g \rangle_{w_\nu} = \int_{-1}^{1} w_\nu(t) f(t) g(t)\, dt$$

and

$$\langle f, g \rangle_{1/w_\nu} = \int_{-1}^{1} [f(t)g(t)/w_\nu(t)]\, dt. \tag{307}$$

Using (306), H_ν can be extended as a bounded linear operator from $L_{w_\nu} \to L_{1/w_\nu}$, generalizing the corresponding result for equations of the first kind. As in Section 8 it follows from (306) that H_0 is invertible and unitary, H_1 has a bounded right inverse, H_{-1} has a bounded left inverse, and H_1 has a one-dimensional nullspace consisting of the constant multiples of χ_0.

To complete our assumptions, we require that $K: L_{w_\nu} \to L_{1/w_\nu}$ be compact, sufficient conditions being that $k(x, t)$ is continuous or that

$$\int_{-1}^{1} \int_{-1}^{1} \frac{w_\nu(t)}{w_\nu(x)} k^2(x, t)\, dx\, dt < \infty. \tag{308}$$

If $\nu = 0$, we assume that (298) has a unique solution $u \in L_{w_\nu}$ for each $f \in L_{1/w_0}$, and for $\nu = 1$, if $l: L_{w_1} \to L_{w_1}$ is a bounded linear functional, then we require that

$$\begin{cases} H_1 u + Ku = f, & (309) \\ l(u) = M, & (310) \end{cases}$$

have a unique solution. For $\nu = -1$ (298) can have a unique solution iff the consistency condition (Ref. 10)

$$\langle Ku - f, \chi_0 \rangle_{1/w_\nu} = 0 \tag{311}$$

is satisfied—a condition which we shall assume holds.

With these assumptions in force we are now ready to develop the standard polynomial Galerkin, collocation, and quadrature methods for solving (298).

17.1.1. Galerkin's Method. For Galerkin's method, following the same pattern as for the GAE, we approximate u by

$$u_N = \sum_{k=0}^{N} a_k \psi_k, \tag{312}$$

form the residual $r_N = H_\nu u_N + K u_N - f$, and set it orthogonal to χ_j, $j = 0, 1, 2, \ldots, N - \nu$, $\nu = 0, \pm 1$, in L_{1/w_ν} in order to determine $\{a_k\}_{k=0}^{N}$. From this and the orthonormality of $\{\chi_j\}_{j=0}^{N-\nu}$, the coefficients $\{a_k\}_{k=0}^{N}$ are obtained by solving the equations

$$a_{j+\nu} + \sum_{k=0}^{N} a_k \langle K\psi_k, \chi_j \rangle_{1/w_\nu} = \langle f, \chi_j \rangle_{1/w_\nu}, \qquad j = 0, 1, 2, \ldots, N - \nu. \tag{313}$$

From (313) if $\nu = 0$, then there are $N + 1$ equations in $N + 1$ unknowns to solve and, as will be shown in Section 19, they have a unique solution for all N sufficiently large.

If $\nu = 1$, then there are only N equations in $N + 1$ unknowns, so that to obtain a square system we append the auxiliary condition $l(u_N) = M$, where in most cases $l(u) = \int_{-1}^{1} w_1(t) u(t)\, dt$ (Ref. 22).

When $\nu = -1$ (adopting the convention that $a_{-1} = 0$), we have an overdetermined system with $N + 2$ equations in $N + 1$ unknowns, and one equation must be dropped to obtain a square system. The argument used in Section 10.4 suggests that we delete the zeroth equation.

As will be shown later, u_N converges to u in L_{w_ν}, and if $k(x, t) \in C^r$ and $f(x) \in C^r$, then $\|u - u_N\|_{w_\nu} = O(N^{-r})$, $r > 0$. Uniform convergence also holds, but with stronger differentiability conditions on the data and at a slightly lower rate (Ref. 16).

Again, for numerical implementation, the inner products in (313) usually have to be computed numerically, and this gives rise to the generalization of the discrete Galerkin method for the GAE. If $k(x, t)$ and $f(x)$ are smooth, then Gaussian integration with respect to $1/w_\nu(x)$ is a good choice for approximating $\langle f, \chi_j \rangle_{1/w_\nu}$, while product Gaussian integration for the weight function $w_\nu(t)/w_\nu(x)$ can be used to approximate $\langle K\psi_k, \chi_j \rangle_{1/w_\nu}$ (Ref. 16). If $Q_L = (\{\sigma_l\}_{l=0}^{L(N)}, \{x_l\}_{l=0}^{N})$ is the quadrature rule used to evaluate $\langle f, \chi_j \rangle_{1/w_\nu}$ and

$$Q_L \times Q_M = (\{w_m \sigma_l\}_{(l,m)=0}^{(L(N), M(N))}, \{(x_l, t_m)\}_{(l,m)=0}^{(L(N), M(N))})$$

is the product rule used for $\langle K\psi_k, \chi_j \rangle_{1/w_\nu}$, then the discrete Galerkin equations are

$$b_{j+\nu} + \sum_{k=0}^{N} \left[\sum_{m=0}^{M(N)} \sum_{l=0}^{L(N)} w_m \sigma_l k(x_l, t_m) \psi_k(t_m) \chi_j(x_l) \right] b_k = \sum_{l=0}^{L(N)} \sigma_l f(x_l) \chi_j(x_l),$$

$$j = 0, 1, 2, \ldots, N - \nu, \quad (314)$$

where

$$v_N = \sum_{k=0}^{N} b_k \psi_k \quad (315)$$

is the discrete Galerkin approximation to u.

If either $k(x, t)$ or $f(x)$ is not continuous, then the numerical integration problem is more difficult. Logarithmically singular kernels can be handled as for equations of the first kind using product integration analogous to (165) (Ref. 29). However, in general, each case must be treated separately. For smooth data and Gaussian rules we have established convergence of the discrete Galerkin method in Ref. 16. For instance, if $\nu = 0$ and at least $N + 1$ integration nodes in the Gaussian rules are used to evaluate integrals of the form $\int_{-1}^{1} w_\nu(t)g(t)\, dt$ and $\int_{-1}^{1} [g(t)/w_\nu(t)]\, dt$, then $\|u - u_N\|_w = O(N^{-r+1})$, $r > 1$, as will be shown in Section 19. This improves on the results of Ref. 16.

17.1.2. Collocation. In the standard collocation method for solving (296), u is again approximated by

$$u_N = \sum_{k=0}^{N} a_k \psi_k, \quad (316)$$

and the residual is set to zero at the zeros $\{x_j\}_{j=0}^{N-\nu}$ of $\chi_{N+1-\nu}(x)$. Using (306), this gives the equations for $\{a_k\}_{k=0}^{N}$ as $(a_{-1} = 0)$

$$\sum_{k=0}^{N} a_k \chi_{k-\nu}(x_j) + \sum_{k=0}^{N} a_k K\psi_k(x_j) = f(x_j), \quad j = 0, 1, \ldots, N - \nu. \quad (317)$$

As for Galerkin's method, we have $N + 1$ equations in $N + 1$ unknowns if $\nu = 0$, N equations if $\nu = 1$ which, along with $l(u_N) = M$, gives a square system. For $\nu = -1$ the system is overdetermined and, if consistent, one equation may be deleted.

In practice the collocation method is usually preferable to Galerkin's method, since only single integrals $K\psi_k(x_j)$ need to be calculated, rather than the double integrals there. For smooth kernels $K\psi_k(x_j)$ can be obtained by Gaussian quadrature with respect to $w_\nu(t)$, otherwise some sort of product quadrature generally has to be employed. Except for the GAE, this technique has apparently found little use in applications, perhaps because as is shown in Section 19, if $k(x, t)$ is smooth and $N + 1$ point Gaussian quadrature with respect to $w_\nu(t)$ is used to compute $K\psi_k(x_j)$, then this method is mathematically equivalent to the Gaussian quadrature method to be discussed next.

17.1.3. Quadrature Methods. As we have already stated, quadrature methods based on (181) for approximating the Cauchy integral in (291) currently appear to be the most popular way of solving CSIES of the second kind. These methods generalize those for the GAE and were apparently first developed by Krenk in Ref. 59, although he did not use (181) directly. As shown in Ref. 61, this method was proved equivalent to using Gaussian quadrature to evaluate (179) and in that paper the derivation was simplified and extended to include Lobatto and Radau rules as well. Krenk also developed algorithms based on these rules, but again his derivation did not use (181) directly (Ref. 83). Although in theory one can consider quadrature methods based on general polynomial interpolatory formulas in (181), apparently little is known about such procedures except in the Gauss, Lobatto and Radau cases (Ref. 3). As shown in Refs. 3, 5, 28 the Gaussian quadrature method can be extended to solve a large class of nonconstant coefficient equations as well, and this topic will be taken up in Chapter 6 by Elliott.

17.1.4. The Gaussian Quadrature Method. As for equations of the first kind, we begin by approximating Ku in (299) by the $N + 1$ point Gaussian quadrature rule for $w_\nu(t)$,

$$(Ku)(x) \simeq \sum_{k=0}^{N} w_k k(x, t_k) u(t_k), \qquad (318)$$

and letting $\int_{-1}^{1} [w_\nu(t)u(t)/(t - x)] \, dt = (S_\nu u)(x)$, we approximate $S_\nu u$ by

$$(S_\nu u)(x) \simeq \sum_{k=0}^{N} \frac{w_k u(t_k)}{t_k - x}$$

$$+ \frac{q_{N+1}(x)u(x)}{\sigma_{N+1}(x)}, \qquad x \neq t_k, \quad k = 0, 1, 2, \ldots, N, \qquad (319)$$

where $\sigma_{N+1}(x)$ and $q_{N+1}(x) = (S_\nu \sigma_{N+1})(x)$ are defined in Section 12.1. Letting $u_N(x)$ denote the approximation to $u(x)$ obtained by substituting (318)-(319) into (299), we find that it satisfies

$$aw_\nu(x)u_N(x) + \frac{b}{\pi}\sum_{k=0}^{N}\frac{w_k u_N(t_k)}{t_k - x} + \frac{b}{\pi}\frac{S_\nu(\sigma_{N+1})u_N(x)}{\sigma_{N+1}(x)}$$

$$+ \sum_{k=0}^{N} w_k k(x, t_k) u_N(t_k) = f(x), \quad x \neq t_k, \quad k = 0, 1, 2, \ldots, N. \tag{320}$$

Now note that $\sigma_{N+1}(x)$ is a multiple c_{N+1} of $\psi_{N+1}(x)$ (since they have the same zeros) so that using (306)

$$aw_\nu(x)u_N(x) + \frac{b}{\pi}\frac{S_\nu(\sigma_{N+1})u_N(x)}{\sigma_{N+1}(x)}$$

$$= \left[aw_\nu(x)\sigma_{N+1}(x) + \frac{b}{\pi}S_\nu(\sigma_{N+1})\right]u_N(x)/\sigma_{N+1}(x)$$

$$= c_{N+1}[(H_\nu\psi_{N+1})(x)]u_N(x)/\sigma_{N+1}(x)$$

$$= c_{N+1}\chi_{N+1-\nu}(x)u_N(x)/c_{N+1}\psi_{N+1}(x)$$

$$= \chi_{N+1-\nu}(x)u_N(x)/\psi_{N+1}(x), \quad x \neq t_k, \quad k = 0, 1, 2, \ldots, N. \tag{321}$$

As for Eq. (190) we can obtain a set of linear equations for $\{u_N(t_k)\}_{k=0}^{N}$ by collocating at the zeros of $\chi_{N+1-\nu}(x)$. Since $\chi_{N+1-\nu}(x)$ is a normalized Jacobi polynomial, it is well known that it has $N + 1 - \nu$ distinct roots x_j, $j = 0, 1, \ldots, N - \nu$, in the open interval $(-1, 1)$ (Ref. 65), and choosing $x = x_j, j = 0, 1, 2, \ldots, N - \nu$ in (320) gives the equations

$$\frac{b}{\pi}\sum_{k=0}^{N}\frac{w_k u_N(t_k)}{t_k - x_j} + \sum_{k=0}^{N} w_k k(x_j, t_k) u_N(t_k) = f(x_j),$$

$$j = 0, 1, 2, \ldots, N - \nu, \tag{322}$$

for $\{u_N(t_k)\}_{k=0}^{N}$. When these equations are solved, (320) serves as a natural or Nyström interpolant as for equations of the first kind.

However, before proceeding with a further discussion of solving these equations, we need to show that they are well defined; that is, that the zeros of $\psi_{N+1}(x)$, which are the integration points $\{t_k\}_{k=0}^N$, are distinct from the zeros of $\chi_{N+1-\nu}(x)$. For this we use the expression (Ref. 30) (see also Section 12.1)

$$w_k = \fint_{-1}^1 [\sigma_{N+1}(t)/\sigma'_{N+1}(t_k)(t-t_k)]\, dt$$

$$= \pi \chi_{N+1-\nu}(t_k)/b\psi'_{N+1}(t_k), \qquad k = 0, 1, 2, \ldots, N, \qquad (323)$$

where (306) has been used in deriving (323). From the theory of Gaussian quadrature $w_k > 0$, so that if $\chi_{N+1-\nu}(t_k) = 0$, then it follows from (323) that $\psi'_{N+1}(t_k) = 0$. But the roots of the Jacobi polynomials are simple, so that $\psi_{N+1}(x)$ and $\chi_{N+1-\nu}(x)$ can have no common zeros. Other proofs of this fact may be found in Ref. 3 or Ref. 61. For instance, using the arguments in Ref. 61 one can also show that exactly one zero of $\chi_{N+1-\nu}(x)$ lies between any two zeros of $\psi_{N+1}(x)$; that is, their zero sets interlace.

Thus for $\nu = 0$ we have a square system; for $\nu = 1$, (322) should be supplemented by $l(u_N) = M$, and for $\nu = -1$ the system is overdetermined.

The convergence of the Gaussian quadrature method will be discussed in Section 19 by relating it to an appropriate case of the discrete Galerkin method.

17.1.5. Lobatto Quadrature: $\nu = 1$.

In applications, equations with index one seem to occur most often in problems in solid mechanics where the values $u(\pm 1)$ are mostly of interest. In this situation, as for equations of the first kind, using Lobatto quadrature in (181) is natural, since approximations to $u(\pm 1)$ can be obtained as in Section 12.3 without interpolation. As for the Gaussian quadrature method, we follow the derivation given in Ref. 61.

Here, if we use an $N+1$ point Lobatto integration rule for $w_1(x) = (1-x)^\alpha (1+x)^\beta$ with ± 1 as nodes, the remaining nodes are the zeros of $P_{N-1}^{(\alpha+1,\beta+1)}(x)$ (Ref. 69). Thus $\sigma_{N+1}(x)$ in (181) may be taken as $d_{N+1}(1-x^2)P_{N-1}^{(\alpha+1,\beta+1)}(x)$, where d_{N+1} is a constant, and the quadrature rule (181) becomes

$$\fint_{-1}^1 \frac{w_1(t)u(t)\,dt}{t-x} \simeq \sum_{k=0}^N \frac{\lambda_k u(t_k)}{t_k - x} + \frac{S_1(\sigma_{N+1})u(x)}{\sigma_{N+1}(x)}, \qquad (324)$$

where $(\{\lambda_k\}_{k=0}^N, \{x_k\}_{k=0}^N)$ are the weights and nodes of the $N+1$ point

Lobatto rule for $w_1(t)$. From the definition of $\sigma_{N+1}(x)$,

$$S_1(\sigma_{N+1}) = d_{N+1} \fint_{-1}^{1} \frac{(1-t^2)(1+t)^\alpha(1+t)^\beta P_{N-1}^{(\alpha+1,\beta+1)}(t)\, dt}{t-x}$$

$$= d_{N+1} \fint_{-1}^{1} \frac{(1-t)^{\alpha+1}(1+t)^{\beta+1} P_{N-1}^{(\alpha+1,\beta+1)}(t)\, dt}{t-x}$$

$$= d_{N+1} s_N(x), \qquad (325)$$

and using (325) in (324) it becomes

$$\fint_{-1}^{1} \frac{w_1(t)u(t)\, dt}{t-x} \simeq \sum_{k=0}^{N} \frac{\lambda_k u(t_k)}{t_k - x} + \frac{s_N(x)u(x)}{(1-x^2)P_{N-1}^{(\alpha+1,\beta+1)}(x)}. \qquad (326)$$

Now if we approximate Ku by $\sum_{k=0}^{N} \lambda_k k(x, t_k) u(t_k)$, substitute (326) into (296), and again denote by $u_N(x)$ the approximation to $u(x)$, we find that $u_N(x)$ satisfies

$$\left[aw_1(x)(1-x^2)P_{N-1}^{(\alpha+1,\beta+1)}(x) + \frac{b}{\pi} s_N(x) \right] u_N(x)/\sigma_{N+1}(x) + \frac{b}{\pi} \sum_{k=0}^{N} \frac{\lambda_k u_N(t_k)}{t_k - x}$$

$$+ \sum_{k=0}^{N} \lambda_k k(x, t_k) u_N(t_k) = f(x), \qquad x \neq t_k, \qquad k = 1, 2, \ldots, N. \qquad (327)$$

Since $w_1(x) = (1-x)^\alpha (1+x)^\beta$, with $\alpha + \beta = -1$ (recall $\nu = 1$), the term in brackets in (327) is

$$a(1-x)^{\alpha+1}(1+x)^{\beta+1} P_{N-1}^{(\alpha+1,\beta+1)}(x)$$

$$+ \frac{b}{\pi} \fint_{-1}^{1} \frac{(1-t)^{\alpha+1}(1+t)^{\beta+1} P_{N-1}^{(\alpha+1,\beta+1)}(t)\, dt}{t-x}$$

$$= c_N P_N^{(-\alpha-1,-\beta-1)}(x), \qquad (328)$$

where we have used (300), the fact that the operator in (328) has index -1, and c_N is a constant. From (328) it follows that u_N satisfies

$$\frac{c_N P_N^{(-\alpha-1,-\beta-1)}(x) u_N(x)}{\sigma_{N+1}(x)} + \frac{b}{\pi} \sum_{k=0}^{N} \frac{\lambda_k u_N(t_k)}{t_k - x} + \sum_{k=0}^{N} \lambda_k k(x, t_k) u_N(t_k)$$

$$= f(x), \qquad x \neq t_k, \qquad k = 0, 1, 2, \ldots, N. \qquad (329)$$

Now $P_N^{(-\alpha-1,-\beta-1)}(x)$ has N distinct roots in $(-1, 1)$ and, collocating at these values $\{x_j\}_{j=0}^{N-1}$ [which are distinct from $\{t_k\}_{k=0}^{N}$ (Ref. 61)], we obtain the N equations

$$\frac{b}{\pi} \sum_{k=0}^{N} \frac{\lambda_k u_N(t_k)}{t_k - x_j} + \sum_{k=0}^{N} \lambda_k k(x_j, t_k) u_N(t_k) = f(x_j),$$

$$j = 0, 1, 2, \ldots, N - 1, \qquad (330)$$

for $\{u_N(t_k)\}_{k=0}^{N}$. Supplementing (330) by $l(u_N) = M$ gives $N + 1$ equations in $N + 1$ unknowns for the determination of $\{u_N(t_k)\}_{k=0}^{N}$. Since $l(u_N) = \int_{-1}^{1} w_1(t) g(t) u_N(t)\, dt$, it should be discretized by the classical Lobatto quadrature rule to give the $(N + 1)$st equation

$$\sum_{k=0}^{N} \lambda_k g(t_k) u_N(t_k) = M, \qquad (331)$$

to get a square system.

18. Other Polynomial Approximation Methods

As we have seen in the previous section the standard polynomial approximation methods for solving (291) are considerably more complex than their counterparts for equations of the first kind, due to the necessity of having to numerically generate roots of Jacobi polynomials, the polynomials themselves, and the weights for the appropriate quadrature rules. In particular, this usually has to be done for each equation separately. Although efficient methods exist for these computations, they make implementing these algorithms more time consuming than those for equations of the first kind. However, as these algorithms can be justified mathematically, there is good reason to use them.

On the other hand, a number of authors have sought to develop procedures along the line of Sections 15 and 16 which do not have these difficulties and here we shall describe three polynomial-based algorithms which have been used in practice in certain circumstances, and appear to give good numerical results, except near $x = \pm 1$. Unfortunately, little mathematical analysis of these methods has been given so, if used, the reader should be conscious of this fact.

The methods we consider are due to Cohen (Ref. 73), who developed an algorithm similar to that of Kalandiya's, and two Chebyshev methods,

Numerical Solution of Cauchy Singular Integral Equations 269

one by Chawla and Kumar (Ref. 84), and another by Hashmi and Delves (Ref. 77). Since the latter is a straightforward generalization of the one in Section 16.2, we leave the details to the reader.

18.1. Cohen's Method. This method is based on a quadrature rule for Cauchy principal value integrals which is not Gaussian (Ref. 73), and leads, as in Kalandiya's method, to an algorithm where the integration and collocation nodes are the same.

Theorem 18.1. Consider the integral

$$\fint_{-1}^{1} \frac{v(t)\,dt}{t-x}. \tag{332}$$

Then it can be approximated by a quadrature rule of the form (Ref. 73)

$$\fint_{-1}^{1} \frac{v(t)\,dt}{t-x} \simeq \sum_{k=1}^{N} \frac{v(t_k)}{P'_N(t_k)} \left[\frac{Q_N(x) - Q_N(t_k)}{x - t_k} \right], \tag{333}$$

where $\{t_k\}_{k=1}^{N}$ are the roots of the Nth Legendre polynomial $P_N(t)$, and

$$Q_N(x) = \fint_{-1}^{1} \frac{P_N(t)\,dt}{t-x} \tag{334}$$

is a Legendre function of the second kind (Ref. 65) (if $x = t_k$ we interpret $[Q_N(x) - Q_N(t_k)]/(x - t)$ as $Q'_N(t_k)$).

Proof. Approximating $v(t)$ by its interpolation polynomial $p_N(t)$ on $\{t_k\}_{k=1}^{N}$,

$$p_N(t) = \sum_{k=1}^{N} \frac{P_N(t) v(t_k)}{(t - t_k) P'_N(t_k)}, \tag{335}$$

and using (335) in (332),

$$\fint_{-1}^{1} \frac{v(t)\,dt}{t-x} \simeq \fint_{-1}^{1} \frac{p_N(t)}{t-x}\,dt = \sum_{k=1}^{N} \frac{v(t_k)}{P'_N(t_k)} \fint_{-1}^{1} \frac{P_N(t)\,dt}{(t-x)(t-t_k)}. \tag{336}$$

But

$$\frac{1}{(t-t_k)(t-x)} = \frac{1}{x-t_k}\left(\frac{1}{t-x} - \frac{1}{t-t_k}\right), \tag{337}$$

and substituting (337) into (336) and carrying out the integrations gives (333). □

For brevity we write the sum in (333) as

$$\sum_{k=1}^{N} q_k(x)v(t_k), \qquad (338)$$

where

$$q_k(x) = \begin{cases} [Q_N(x) - Q_N(t_k)]/P'_N(t_k)(x - t_k), & x \neq t_k, \\ Q'_N(t_k)/P'_N(t_k), & x = t_k. \end{cases} \qquad (339)$$

In Cohen's algorithm he uses (291) without consideration of the possible endpoint singularities, approximates the Cauchy integral by (333), and $\int_{-1}^{1} k(x, t)v(t)\, dt$ by $\sum_{k=1}^{N} w_k k(x, t_k)v_N(t_k)$ where $\{w_k\}_{k=1}^{N}$ are the weights of the N point Gaussian quadrature rule for $w(t) = 1$. Then using these in (291) and collocating at $\{t_j\}_{j=1}^{N}$ gives the N equations for $v_N(t_j) \simeq v(t_j)$, $j = 1, 2, \ldots, N$:

$$av_N(t_j) + \frac{b}{\pi} \sum_{k=1}^{N} q_k(t_j)v_N(t_k)$$

$$+ \sum_{k=1}^{N} w_k k(t_j, t_k)v_N(t_k) = f(t_j), \qquad j = 1, 2, \ldots, N. \qquad (340)$$

In Ref. 73 Cohen extends this procedure to solve equations with nonconstant coefficients, and solves a number of problems occurring in high energy physics with quite good accuracy. He also considers the possibility that the homogeneous equation may have nonzero solutions, and applies the above procedure to (338) with $f(x) = 0$ to obtain an approximate solution $y_N(t)$ to that equation. Then the linear combination $v_N(t) + cy_N(t)$ may be used to satisfy an auxiliary condition.

Obviously in such a method we are free to choose the quadrature nodes $\{t_k\}_{k=0}^{N}$ [as Cohen points out (Ref. 73)], and perhaps using the Chebyshev nodes they might do just as well (and be easier to implement) since they are known explicitly.

18.2. The Method of Chawla and Kumar. In Ref. 84 Chawla and Kumar consider a general CSIE in the form

$$u(x) = f(x) + \lambda \int_{-1}^{1} \frac{k(x, t)}{t - x} u(t)\, dt. \qquad (341)$$

To solve this equation they approximate $k(x, t)$ by a degenerate kernel of the form

$$k_M(x, t) = \sum_{i=0}^{M} b_i(x) T_i(t), \qquad (342)$$

and simultaneously approximate u by a finite Chebyshev series

$$u_N(x) = \sum_{k=0}^{N} a_k T_k(x). \qquad (343)$$

Substituting (342) and (343) into (341) and collocating at the zeros $\{x_j\}_{j=0}^{N}$ of $T_{N+1}(x)$ they get the following equations for $\{a_k\}_{k=0}^{N}$:

$$\sum_{k=0}^{N} a_k T_k(x_j) + \lambda \sum_{k=0}^{N} a_k \fint_{-1}^{1} \frac{k_M(x_j, t) T_k(t)\, dt}{t - x_j} = f(x_j),$$

$$j = 0, 1, \ldots, N, \qquad (344)$$

where

$$\fint_{-1}^{1} \frac{k_M(x_j, t) T_k(t)\, dt}{t - x_j} = \sum_{i=0}^{M} b_i(x_j) \fint_{-1}^{1} \frac{T_i(t) T_k(t)\, dt}{t - x}. \qquad (345)$$

Thus to implement the algorithm one must be able to calculate

$$\fint_{-1}^{1} [T_i(t) T_k(t)/(t - x_j)]\, dt, \qquad i = 0, \ldots, M, \qquad k = 0, 1, 2, \ldots, N.$$

For this they use the identity

$$T_i(t) T_k(t) = [T_{i+k}(t) + T_{|i-k|}(t)]/2, \qquad (346)$$

and then

$$\fint_{-1}^{1} \frac{T_i(t) T_k(t)\, dt}{t - x_j} = \tfrac{1}{2}[\lambda_{i+k}(x_j) + \lambda_{|i-k|}(x_j)], \qquad (347)$$

where $\lambda_i(x) = \fint_{-1}^{1} [T_i(t)/(t - x)]\, dt$. As in Section 16.2, these can be obtained recursively by Eqs. (283)–(285).

Last one requires $b_i(x_j)$, and these are obtained in Ref. 84 by interpolating $k(x_j, t)$ at the Chebyshev nodes $\{t_j\}_{j=0}^{N}$. These coefficients can be obtained explicitly (Ref. 84) and the algorithm is complete.

As with Cohen's algorithm, the authors ignore the endpoint singularities, and also the possible nonuniqueness of the solution.

In Ref. 84 this method is used to solve several test problems with known unique solutions and no endpoint singularities. Here the algorithm works well, with 7-9 figure accuracy being obtained for $N = M = 10$.

No convergence proof is given, and no problems are solved where the solution is not unique and/or has endpoint singularities.

18.3. Hashmi and Delves' Algorithm.

In Ref. 77 Hashmi and Delves consider the algorithm of Section 16.2 applied to equations of the second kind in the form (338). As for equations of the first kind they approximate the solution by a finite Chebyshev series $\sum_{k=0}^{N} a_k T_k(t)$ whose coefficients are determined by Galerkin's method as in Section 16.2. Since the details are basically the same as there, we refer the reader to Ref. 77 for more information and numerical examples.

18.4. Piecewise Polynomial Methods.

At present a number of papers have been published using piecewise-polynomial approximations for solving Eq. (2). In Ref. 44 Thomas developed a Galerkin method for equations whose solutions are in $L_2[-1, 1]$, thereby ruling out ones with inverse square-root singularities. This work has been extended by Prössdorf and Elschner when the operator is strongly elliptic (Ref. 6), and collocation approximations have been considered under the same conditions. However, equations of the first kind generally do not satisfy this condition (Ref. 6), so that while these methods can be quite useful in practice, as shown in Refs. 7-8, the full theory (at least to our knowledge) has not been completely worked out.

19. Convergence

In this section we prove a number of the convergence theorems that have been stated previously in the chapter. The convergence problem is one that has attracted much attention in the past 15 years, and for the polynomial-based Galerkin, collocation, and Gaussian quadrature methods developed in Sections 10, 11, 12, and 17, the theory is essentially complete, at least if the data are sufficiently smooth. As for Fredholm equations, a variety of approaches have been employed to prove convergence, with slight differences in assumptions being required depending on the method used. For example, Linz (Ref. 37), Ioakimidis (Ref. 22), Elliott (Refs. 5, 15), and

Numerical Solution of Cauchy Singular Integral Equations

Gerasoulis (Ref. 68) use techniques based on regularization, while Duskov and Gabdulhaev (Ref. 85), Dziskhariani (Ref. 86), Fromme and Golberg (Ref. 13), Junghanns and Silbermann (Ref. 3), and Miel (Ref. 42) use projection methods. As the regularization approach will be treated by Elliott in Chapter 6 we will consider only projection methods here. And since a comprehensive development would probably warrant a separate chapter, we will confine ourselves to proving the convergence of the following algorithms:

1. The polynomial-based Galerkin methods in Section 11 and 17 for $\nu = 0$ and $\nu = 1$.
2. The polynomial-based collocation methods in Sections 11 and 17 for $\nu = 0$ and $\nu = 1$.
3. The discrete Galerkin method in Sections 10 and 17 where $\nu = 0$, 1 and the data are sufficiently smooth.
4. The Gaussian quadrature method in Sections 12 and 17 for $\nu = 0$ and $\nu = 1$.

To avoid repetition in stating theorems, the terms Galerkin's method, collocation method, etc., will refer to those in algorithms (1)-(4) for approximating the solution u of Eq. (96) and/or Eq. (296). The approximations themselves will be denoted by u_N or v_N as before.

19.1. Mean-Square Convergence of Galerkin's Method: $\nu = 0$.

Just as Galerkin's method can be derived in both a direct and indirect fashion, its mean-square convergence can be proved in both ways as well. As both of these approaches have been used in the literature, we present both procedures here and, as observed by Ioakimidis (Ref. 22), while the two methods do not always yield the same convergence rates [the indirect approach sometimes does better (Ref. 22)], in this case it does, and it is instructive to compare the two. From an algebraic point of view, the direct approach usually leads to simpler manipulations—a viewpoint also taken by Junghanns and Silbermann in Ref. 3.

19.1.1. The Indirect Method. As shown in Sections 10.1 and 17.1.1, the CSIE with $\nu = 0$ can be considered as an operator equation from $L_w \to L_{1/w}$ of the form

$$H_0 u + Ku = f, \tag{348}$$

where H_0 is unitary and K is compact. To begin our convergence analysis we write (348) in the form (recall $H_0^{-1} = H_0^*$)

$$u + H_0^* K u = H_0^* f, \tag{349}$$

and then show that the Galerkin equations have a similar form. For this, let $Y_N = \text{span}\{\chi_k\}_{k=0}^N$ and denote by P_N the operator of orthogonal projection onto Y_N. Then, as for Fredholm equations, we find that the Galerkin approximation u_N satisfies the operator equation (Ref. 1)

$$P_N H_0 u_N + P_N K u_N = P_N f. \tag{350}$$

Since u_N is a polynomial of degree $\leq N$, $H_0 u_N \in Y_N$ and since $P_N Y_N = Y_N$, $P_N H_0 u_N = H_0 u_N$, so that (350) becomes

$$H_0 u_N + P_N K u_N = P_N f, \tag{351}$$

and then

$$u_N + H_0^* P_N K u_N = H_0^* P_N f. \tag{352}$$

Since $H_0 H_0^* = I$, the identity operator on $L_{1/w}$,

$$u_N + (H_0^* P_N H_0) H_0^* K u_N = (H_0^* P_N H) H^* f. \tag{353}$$

It can now be verified that $Q_N = H_0^* P_N H_0$ is the operator of orthogonal projection onto $X_N = \text{span}\{\psi_k\}_{k=0}^N$, so that u_N satisfies

$$u_N + Q_N L u_N = Q_N g, \tag{354}$$

where $L = H_0^* K$ and $g = H_0^* f$. Since K is compact and H_0^* is bounded, L is compact (Ref. 13), and u_N satisfies a Galerkin approximating equation for

$$u + Lu = g,$$

which is just (349). To prove the convergence of u_N in L_w we can now use the following well-known convergence theorem for operator equations of the second kind (Ref. 19).

Theorem 19.1. See Ref. 19. Suppose that X is a Hilbert space, $L: X \to X$ is compact, and assume that $u + Lu = g$ has a unique solution for each $g \in X$. If $\{\phi_k\}_{k=0}^\infty$ is an orthonormal basis for X, and Q_N is the operator of orthogonal projection onto $\text{span}\{\phi_k\}_{k=0}^N$, then for all N sufficiently large, $N \geq N_0$, $(I + Q_N L)^{-1}$ exists and the norms $\|(I + Q_N L)^{-1}\|$ are uniformly bounded. From this it follows for $N \geq N_0$ that $u_N + Q_N L u_N = Q_N g$ has a unique solution in X, and

$$\|u - Q_N u\|_X \leq \|u - u_N\|_X \leq [1 + \gamma(N)]\|u - Q_N u\|_X, \qquad (355)$$

where $\gamma(N) \to 0$ as $N \to \infty$.

Proof. By assumption, $Q_N f \to f$ for all $f \in X$, and since L is compact, it follows by a standard theorem that $\|L - Q_N L\| \to 0$ (Ref. 19). Hence, for all $N \geq N_0$, the operators $(I + Q_N L)^{-1}$ exist and have uniformly bounded norms (Ref. 19). Thus for $N \geq N_0$, $u_N = (I + Q_N L)^{-1} Q_N g$ is the unique solution to $u_N + Q_N L u_N = Q_N g$.

To get (355) we first observe that $Q_N u$ is the best approximation to u in the norm of X by an element in X_N (Ref. 19), so that

$$\|u - Q_N u\|_X \leq \|u - u_N\|_X. \qquad (356)$$

For the right-hand side of (355), use

$$u - u_N = (I + L)^{-1} g - (I + Q_N L)^{-1} Q_N g = (I + Q_N L)^{-1} (I - Q_N) u \qquad (357)$$

and

$$(I + Q_N L)^{-1} = I - (I + Q_N L)^{-1} Q_N L. \qquad (358)$$

Since $I - Q_N$ is a projection, $(I - Q_N) = (I - Q_N)^2$, and using this and (358) in (357) gives

$$u - u_N = (u - Q_N u) - (I + Q_N L)^{-1} Q_N (L - L Q_N)(u - Q_N u). \qquad (359)$$

But Q_N is an orthogonal projection, so that $\|L - L Q_N\| \to 0$ (Ref. 13), and taking norms in (359) ($\|Q_N\| = 1$)

$$\|u - u_N\|_X \leq \|u - Q_N u\|_X + \|(I + Q_N L)^{-1}\| \|L - L Q_N\| \|u - Q_N u\|_X,$$

which shows that

$$\|u - u_N\|_X \leq [1 + \gamma(N)]\|u - Q_N u\|_X, \qquad N \geq N_0,$$

where $\gamma(N) = \|(I + Q_N L)^{-1}\| \|L - L Q_N\| \to 0$ as $N \to \infty$. □

Theorem 19.2. Assume that (98) or (298) with $\nu = 0$ is solved by the Galerkin method of Section 10.2 or 17.1.1. Then u_N converges in L_w to u,

$$\|u - Q_N u\|_w \leq \|u - u_N\|_w \leq [1 + \gamma(N)]\|u - Q_N u\|_w \tag{360}$$

for all N sufficiently large, and $\gamma(N) \to 0$ as $N \to \infty$.

Proof. Apply Theorem 19.1 with $Q_N = H_0^* P_N H_0$, $L = H_0^* K$, $g = H_0^* f$, and $\phi_k = \psi_k$, $k = 0, 1, 2, \ldots$. □

To get more precise convergence rates for u_N from (360) we apparently need to know something about the smoothness of u. However, this problem can be overcome if we observe that $\|u - Q_N u\|_w = \|u - H_0^{-1} P_N H_0\|_w = \|H_0^{-1}(H_0 - P_N H_0)\|_w$ so that

$$\|u - u_N\|_w \leq [1 + \gamma(N)]\|H_0^{-1}\|\|(I - P_N)H_0 u\|_{1/w}$$

$$\leq [1 + \gamma(N)]\|(I - P_N)H_0 u\|_{1/w},$$

since $\|H_0^{-1}\| = 1$. Now if $k(x, t) \in C^r$ and $f(x) \in C^r$, then $H_0 u = -Ku + f \in C^r$, so that a standard argument using Jackson's theorem (Ref. 72) gives $\|(I - P_N)H_0 u\|_{1/w} = O(N^{-r})$, and

$$\|u - u_N\|_w = O(N^{-r}), \quad r > 0. \tag{361}$$

19.1.2. The Direct Method. Here we observe that the previous results can be obtained directly from (348) and (351) without passing to the equivalent equations of the second kind. Using essentially the same arguments as in Theorem 19.1, we find that for $N \geq N_0$, $\{(H_0 + P_N K)^{-1}\}$ exist and have uniformly bounded norms. Thus for $N \geq N_0$,

$$u - u_N = (H_0 + K)^{-1} f - (H_0 + P_N K)^{-1} P_N f$$

$$= (H_0 + P_N K)^{-1}[(H_0 + P_N K) - P_N (H_0 + K)] u,$$

which gives

$$u - u_N = (H_0 + P_N K)^{-1}[(I - P_N) H_0 u].$$

But $(H_0 + P_N K)^{-1} = H_0^{-1} - (H_0 + P_N K)^{-1} P_N K H_0^{-1}$, and using this and $(I - P_N)^2 = (I - P_N)$,

$$u - u_N = H_0^{-1}(I - P_N) H_0 u$$

$$- (H_0 + P_N K)^{-1} P_N [K H_0^{-1}(I - P_N)](I - P_N) H_0 u.$$

From this, we get on taking norms and using $\|H_0^{-1}\| = 1$ that

$$\|u - u_N\|_w \le [1 + \beta(N)]\|(I - P_N)H_0 u\|_{1/w}, \tag{362}$$

where $\beta(N) = \|(H_0 + P_N K)^{-1}\| \|P_N\| \|KH_0^{-1} - KH_0^{-1} P_N\|$. Since P_N is orthogonal, $\|KH_0^{-1} - KH_0^{-1} P_N\| \to 0$, so that $\beta(N) \to 0$, $N \to \infty$ and $\|u - u_N\|_w \to 0$. Also if $k(x, t) \in C^r$ and $f(x) \in C^r$, $r > 0$, $\|(I - P_N)H_0 u\|_{1/w} = O(N^{-r})$ so that $\|u - u_N\|_w = O(N^{-r})$, $r > 0$.

Before leaving this section, we make several comments regarding the mean-square convergence rate of Galerkin's method. From (355) we see that if we know something about the smoothness of u, then convergence rates follow, while from (362) we need only know something about the smoothness of the data $k(x, t)$ and $f(x)$, since this gives us immediate information about the smoothness of $H_0 u$. Because it is this latter information that is usually available in most problems, direct error estimates of the form (362) are usually preferable. However, there may be cases where the kernel is not smooth, but the solution is, and then (360) is preferable. For instance, this may happen when $k(x, t) = a(x, t)\log(|x - t|) + b(x, t)$ and $\nu = 0$ in the generalized airfoil equation. As an example, the exact solution formula for the Küssner-Schwarz kernel shows that if $f(x)$ is a polynomial, then so is the solution $u(x)$ (Ref. 13). Although little formal work seems to have been done on this problem, let us give a heuristic argument as to why logarithmically singular kernels may produce smooth solutions in this case, and so numerical algorithms can have rapid rates of convergence.

For this, consider the equation

$$\frac{1}{\pi} \int_{-1}^{1} \frac{w_0(t) u(t)\, dt}{t - x} + \int_{-1}^{1} w_0(t) \log(|x - t|) u(t)\, dt = f(x),$$

where $f(x)$ is smooth, and let

$$v(x) = \int_{-1}^{1} w_0(t) \log(|x - t|) u(t)\, dt.$$

Then formally,

$$v'(x) = -\int_{-1}^{1} \frac{w_0 u(t)\, dt}{t - x},$$

so that $v(x)$ satisfies the differential equation

$$-\frac{1}{\pi} v'(x) + v(x) = f(x),$$

which shows that $v(x)$ is C^{r+1} if $f(x)$ is C^r, and so $v'(x)$ is C^r. Thus

$$\frac{1}{\pi}\int_{-1}^{1}\frac{w_0(t)u(t)\,dt}{t-x} = g(x), \tag{363}$$

where $g(x) \in C^r$. Using the Söhngen inversion formula in (363), we find that $u(x)$ is differentiable if r is sufficiently large (for more precise results consult Ref. 10).

Finally, as we mentioned in Section 9, Galerkin's method may be regarded as a degenerate kernel method. In fact, by explicitly computing $P_N K u_N$ in (350), we find that $P_N K u_N = K_N u_N$ where

$$K_N u_N(x) = \int_{-1}^{1} w_0(t) k_N(x,t) u_N(t)\,dt,$$

and

$$k_N(x,t) = \sum_{k=1}^{N} \chi_k(x)\beta_k(t),$$

with

$$\beta_k(t) = \int_{-1}^{1} [k(x,t)\chi_k(x)/w_0(x)]\,dx.$$

19.2. Mean-Square Convergence of Galerkin's Method: $\nu = 1$.

Our approach to proving the convergence of Galerkin's method when $\nu = 1$ will follow that of Ref. 16, where using a suggestion of Linz (Ref. 45), (104)–(114) or (309)–(310) are converted to an equivalent problem of index zero, and then the results of the previous subsection can be used.

Thus we consider the equations

$$\begin{cases} H_1 u + K u = f, & (364) \\ l(u) = M, & (365) \end{cases}$$

where H_1 is given by either (63) or (297) and has index one, and so is not invertible. We shall assume, however, that (364)–(365) have a unique solution $u \in L_w$ for each $f \in L_{1/w}$ and that $l(u)$ is bounded, so that it has the representation

$$l(u) = \int_{-1}^{1} w_1(t)g(t)u(t)\,dt, \qquad g(t) \in L_w.$$

Now let $\hat{L}_{1/w}$ be the Cartesian product of $L_{1/w}$ and the real numbers R. That is,

$$\hat{L}_{1/w} = \{(u, x): u \in L_{1/w}, x \in R\},$$

with the inner product

$$\langle (u_1, x_1), (u_2, x_2) \rangle = \langle u_1, u_2 \rangle_{1/w} + x_1 x_2,$$

and the norm

$$\|(u, x)\| = (\|u\|_{1/w} + x^2)^{1/2}.$$

Also define $\hat{H}_1: L_w \to \hat{L}_{1/w}$ by

$$\hat{H}_1 u = (H_1 u, l(u)), \tag{366}$$

and let $g = \sum_{k=0}^{\infty} \langle g, \psi_k \rangle_w \psi_k = \sum_{k=0}^{\infty} g_k \psi_k$. In Ref. 16 it was shown that if $g_0 \neq 0$ (which will certainly be true in the most common case where $g = 1$), then $\hat{H}_1$ is bounded with a bounded inverse, so (364)-(365) can be written as the operator equation

$$\hat{H}_1 u + \hat{K} u = \hat{f}, \tag{367}$$

where $\hat{K}: L_w \to \hat{L}_{1/w}$ is defined by $\hat{K} u = (Ku, 0)$ and $\hat{f} = (f, M)$.

If K is compact, so is $\hat{K}$, and (366) now has the form of an equation of index zero, since $\hat{H}_1$ is invertible (but not necessarily unitary). To analyze the convergence of Galerkin's method we formulate it as a projection method. For this let P_N be the operator of orthogonal projection of $L_{1/w}$ onto $\text{span}\{\chi_k\}_{k=0}^{N-1}$ and define $\hat{P}_N: \hat{L}_{1/w} \to \hat{L}_{1/w}$ by

$$\hat{P}_N(u, x) = (P_N u, x), \quad u \in L_{1/w}, \quad x \in R. \tag{368}$$

Lemma 19.1. $\hat{P}_N$ is an orthogonal projection on $\hat{L}_{1/w}$ and $\hat{P}_N f \to f$ for all $f \in \hat{L}_{1/w}$.

Proof. To prove that $\hat{P}_N$ is a projection we need to show that $\hat{P}_N^2 = \hat{P}_N$. But $\hat{P}_N[\hat{P}_N(u, x)] = \hat{P}_N(P_N u, x) = (P_N^2 u, x) = (P_N u, x) = \hat{P}_N(u, x)$, since $P_N^2 = P_N$.

To prove that $\hat{P}_N$ is orthogonal we have to show for every two elements (u, x) and (v, y) in $\hat{L}_{1/w}$ that

$$\langle \hat{P}_N(u, x), (v, y) \rangle = \langle (u, x), \hat{P}_N(v, y) \rangle,$$

which will establish that $\hat{P}_N^* = \hat{P}_N$. But

$$\langle \hat{P}_N(u, x), (v, y) \rangle = \langle (P_N u, v) \rangle_{1/w} + xy = \langle u, P_N v \rangle_{1/w} + xy,$$

since P_N is orthogonal on $L_{1/w}$. Also $\langle (u, x), \hat{P}_N(v, y) \rangle = \langle u, P_N v \rangle_{1/w} + xy$ and orthogonality holds.

Because $P_N u \to u$ for all $u \in L_{1/w}$, it follows immediately that $\hat{P}_N f = \hat{P}_N(u, x) = (P_N u, x) \to (u, x) = f$ for all $f \in \hat{L}_{1/w}$. □

Lemma 19.2. The Galerkin approximation u_N satisfies the operator equation

$$\hat{H}_1 u_N + \hat{P}_N \hat{K} u_N = \hat{P}_N \hat{f}. \tag{369}$$

Proof. We note, using the definition of P_N, that u_N can be shown to satisfy (Ref. 16)

$$\begin{cases} P_N H_1 u_N + P_N K u_N = P_N f, & (370) \\ 1(u_N) = M. & (371) \end{cases}$$

Since $H_1 u_N = \sum_{k=0}^{N} a_k H_1 \psi_k = \sum_{k=1}^{N} a_k \chi_{k-1} = \sum_{k=0}^{N-1} a_{k+1} \chi_k$, $P_N H_1 u_N = H_1 u_N$, so that (370)-(371) become

$$\begin{cases} H_1 u_N + P_N K u_N = P_N f, & (372) \\ l(u_N) = M. & (373) \end{cases}$$

Using the definition of $\hat{H}_1$, (372)-(373) are equivalent to

$$\hat{H}_1 u_N + \hat{P}_N \hat{K}_N u_N = \hat{f}_N, \tag{374}$$

where $\hat{K}_N u_N = (P_N K u_N, 0) = \hat{P}_N(K u_N, 0) = \hat{P}_N \hat{K} u_N$, and $f_N = (P_N f, M) = \hat{P}_N(f, M) = \hat{P}_N \hat{f}$, so that

$$\hat{H}_1 u_N + \hat{P}_N \hat{K} u_N = \hat{P}_N \hat{f}, \tag{375}$$

which proves the lemma. □

Using Lemma 19.2 and the argument in the previous subsection we arrive at the following convergence theorem for Galerkin's method when $\nu = 1$.

Theorem 19.3. In (364)-(365) assume that $K: L_w \to L_{1/w}$ is compact, and $l(u) = \int_{-1}^{1} w_1 g(t) u(t) \, dt$, with $g_0 = \langle g, \psi_0 \rangle_w \neq 0$. Then for all N

sufficiently large u_N exists, and

$$\|u - Q_N u\|_w \le \|u - u_N\|_w \le [\|\hat{H}_1^{-1}\| + \gamma(N)]\|(I - \hat{P}_N)\hat{H}_1 u\|, \quad (376)$$

where Q_N is the operator of orthogonal projection onto $\operatorname{span}\{\psi_k\}_{k=0}^N$ and $\gamma(N) \to 0$ as $N \to \infty$.

Proof. This is a straightforward application of the result in Section 19.1.2, with H_0 being replaced by $\hat{H}_1$, K by $\hat{K}$, P_N by $\hat{P}_N$, and f by $\hat{f}$ [note that the unitarity of H_0 does not play an essential role in deriving (362), the boundedness of H_0 and H_0^{-1} are sufficient]. □

Corollary 19.1. $\|u - u_N\|_w \to 0$, as $N \to \infty$.

Proof. From (366) and (368) $\|(I - \hat{P}_N)(\hat{H}_1 u)\| = \|(I - P_N)H_1 u\|_{1/w}$, so that

$$\|u - u_N\|_w \le c \|(I - P_N)H_1 u\|_{1/w}. \quad (377)$$

Since $P_N g \to g$ for all $g \in L_{1/w}$, $\|(I - P_N)H_1 u\|_{1/w} \to 0$ and $\|u - u_N\|_w \to 0$. □

Corollary 19.2. If $k(x, t) \in C^r$ and $f(x) \in C^r$, then

$$\|u - u_N\|_w = O(N^{-r}), \quad r > 0.$$

Proof. Since $H_1 u = -Ku + f$, $H_1 u \in C^r$, and using Jackson's theorem to estimate $\|(I - P_N)H_1 u\|_{1/w}$ the result follows from (377). □

19.3. Uniform Convergence of Galerkin's Method: $\nu = 0, 1$.

To prove the uniform convergence of Galerkin's method we need some bounds on the uniform norms of the normalized Jacobi polynomials $\{\psi_k\}_{k=0}^\infty$. These bounds may be found, for instance, in Szëgo's classical treatise on orthogonal polynomials (Ref. 65).

Changing our notation slightly from Section 17, let $p_k^{(\alpha,\beta)}(x)$ denote the kth Jacobi polynomial normalized so that $\int_{-1}^{1} (1-t)^\alpha (1+t)^\beta [p_k^{(\alpha,\beta)}(t)]^2 \, dt = 1$ (these are just the polynomials that we have referred to previously as ψ_k). Then $\|p_k^{(\alpha,\beta)}\|_\infty = O(k^{\mu+1/2})$, where $\mu = \max(\alpha, \beta, -\tfrac{1}{2})$.

Theorem 19.4. If u_N is the Galerkin approximation to the solution u of (296) with $\nu = 0$ or $\nu = 1$, $k(x, t) \in C^r$, $f(x) \in C^r$, and $r - \mu - \tfrac{3}{2} > 0$, then u_N converges uniformly to u and

$$\|u - u_N\|_\infty = O(N^{-r+\mu+3/2}). \tag{378}$$

Proof. To obtain (378) we use the mean-square convergence results of the previous subsections and arguments analogous to those of Linz (Ref. 37) and Ioakimidis (Ref. 22).

Since $u \in L_w$, we can expand it as $u = \sum_{k=0}^\infty \langle u, \psi_k \rangle_w \psi_k$ so that

$$u - u_N = \sum_{k=0}^N [\langle u, \psi_k \rangle_w - \langle u_N, \psi_k \rangle_w]\psi_k + \sum_{k=N+1}^\infty \langle u, \psi_k \rangle_w \psi_k, \tag{379}$$

and to bound $\|u - u_N\|_\infty$ we only need to estimate $|\langle u, \psi_k \rangle_w - \langle u_N, \psi_k \rangle_w|$, and $\langle u, \psi_k \rangle_w$, since $\|\psi_k\|_\infty = O(k^{\mu+1/2})$. Now from our estimates of $\|u - u_N\|_w$ and the Cauchy-Schwarz inequality

$$|\langle u, \psi_k \rangle_w - \langle u_N, \psi_k \rangle_w| \le \|u - u_N\|_w \|\psi_k\|_w = \|u - u_N\|_w = O(N^{-r}). \tag{380}$$

To estimate $|\langle u, \psi_k \rangle_w|$ we use the fact that $\langle u, \psi_k \rangle_w = \langle H_0 u, \chi_k \rangle_{1/w}$ if $\nu = 0$, and $\langle u, \psi_k \rangle_w = \langle H_1 u, \chi_{k-1} \rangle_{1/w}$ if $\nu = 1$, and show in both cases that $|\langle u, \psi_k \rangle_w| = O(k^{-r})$. Since the proofs are essentially the same, we consider only $\nu = 0$. For this let e_{k-1} be the polynomial of degree $\le k - 1$ of best uniform approximation to $H_0 u$, then

$$\langle u, \psi_k \rangle_w = \langle H_0 u, \chi_k \rangle_{1/w} = \langle H_0 u - e_{k-1}, \chi_k \rangle_{1/w}, \tag{381}$$

since $\langle e_{k-1}, \chi_k \rangle_{1/w} = 0$. Then using the Cauchy-Schwarz inequality and Jackson's theorem again

$$|\langle H_0 u - e_{k-1}, \chi_k \rangle_{1/w}| \le \|H_0 u - e_{k-1}\|_{1/w} \|\chi_k\|_{1/w}$$

$$= \|H_0 u - e_{k-1}\|_{1/w} = O(k^{-r}). \tag{382}$$

Now taking norms on both sides of (379) and using (380) and (382) gives (where the quantities c are constants independent of k and N)

$$\|u - u_N\|_\infty \le c_1 \sum_{k=0}^N N^{-r} k^{\mu+1/2} + c_2 \sum_{k=N+1}^\infty k^{-r+\mu+1/2} = O(N^{-r+\mu+3/2}). \quad \square$$

This result agrees with that found by Ioakimidis (Ref. 22) who obtained it by using regularization. For equations of the first kind with $\nu = 1$, $\|u - u_N\|_\infty = O(N^{-r+1})$, which agrees with the result of Linz (Ref. 37). For

$\nu = 0$, and equations of the first kind, (378) gives $\|u - u_N\|_\infty = O(N^{-r+2})$, $r > 2$.

On the other hand, these results are slightly weaker than those of Junghanns who obtained $O(N^{-r+\mu+1})$ convergence in this case (Ref. 41). The discrepancy arises from the different methods used in going from mean-square to uniform convergence. However, for practical purposes these differences are not great.

19.4. The Sloan Iterate. As shown in Chapter 2 by Sloan, iterating a Galerkin approximation for equations of the second kind has quite remarkable superconvergence properties, and some of these results can be generalized to the Hilbert space setting for equations of the form $Hu + Ku = f$, where H is invertible and K is compact. Thus they apply to CSIES with $\nu = 0$ or 1 when the polynomial-based Galerkin method is used. This generalization seems to have been first proposed in Ref. 26 for the GAE with $\nu = 0$, and then extended by Elliott in Ref. 40. However, in neither of these papers were numerical results given, and the method appears to be more difficult to implement than for Fredholm equations (a recent calculation is given, however, in Ref. 27). In light of this, our discussion will be brief, although we feel that this idea warrants further investigation.

Suppose now that we have computed a Galerkin approximation u_N when $\nu = 0$ or $\nu = 1$. Then, as we have seen in the previous subsection, u_N satisfies an operator equation of the form

$$Hu_N + P_N K u_N = P_N f, \tag{383}$$

where H is invertible, K is compact, and P_N is an orthogonal projection. The Sloan iterate u_N^s is then given by solving

$$Hu_N^s = -Ku_N + f, \tag{384}$$

where u_N^s is well defined, since H is invertible.

Theorem 19.5. u_N^s converges in L_w to u and

$$\|u - u_N^s\|_w \leq c \|K - KQ_N\| \|u - Q_N u\|_w, \tag{385}$$

where $Q_N = H^{-1} P_N H$.

Proof. We first show that u_N^s satisfies

$$Hu_N^s + KQ_N u_N^s = f. \tag{386}$$

To see this, operate on both sides of (384) with P_N giving

$$P_N H u_N^s = -P_N K u_N + P_N f = H u_N,$$

where the last step follows from (383). Thus $u_N = H^{-1} P_N H u_N^s = Q_N u_N^s$ and substituting this into (383) gives (386).

Since $Q_N = H^{-1} P_N H$, where P_N is an orthogonal projection, it follows that $\|K - KQ_N\| \to 0$. Thus for all N sufficiently large, $(H + KQ_N)^{-1}$ exists, and (386) has a unique solution $u_N = (H + KQ_N)^{-1} f$. Hence

$$u - u_N^s = (H + K)^{-1} f - (H + KQ_N)^{-1} f$$

$$= -(H + KQ_N)^{-1}[K(I - Q_N)u], \quad (387)$$

and since $(I - Q_N)^2 = I - Q_N$, (387) becomes

$$u - u_N^s = -(H + KQ_N)^{-1} K(I - Q_N)(I - Q_N)u. \quad (388)$$

Taking norms in (388) gives

$$\|u - u_N^s\|_w \le c \|K - KQ_N\| \|u - Q_N u\|_w,$$

which establishes the convergence of u_N^s to u. □

Letting $\beta(N) = c\|K - KQ_N\|$, we see that $\beta(N) \to 0$ as $N \to \infty$ so that $\|u - u_N^s\|_w \le \beta(N) \|u - Q_N u\|_w$ which shows that u_N^s converges more rapidly in L_w than u_N.

If $\nu = 0$, and we have smoothness conditions on $k(x, t)$ and $f(x)$, then it follows that u_N^s converges in L_w at twice the rate of u_N. This is analogous to the superconvergence results in Chapter 2 for Fredholm equations.

Theorem 19.6. Suppose that $\nu = 0$, $k(x, t) \in C^r$ and $f(x) \in C^r$, then

$$\|u - u_N^s\|_w = O(N^{-2r}), \quad r > 0. \quad (389)$$

Proof. Since $\nu = 0$, Q_N is an orthogonal projection, so that $\|K - KQ_N\| = O(N^{-r})$, and $\|u - Q_N u\|_w = O(N^{-r})$. Thus (389) follows immediately from (385). □

19.5. Convergence of the Collocation Method. As for Galerkin's method, the collocation methods developed in Sections 11 and 17 can also be analyzed as projection methods, the one major difference being the fact that the projections involved are not bounded in $L_{1/w}$. Since the details, as shown for Galerkin's method, are essentially the same for indices zero and one, we will carry out the calculations for $\nu = 0$, as those for $\nu = 1$ can be reduced to this case.

We begin by showing that the algorithm given by (316) and (317) can be formulated as a projection method, establish its mean-square convergence, and then obtain uniform convergence in the same manner as for Galerkin's method. To prove these theorems we need a classical result in interpolation theory—the Erdös-Turan theorem (Ref. 65) which we state as Theorem 19.7.

Theorem 19.7. See Ref. 65. Let $w(t)$ be a nonnegative integrable weight function such that $\int_a^b t^n w(t)\,dt < \infty$, $n \geq 0$, and let $\{\psi_k\}_{k=0}^\infty$ be a set of orthogonal polynomials associated with $w(t)$ (Ref. 65). If $p_N(t)$ is the unique polynomial of degree $\leq N$ obtained by interpolating the continuous function $f(t)$ at the zeros of $\psi_{N+1}(t)$, then

$$\lim_{N\to\infty} \int_a^b w(t)[f(t) - p_N(t)]^2\,dt = 0. \tag{390}$$

In functional analytical terms we can state Theorem 19.7 in the following way. Let $P_N : C^0 \to L_w$ be the operator which maps $f(t)$ onto $p_N(t)$ [P_N is well defined because the interpolant $p_N(t)$ is unique]. P_N is a projection operator, and for $f \in C^0$

$$\|f - P_N f\|_w \to 0. \tag{391}$$

Also, P_N considered as an operator from C^0 to L_w is bounded, and (Ref. 1)

$$\|P_N\|_{C^0 \to L_w} = \left[\int_a^b w(t)\,dt\right]^{1/2}. \tag{392}$$

Now let $w(t)$ be the weight function when $\nu = 0$ and let P_N be the operator which maps $g(t) \in C^0$ onto its polynomial interpolant at the zeros of χ_{N+1}. Then it follows easily from this (Ref. 1) that u_N, the collocation approximation, satisfies

$$P_N H_0 u_N + P_N K u_N = P_N f. \tag{393}$$

Since $H_0 u_N$ is a polynomial of degree $\leq N$, $P_N H_0 u_N = H_0 u_N$ so that (393) becomes

$$H_0 u_N + P_N K u_N = P_N f. \tag{394}$$

With these preliminaries we are now ready to establish the mean-square convergence of u_N.

Theorem 19.8. Let $k(x, t)$ be such that K defines a compact operator from $L_w \to C^0$ and assume that $f(x)$ is continuous. Then u_N converges to u in L_w and

$$\|u - u_N\|_w \leq c\|H_0 u - P_N H_0 u\|_{1/w}. \tag{395}$$

Proof. Since K is compact from $L_w \to C^0$, P_N is bounded from $C^0 \to L_{1/w}$ and $P_N g \to g$, for all $g \in C^0$, it follows that $\|K - P_N K\| \to 0$ (Ref. 19). From this point on the argument is essentially the same as in Section 19.1.2. That is, for all N sufficiently large, $N \geq N_0$, $(H_0 + P_N K)^{-1}$ exists and their norms are uniformly bounded. Thus for all $N \geq N_0$, $u_N = (H_0 + P_N K)^{-1} P_N f$ exists,

$$u - u_N = (H_0 + P_N K)^{-1}(I - P_N)H_0 u, \tag{396}$$

and

$$\|u - u_N\|_{1/w} \leq c\|H_0 u - P_N H_0 u\|_{1/w}.$$

Because $H_0 u = -Ku + f$ is continuous, it follows from the Erdös–Turan theorem that $\|H_0 u - P_N H_0 u\|_{1/w} \to 0$. □

Note. Using a generalization of the Erdös–Turan theorem (Ref. 65), it suffices to have $f(x)$ Riemann integrable for Theorem 19.8 to hold. Thus the standard collocation method converges in L_w if $f(x)$ is a step-function.

With a little extra work we can get convergence rates identical to those for Galerkin's method.

To see this, let e_N be the polynomial of degree $\leq N$ of best uniform approximation to $H_0 u$. Then

$$\|H_0 u - P_N H_0 u\|_{1/w} = \|(H_0 u - e_N) + P_N(e_N - H_0 u)\|_{1/w}$$
$$\leq \|H_0 u - e_N\|_{1/w} + \|P_N\|_{C^0 \to L_{1/w}} \|H_0 u - e_N\|_\infty, \tag{397}$$

and if $k(x, t) \in C^r$ and $f(x) \in C^r$, $r > 0$, it again follows from Jackson's theorem that both terms on the right in (397) are $O(N^{-r})$. Thus u_N converges to u in L_w and

$$\|u - u_N\|_w = O(N^{-r}). \tag{398}$$

If one examines the proof used to establish uniform convergence in Theorem 19.4, then it can be seen that only the estimate (398) is used along with the uniform bounds on $\{\psi_k\}_{k=0}^{\infty}$. Thus under the same conditions as in Theorem 19.4 we find that u_N converges uniformly to u and $\|u - u_N\|_{\infty} = O(N^{-r+\mu+3/2})$.

19.6. Convergence of the Discrete Galerkin Method. In Sections 19.1–19.3 we established the convergence of Galerkin's method assuming that all the integrals $\{\langle K\psi_k, \chi_j\rangle_{1/w}\}$ and $\{\langle f, \chi_j\rangle_{1/w}\}$ were calculated exactly. However, as we noted in Sections 10 and 17 this is rarely possible in practice, and usually some numerical method must be used. Here we examine the convergence of the discrete Galerkin method proposed in Sections 10 and 17 when $\nu = 0$ and 1 and prove uniform convergence if $k(x, t)$ and $f(x)$ are sufficiently differentiable.

The approach taken to analyze convergence in this case is through the use of perturbed projection methods for equations of the form $Hu + Ku = f$, where H is a bounded invertible operator from a Hilbert space X to another Hilbert space Y, and K is compact (Ref. 42). We shall only outline the basic details of the theory; for more information the reader should consult Ref. 42 or Ref. 87.

19.6.1. Perturbed Projection Methods. Let X and Y be Hilbert spaces with inner products $\langle \cdot, \cdot \rangle_X$ and $\langle \cdot, \cdot \rangle_Y$ respectively, and let X_N be a finite-dimensional subspace of X. Assume that $H: X \to Y$ is bounded with a bounded inverse and let

$$Y_N = HX_N = \{y_N \in Y: y = Hx_N, x_N \in X_N\}.$$

Also assume that $P_N: Y \to Y_N$ is a bounded linear projection [this condition can be relaxed (Ref. 88)], let $K: X \to Y$ be compact, and assume that the equation

$$(H + K)u = f \tag{399}$$

has a unique solution $u \in X$ for each $f \in Y$.

To approximate u, we consider the sequence $v_N \in X_N$ which is defined by

$$(H + P_N K + R_N)v_N = P_N f + r_N, \tag{400}$$

where $R_N: X_N \to Y_N$ is a linear operator and $r_N \in Y_N$. Let $Q_N = H^{-1}P_N H$, which is a bounded projection from $X \to X_N$, and denote by $\|R_N\|_N$ the operator norm of R_N induced by the norms on X and Y. Then we have the following theorem (Ref. 88).

Theorem 19.9. Given all the definitions above, assume that $\|f - P_N f\|_Y \to 0$, for all $f \in Y$, $\|R_N\|_N \to 0$, and $\|r_N\|_Y \to 0$. Then $\|K - P_N K\| \to 0$, and for all N sufficiently large, say $N \geq N_0$, the operator $A_N = H + P_N K + R_N$ restricted to X_N has an inverse $A_N^{-1} : Y_N \to X_N$, so that (400) has a unique solution

$$v_N = A_N^{-1}(P_N f + r_N). \tag{401}$$

Moreover, the norms $\|A_N^{-1}\|_N$ are uniformly bounded, and

$$\|u - v_N\|_X = c(\|u - Q_N u\|_X + \|R_N\|_N + \|r_N\|_Y). \tag{402}$$

Proof. See Ref. 88. □

Since $\|u - Q_N u\|_X = \|H^{-1}(Hu - P_N Hu)\|_X \leq \|H^{-1}\| \|Hu - P_N Hu\|_Y$, $\|u - Q_N u\|_X \to 0$, and it follows from (402) that v_N converges to u.

To apply Theorem 19.9 to prove the convergence of the discrete Galerkin method, we need to show that the discrete Galerkin approximation defined in Sections 10 and 17 satisfies an equation of the form in (400) for properly chosen values of R_N and r_N. For conciseness we consider only the case $\nu = 0$, as the details for $\nu = 1$ follow in a similar fashion by using the operator formulation (367) (Ref. 16).

Now define $\tilde{P}_N : L_{1/w} \to \text{span}\{\chi_k\}_{k=0}^N = Y_N$ by

$$\tilde{P}_N u = \sum_{j=0}^{N} \left[\sum_{l=0}^{L(N)} \sigma_l \chi_k(x_l) u(x_l) \right] \chi_j(x), \tag{403}$$

and $\tilde{K} : L_w \to L_{1/w}$ by

$$\tilde{K} u = \sum_{m=0}^{M(N)} w_m k(x, t_m) u(t_m), \tag{404}$$

where the quadrature rules are defined as in Sections 10 and 17.1.1. Then with some tedious algebra, it can be shown that the discrete Galerkin approximation v_N satisfies (Ref. 16)

$$H_0 v_N + \tilde{P}_N \tilde{K} v_N = \tilde{P}_N f. \tag{405}$$

Rewriting (405) as

$$H_0 v_N + P_N K v_N + (\tilde{P}_N \tilde{K} - P_N K) v_N = P_N f + (\tilde{P}_N f - P_N f), \tag{406}$$

(406) is equivalent to

$$H_0 v_N + P_N K v_N + R_N v_N = P_N f + r_N, \tag{407}$$

where

$$R_N = \tilde{P}_N \tilde{K} - P_N K \quad \text{and} \quad r_N = \tilde{P}_N f - P_N f. \tag{408}$$

In Ref. 16 it was shown that

$$R_N v_N = -\sum_{k=0}^{N} b_k \left(\sum_{j=0}^{N} E_{kj} \chi_j \right), \tag{409}$$

and

$$r_N = -\sum_{j=0}^{N} e_j \chi_j, \tag{410}$$

where e_j is the error in approximating $\langle f, \chi_j \rangle_{1/w}$ by the quadrature rule $Q_L(f\chi_j)$ and E_{kj} is the error in evaluating $\langle K\psi_k, \chi_j \rangle_{1/w}$ by the product quadrature rule $Q_M \times Q_L(k\psi_j\chi_k)$ (Ref. 16). From (409)–(410) it follows that (Ref. 16)

$$\|R_N\|_N \leq \left(\sum_{j=0}^{N} \sum_{k=0}^{N} E_{kj}^2 \right)^{1/2} \tag{411}$$

and

$$\|r_N\|_{1/w} = \left(\sum_{j=0}^{N} e_j^2 \right)^{1/2}, \tag{412}$$

so we now need to estimate $\{e_j\}_{j=0}^{N}$ and $\{E_{kj}\}_{(j,k)=0}^{N}$.

Theorem 19.10. Suppose that the quadrature rules Q_L and Q_M have polynomial precision $\geq 2N$, and that the weights $\{\sigma_l\}_{l=0}^{L(N)}$ and $\{w_m\}_{m=0}^{M(N)}$ are positive for all $N \geq 0$. If $f \in C^r$, $r > 0$, then

$$|e_j| = O(N^{-r}), \quad 0 \leq j \leq N, \tag{413}$$

and

$$|E_{kj}| = O(N^{-r}), \quad 0 \leq (j, k) \leq N. \tag{414}$$

Proof. By the definition of e_j,

$$e_j = \int_{-1}^{1} [f(t)\chi_j(t)/w(t)]\, dt - \sum_{l=0}^{L(N)} \sigma_l f(t_l)\chi_j(t_l), \qquad (415)$$

and letting $p_N(t)$ be the polynomial of best uniform approximation to $f(t)$ of degree $\leq N$,

$$e_j = \int_{-1}^{1} [(f - p_N)(t)\chi_j(t)/w(t)]\, dt$$

$$+ \left[\int_{-1}^{1} [p_N(t)\chi_j(t)/w(t)]\, dt - \sum_{l=0}^{L(N)} \sigma_l p_N(t_l)\chi_j(t_l) \right]$$

$$+ \left[\sum_{l=0}^{L(N)} \sigma_l p_N(t_l)\chi_j(t_l) - \sum_{l=0}^{L(N)} \sigma_l f(t_l)\chi_j(t_l) \right], \qquad 0 \leq j \leq N. \quad (416)$$

Since Q_L has precision $\geq 2N$, the first term in brackets is zero, because it is the quadrature error in integrating $p_N(t)\chi_j(t)$, which is a polynomial of degree $\leq 2N$ for $0 \leq j \leq N$. Thus,

$$|e_j| \leq \|f - p_N\|_\infty \left\{ \int_{-1}^{1} [|\chi_j(t)|/w(t)]\, dt + \sum_{l=0}^{L(N)} \sigma_l |\chi_j(t_l)| \right\}. \qquad (417)$$

By the Cauchy–Schwarz inequality for integrals and sums,

$$\int_{-1}^{1} [|\chi_j(t)|/w(t)]\, dt \leq \left[\int_{-1}^{1} dt/w(t) \right]^{1/2} \|\chi_j\|_{1/w}$$

$$= \left[\int_{-1}^{1} dt/w(t) \right]^{1/2}, \qquad (418)$$

and

$$\sum_{l=0}^{L(N)} \sigma_l |\chi_j(t_l)| \leq \left(\sum_{l=0}^{L(N)} \sigma_l \right)^{1/2} \left[\sum_{l=0}^{L(N)} \sigma_l \chi_j^2(t_l) \right]^{1/2}$$

$$= \left[\int_{-1}^{1} dt/w(t) \right]^{1/2} \|\chi_j\|_{1/w} = \left[\int_{-1}^{1} dt/w(t) \right]^{1/2}. \qquad (419)$$

Thus,
$$|e_j| \le c\|f - p_N\|_\infty = O(N^{-r}),$$

where the last estimate follows from Jackson's theorem.

A similar proof shows that the error

$$\left| \int_{-1}^{1} w(t)f(t)\psi_j(t)\, dt - \sum_{m=0}^{M(N)} w_m f(t_m)\psi_j(t_m) \right| = O(N^{-r}), \quad 0 \le j \le N, \tag{420}$$

as well.

To estimate E_{kj} we have

$$E_{kj} = \int_{-1}^{1}\int_{-1}^{1} [w(y)/w(x)]k(x,y)\psi_k(y)\chi_j(x)\, dx\, dy$$

$$- \sum_{l=0}^{L(N)} \sum_{m=0}^{M(N)} \sigma_l w_m k(x_l, y_m)\psi_k(y_m)\chi_j(x_l), \tag{421}$$

and letting

$$g(y) = \int_{-1}^{1} [k(x,y)\chi_j(x)/w(x)]\, dx, \tag{422}$$

$$E_{kj} = \int_{-1}^{1} w(y)g(y)\psi_k(y)\, dy$$

$$- \sum_{m=0}^{M(N)} \left[\sum_{l=0}^{L(N)} \sigma_l \chi_j(x_l) k(x_l, y_m) \right] w_m \psi_k(y_m). \tag{423}$$

Defining $k_m(x) = k(x, y_m)$, the sum in (423) is

$$\sum_{m=0}^{M(N)} w_m \psi_k(y_m) Q_L(k_m \chi_j), \tag{424}$$

and

$$Q_L(k_m \chi_j) = \int_{-1}^{1} [k(x, y_m)\chi_j(x)/w(x)]\, dx - e_L(k_m \chi_j)$$

$$= g(y_m) - e_L(k_m \chi_j), \tag{425}$$

where $e_L(k_m \chi_j)$ is the error in integrating $k_m \chi_j$ by Q_L.

Substituting (425) into (424) gives

$$E_{kj} = \left[\int_{-1}^{1} w(y)g(y)\psi_k(y)\,dy - \sum_{m=0}^{M(N)} w_m\psi_k(y_m)g(y_m) \right]$$

$$+ \sum_{m=0}^{M(N)} w_m\psi_k(y_m)e_L(k_m\chi_j). \tag{426}$$

By the first part of the theorem,

$$|e_L(k_m\chi_j)| = O(N^{-r}), \qquad 0 \le j \le N,$$

uniformly in m, and the first term in (426) is $O(N^{-r})$ as well. Thus

$$|E_{kj}| \le c_1 N^{-r} + c_2 N^{-r} \left(\sum_{m=0}^{M(N)} w_m |\psi_k(y_m)| \right). \tag{427}$$

But the last sum in (427) is bounded by a constant [using the same argument as for (419)] giving $|E_{kj}| = O(N^{-r})$, $0 \le (j, k) \le N$. □

Theorem 19.11. ($\nu = 0$). Let v_N be the discrete Galerkin approximation defined by (130)-(131) or (314)-(315), and assume that Q_L and Q_m satisfy the conditions of Theorem 19.10. Then, if $k(x, t) \in C^r$, and $f(x) \in C^r$, $r > 1$, v_N converges in L_w and

$$\|u - v_N\|_w = O(N^{-r+1}). \tag{428}$$

Proof. That $\|R_N\|_N = O(N^{-r+1})$ and $\|r_N\|_{1/w} = O(N^{-r+1/2})$ follow from (413)-(414) and Theorem 19.10. In Section 19.1.1 it was shown that $\|u - Q_N u\|_w$ is $O(N^{-r})$, so the error estimate (402) gives

$$\|u - v_N\|_w = O(N^{-r}) + O(N^{-r+1}) + O(N^{-r+1/2})$$

$$= O(N^{-r+1}), \qquad r > 1. \qquad \square$$

Corollary 19.3. Under the conditions of the theorem, the discrete Galerkin method converges uniformly to u if $r - \mu + \frac{5}{2} > 0$, and then

$$\|u - v_N\|_\infty = O(N^{-r+\mu+5/2}). \tag{429}$$

Proof. The proof follows in exactly the same manner as going from Theorems 19.2-19.3 to Theorem 19.4, except now the estimate $\|u - u_N\|_w =$

$O(N^{-r})$ is replaced by $\|u - u_N\|_w = O(N^{-r+1})$ in (379), and this leads directly to (429) and the corresponding convergence result. □

Corollary 19.4. If $\nu = 0$, and Q_M and Q_L are the $N + 1$ point Gaussian quadrature rules with respect to $w(t)$ and $1/w(t)$ in Theorem 19.11, then the results of both Theorem 19.11 and Corollary 19.3 are valid.

Proof. As is well known, Gaussian quadrature with $N + 1$ nodes has precision $2N + 1 \geq 2N$, and positive weights, and so the results follow. □

To get similar results for $\nu = 1$ we have to make some minor changes in our arguments for $\nu = 0$. First, we assume that the auxiliary condition $l(u_N)$ is evaluated exactly, so that the mean-square convergence result of Theorem 19.3 is valid. Also (Ref. 16)

$$r_N = -\sum_{j=0}^{N-1} e_j \chi_j \tag{430}$$

and

$$R_N v_N = -\sum_{k=0}^{N} b_k \left(\sum_{j=0}^{N-1} E_{kj} \chi_j \right), \tag{431}$$

where e_j is the error in numerically integrating $\langle f, \chi_j \rangle_{1/w}$ by Q_L, and E_{kj} is the corresponding error for $\langle K\psi_k, \chi_j \rangle_{1/w}$. In this case, since $\deg(\chi_j) \leq N - 1$, $0 \leq j \leq N - 1$, we only need Q_L to have precision $2N - 2$ for Theorem 19.10 to hold. Thus, in this situation, if Q_M is the Gaussian quadrature rule for $w_1(t)$ with $N + 1$ nodes, and Q_L is the Gaussian quadrature rule with N nodes, the results in Theorem 19.11 and Corollary 19.3 are true. □

Last we note that if Gaussian quadrature rules with $N + 1$ nodes are used when $\nu = 0$, and those with $N + 1$ and N nodes are used if $\nu = 1$, then the discrete Galerkin method is mathematically equivalent to collocation (Ref. 1), and so Theorem 19.11 and the following corollaries and comments can be used to give convergence proofs for that method, but under slightly stronger differentiability conditions on $k(x, t)$ and $f(x)$ than in Section 19.5.

Following a similar line of reasoning we can also prove convergence of the Gaussian quadrature method when polynomial interpolation is used to obtain a global solution.

19.7. Convergence of the Gaussian Quadrature Method. Here we will show that the Gaussian quadrature method for $\nu = 0$ or $\nu = 1$ using a polynomial interpolant is equivalent to a perturbed Galerkin method, and so we can obtain a convergence proof of that method from the results of the previous subsection. Again we concentrate on the details for $\nu = 0$.

Lemma 19.3. Suppose $\nu = 0$ in (313), and the inner products are evaluated by using Gaussian quadrature rules with $N + 1$ nodes. Then $\{v_N(t_m)\}_{m=0}^N$ satisfy the quadrature equations (322).

Proof. Because Q_L has precision $2N + 1$, $\langle \chi_k, \chi_j \rangle_{1/w} = \sum_{l=0}^N \sigma_l \chi_k(x_l) \times \chi_j(x_l)$ so that

$$b_j = \sum_{k=0}^N \left[\sum_{l=0}^N \sigma_l \chi_k(x_l) \chi_j(x_l) \right] b_k,$$

and substituting this back into the discrete Galerkin equations (314) gives

$$\sum_{k=0}^N \left[\sum_{l=0}^N \sigma_l \chi_k(x_l) \chi_j(x_l) \right] b_k + \sum_{k=0}^N b_k \left[\sum_{m=0}^N \sum_{l=0}^N w_m \sigma_l k(x_l, t_m) \psi_k(t_m) \chi_j(x_l) \right]$$

$$= \sum_{l=0}^N \sigma_l f(x_l) \chi_j(x_l), \qquad 0 \leq j \leq N. \tag{432}$$

But

$$\sum_{k=0}^N \left[\sum_{l=0}^N \sigma_l \chi_k(x_l) \chi_j(x_l) \right] b_k = \sum_{l=0}^N \sigma_l y_N(x_l) \chi_j(x_l), \tag{433}$$

where

$$y_N(x_l) = \sum_{k=0}^N b_k \chi_k(x_l) = H_0 v_N(x_l), \tag{434}$$

so that (432) simplifies to

$$\sum_{l=0}^N \sigma_l H_0 v_N(x_l) \chi_j(x_l) + \sum_{l=0}^N \sum_{m=0}^N \sigma_l w_m k(x_l, t_m) v_N(t_m) \chi_j(x_l)$$

$$= \sum_{l=0}^N \sigma_m f(x_l) \chi_j(x_l), \qquad j = 0, 1, 2, \ldots, N. \tag{435}$$

Since the matrix $[\chi_j(x_l)]$, $(j, l) = 0, 1, 2, \ldots, N$, is invertible, and $\sigma_l > 0$,

(435) are equivalent to (Ref. 16)

$$H_0 v_N(x_l) + \sum_{m=0}^{N} w_m k(x_l, t_m) v_N(t_m) = f(x_l), \quad l = 0, 1, 2, \ldots, N. \quad (436)$$

Since the Gaussian quadrature rule (181) with $N+1$ nodes is exact for all polynomials of degree $\leq 2N+2$,

$$H_0 v_N(x_l) = \frac{b}{\pi} \sum_{m=0}^{N} \frac{w_m v_N(t_m)}{t_m - x_l}, \quad (437)$$

so that

$$\frac{b}{\pi} \sum_{m=0}^{N} \frac{w_m v_N(t_m)}{t_m - x_l} + \sum_{m=0}^{N} w_m k(x_l, t_m) v_N(t_m) = f(x_l), \quad l = 0, 1, 2, \ldots, N,$$

which are identical to (322). □

A similar calculation using Gaussian quadrature with N nodes to evaluate $\langle f, \chi_j \rangle_{1/w}$ and $\langle \chi_k, \chi_j \rangle_{1/w}$, $(j, k) = 0, 1, \ldots, N-1$, shows that the discrete Galerkin system for $\nu = 1$ is the same as the quadrature system if $l(v_N)$ is evaluated exactly. From these observations we arrive at the following theorem.

Theorem 19.12. Assume that $k(x, t) \in C^r$ and $f(x) \in C^r$, where $r - \mu - \frac{5}{2} > 0$. If $\nu = 0$ or $\nu = 1$, then the Gaussian quadrature method using the Lagrange polynomial interpolant v_N with $l(v_N)$ computed exactly if $\nu = 1$ converges uniformly, and

$$\|u - v_N\|_\infty = O(N^{-r+\mu+5/2}). \quad (438)$$

Proof. The proof follows from the convergence results of the previous subsection, Lemma 19.3, and the comments following it. □

It is interesting to compare the conditions for convergence obtained for the polynomial interpolant v_N with those for the Nyström interpolant u_N when $\nu = 1$ and (296) is an equation of the first kind. In that case $\mu = -\frac{1}{2}$ so that (438) gives $\|u - v_N\|_\infty = O(N^{-r+2})$, $r > 2$. Thus, in our proof, the data are required to have Hölder continuous *second derivatives*, while Gerasoulis in Ref. 68 requires only C^1 data for the convergence of the Nyström interpolant. On the other hand, Junghanns gets $O(N^{-r+1/2})$ convergence for the polynomial interpolant, which requires only that the data be Hölder continuous of order $\mu > \frac{1}{2}$ (Ref. 41), and this seems to be the best result available in this case (Elliott in Ref. 15 has given a similar result).

It is not clear whether the conditions for convergence of the discrete Galerkin method can be weakened. However, we have chosen it as a way of presenting the convergence proofs for the standard polynomial approximation methods primarily because it unifies the presentation, and we find it somewhat less complicated analytically than some of the other proofs in the literature (Refs. 15, 41, 68).

20. Conclusions

We have given an overview of many of the polynomial-based algorithms that have been developed for solving Cauchy singular equations with constant coefficients. As one can see, the basic theory is essentially complete for this class of problems, but many practical problems of implementation and application remain. Extensions to equations with nonconstant coefficients have also been made, and this will be taken up by Elliott in the next chapter, an area where many important advances can be expected to occur in the near future. We hope that the reader will find this and subsequent chapters a useful reference to what has become a substantial body of knowledge.

References

1. GOLBERG, M. A., *Numerical Solution of Cauchy Singular Integral Equations with Constant Coefficients*, Journal of Integral Equations, Vol. 9, pp. 127–151, 1985.
2. GERASOULIS, A., AND VICHNEVETSKY, R., Editors, *Numerical Solution of Singular Integral Equations*, IMACS, 1984.
3. JUNGHANNS, V. P., AND SILBERMANN, B., *Zur Theorie der Naherungsverfahrens für Singulare Integralgleichungen auf Intervallen*, Mathematische Nachrichten, Vol. 103, pp. 199–244, 1981.
4. TUCK, E. O., *Application and Solution of Cauchy Singular Equations*, Numerical Solution of Integral Equations, Edited by R. S. Anderssen, F. R. de Hoog, and M. A. Lukas, Sijthoff and Noordhoff, Leyden, Holland, 1980.
5. ELLIOTT, D., *The Classical Collocation Method for Singular Integral Equations with Cauchy Kernels*, SIAM Journal on Numerical Analysis, Vol. 19, pp. 816–832, 1982.
6. PRÖSSDORF, S., AND ELSCHNER, J., *Finite Element Methods for Singular Integral Equations on an Interval*, Engineering Analysis, Vol. 1, pp. 83–87, 1984.
7. GERASOULIS, A., *Singular Integral Equations: Direct and Iterative Methods*, Numerical Solution of Singular Integral Equations, Edited by A. Gerasoulis and R. Vichnevetsky, IMACS, 1984.

8. JEN, E., AND SRIVASTAV, R. P., *Cubic Splines and Approximate Solution of Singular Integral Equations*, Mathematics of Computation, Vol. 37, pp. 417-423, 1981.
9. TRICOMI, F. G., *Integral Equations*, Interscience, New York, New York, 1957.
10. MUSKHELISHVILI, N. I., *Singular Integral Equations*, Noordhoff, Groningen, Holland, 1953.
11. PETERS, A. J., *A Note on the Integral Equation of the First Kind with a Cauchy Kernel*, Communications on Pure and Applied Mathematics, Vol. 16, pp. 57-61, 1963.
12. LINZ, P., *Analytical and Numerical Methods for Volterra Equations*, Society for Industrial and Applied Mathematics, Philadelphia, Pennsylvania, 1985.
13. FROMME, J. A., AND GOLBERG, M. A., *Numerical Solution of a Class of Integral Equations Arising in Two Dimensional Aerodynamics*, Solution Methods for Integral Equations: Theory and Applications, Edited by M. A. Golberg, Plenum Press, New York, New York, 1979.
14. FROMME, J. A., AND GOLBERG, M. A., *Unsteady Two Dimensional Airloads Acting on Oscillating Thin Airfoils in Subsonic Vertilated Wind Tunnels*, NASA Contractor's Report, No. 2967, Washington DC, 1978.
15. ELLIOTT, D., *Convergence Theorems for Singular Integral Equations*, Chapter 6, this volume.
16. GOLBERG, M. A., *The Perturbed Galerkin Method for Cauchy Singular Integral Equations with Constant Coefficients*, Applied Mathematics and Computation, Vol. 26, pp. 1-33, 1988.
17. JEN, E., AND SRIVASTAV, R. P., *Solving Singular Integral Equations Using Gaussian Quadrature and Overdetermined Systems*, Computers and Mathematics with Applications, Vol. 9, pp. 625-633, 1983.
18. GOLBERG, M. A., Editor, *Solution Methods for Integral Equations: Theory and Applications*, Plenum Press, New York, New York, 1979.
19. ATKINSON, K. E., *A Survey of Numerical Methods for the Solution of Fredholm Integral Equations of the Second Kind*, Society for Industrial and Applied Mathematics, Philadelphia, Pennsylvania, 1976.
20. SESKO, M. A., *On the Numerical Solution of Singular Integral Equations*, Differential Equations, Vol. 13, pp. 1039-1045, 1977.
21. CHATELIN, F., AND GUESSOUS, N., *Iterative Refinement for the Solution of Cauchy Singular Integral Equations*, Numerical Solution of Singular Integral Equations, Edited by A. Gerasoulis and R. Vichnevetsky, IMACS, 1984.
22. IOAKIMIDIS, N. I., *Further Convergence Results for the Weighted Galerkin Method of Numerical Solution of Cauchy-Type Singular Integral Equations*, Mathematics of Computation, Vol. 41, pp. 79-85, 1983.
23. IOAKIMIDIS, N. I., *A Strange Convergence Property of the Lobatto-Chebyshev Method for Numerical Determination of Stress Intensity Factors*, Computers and Structures, Vol. 17, pp. 206-209, 1983.
24. IOAKIMIDIS, N. I., *The Numerical Solution of Crack Problems in Plane Elasticity in the Case of Loading Discontinuities*, Engineering Fracture Mechanics, Vol. 13, pp. 709-716, 1980.

25. IOAKIMIDIS, N. I., AND THEOCARIS, P. S., *A Remark on the Numerical Evaluation of Stress Intensity Factors by the Method of Singular Integral Equations*, Inernational Journal of Numerical Methods in Engineering, Vol. 14, pp. 1710-1714, 1979.
26. FROMME, J. A., GOLBERG, M. A., AND WERTH, J., *Two-Dimensional Aerodynamic Interference Effects on Oscillating Airfoils with Flaps in Ventilated Subsonic Wind Tunnels*, NASA Contractor's Report, No. 2140, Washington, DC, 1979.
27. FAATH, J., *Diskrete Galerkin-änliche Verfahren zur Lösung von Operatorgleichungen*, University of Kaiserslautern, PhD Thesis, 1986.
28. ELLIOTT, D., *Orthogonal Polynomials Associated with Singular Integral Equations Having a Cauchy Kernel*, SIAM Journal on Numerical Analysis, Vol. 13, pp. 1041-1052, 1982.
29. KRENK, S., *Polynomial Solutions to Singular Integral Equations with Applications to Elasticity Theory*, RISØ National Laboratory, Roskilde, Denmark, 1981.
30. WELSTEAD, S., *Orthogonal Polynomials Applied to the Solution of Singular Integral Equations*, Purdue University, PhD Thesis, 1982.
31. TRIOMI, F. G., *On the Finite Hilbert Transformation*, Quarterly Journal of Mathematics, Vol. 2, 1951.
32. FOX, L., AND PARKER, I., *Chebyshev Polynomials in Numerical Analysis*, Oxford Press, London, England, 1968.
33. ELLIOTT, D., AND DOW, M. L., *The Numerical Solution of Singular Integral Equations*, SIAM Journal on Numerical Analysis, Vol. 16, pp. 115-134, 1979.
34. KARPENKO, L. N., *Approximate Solution of Singular Integral Equations by Means of Jacobi Polynomials*, Journal of Applied Mathematics and Mechanics, Vol. 30, pp. 668-675, 1966.
35. ERDOGAN, F., *Approximate Solution of Systems of Singular Integral Equations*, SIAM Journal on Applied Mathematics, Vol. 17, pp. 1041-1059, 1969.
36. ERDOGAN, F., GUPTA, G. D., AND COOK, T. S., *Numerical Solution of Singular Integral Equations*, Methods of Analysis and Solution of Crack Problems, Edited by G. C. Sih, Noordhoff, Leyden, Holland, 1973.
37. LINZ, P., *An Analysis of a Method for Solving Singular Integral Equations*, BIT, Vol. 17, pp. 429-437, 1977.
38. IOAKIMIDIS, N. I., *On the Weighted Galerkin Method of Numerical Analysis of Cauchy Type Singular Integral Equations*, SIAM Journal on Numerical Analysis, Vol. 18, pp. 1120-1127, 1981.
39. VENTURINO, E., *The Convergence of the Galerkin Method for Singular Integral Equations of the Second Kind*, Numerical Solution of Singular Integral Equations, Edited by A. Gerasoulis and R. Vichnevetsky, IMACS, 1984.
40. ELLIOTT, D., *A Galerkin-Petrov Method for Singular Integral Equations*, Journal of the Australian Mathematical Society, Series B, Vol. 25, pp. 261-275, 1983.
41. JUNGHANNS, P., *Uniform Convergence of Approximate Methods for Cauchy Type Singular Equations over $(-1, 1)$*, Wissenschaftliche Zeitschrift Technische Hocschule, Karl-Marx Stadt, Vol. 26, pp. 250-256, 1984.
42. MIEL, G., *On the Galerkin and Collocation Methods for a Cauchy Singular Integral Equation*, SIAM Journal on Numerical Analysis, Vol. 23, pp. 135-143, 1986.

43. GOLBERG, M. A., *A Note on a Superconvergence Result for the Generalized Airfoil Equation*, Applied Mathematics and Computation, to appear.
44. THOMAS, K. S., *Galerkin Methods for Singular Integral Equations*, Mathematics of Computation, Vol. 36, pp. 193–207, 1981.
45. LINZ, P., *Stability Analysis for the Numerical Solution of Singular Integral Equations*, Numerical Solution of Singular Integral Equations, Edited by A. Gerasoulis and R. Vichnevetsky, IMACS, 1984.
46. FROMME, J. A., AND GOLBERG, M. A., *Numerical Solution of a Class of Integral Equations Arising in Two Dimensional Aerodynamics—The Problem of Flaps*, Solution Methods for Integral Equations: Theory and Applications, Edited by M. A. Golberg, Plenum Press, New York, New York, 1979.
47. GOLBERG, M. A., LEA, M., AND MIEL, G., *A Superconvergence Result for the Generalized Airfoil Equation with Application to the Flap Problem*, Journal of Integral Equations, Vol. 5, pp. 175–185, 1983.
48. GOLBERG, M. A., *Galerkin's Method for CSIES of the First Kind*, Numerical Solution of Singular Integral Equations, Edited by A. Gerasoulis and R. Vichnevetsky, IMACS, 1984.
49. MCKENNA, M., *The Numerical Evaluation of a Class of Logarithmically Singular Integral Transforms*, University of Nevada, Las Vegas, Masters Thesis, 1987.
50. FROMME, J. A., AND GOLBERG, M. A., *Reformulation of Possio's Kernel with Application to Unsteady Wind Tunnel Interference*, AIAA Journal, Vol. 18, pp. 417–426, 1980.
51. ELLIOTT, D., AND PAGET, D. F., *Product Integration Rules and Their Convergence*, BIT, Vol. 16, pp. 32–40, 1976.
52. ATKINSON, K., *An Introduction to Numerical Analysis*, John Wiley and Sons, New York, New York, 1978.
53. CUMINATO, J. A., *On the Uniform Convergence of a Collocation Method for a Class of Singular Integral Equations*, BIT, Vol. 27, pp. 190–202, 1987.
54. ERDOGAN, F., AND GUPTA, G. D., *On the Numerical Solution of Singular Integral Equations*, Quarterly of Applied Mathematics, Vol. 24, pp. 525–534, 1974.
55. KALANDIYA, A. I., *Approximate Solution of a Class of Singular Integral Equations*, Doklady Akademi Nauk SSSR, Vol. 125, pp. 715–718, 1959.
56. MONEGATO, G., *The Numerical Evaluation of One-Dimensional Cauchy Principal Value Integrals*, Computing, Vol. 29, pp. 337–354, 1982.
57. HUNTER, D. B., *Some Gauss Type Formulae for the Evaluation of Cauchy Principal Value Integrals*, Numerische Mathematik, Vol. 11, pp. 419–424, 1972.
58. CHAWLA, M. M., AND RAMAKRISHNAN, T. R., *Modified Gauss Jacobi Quadrature Formulas for the Numerical Evaluation of Cauchy Type Singular Integrals*, BIT, Vol. 14, pp. 14–21, 1974.
59. KRENK, S., *On Quadrature Formulas for Singular Integral Equations of the First and Second Kinds*, Quarterly of Applied Mathematics, Vol. 33, pp. 225–233, 1972.
60. IOAKIMIDIS, N. I., AND THEOCARIS, P. S., *On the Numerical Evaluation of Cauchy Principal Value Integrals*, Revue Roumaine des Sciences Techniques, Série de Mécanique Appliquée, Vol. 22, pp. 803–810, 1977.

61. IOAKIMIDIS, N. I., AND THEOCARIS, P. S., *On the Numerical Solution of a Class of Singular Integral Equations*, Journal of Mathematical and Physical Sciences, Vol. 11, pp. 219-235, 1977.
62. ELLIOTT, D., AND PAGET, D. F., *On the Convergence of a Quadrature Rule for Evaluating Certain Cauchy Principal Value Integrals*, Numerische Mathematik, Vol. 23, pp. 311-319, 1979.
63. THEOCARIS, P. S., AND IOAKIMIDIS, N. I., *Numerical Integration Methods for the Solution of Singular Integral Equations*, Quarterly of Applied Mathematics, Vol. 35, pp. 173-183, 1977.
64. THEOCARIS, P. S., AND IOAKIMIDIS, N. I., *Numerical Solution of Cauchy Type Singular Equations*, Transactions of the Academy of Athens, Vol. 40, pp. 1-39, 1977.
65. SZËGO, G., *Orthogonal Polynomials*, American Mathematical Society, Providence, Rhode Island, 1975.
66. IOAKIMIDIS, N. I., *Some Remarks on the Numerical Solution of Singular Integral Equations with Index Equal to* -1, Computers and Structures, Vol. 14, pp. 403-407, 1981.
67. IOAKIMIDIS, N. I., *On the Natural Interpolation Formula for Cauchy Type Singular Integral Equations of the First Kind*, Computing, Vol. 26, pp. 73-77, 1981.
68. GERASOULIS, A., *Singular Integral Equations—The Convergence of the Nyström Interpolant of the Gauss-Chebyshev Method*, BIT, Vol. 22, pp. 200-210, 1982.
69. DAVIS, P. J., AND RABINOWITZ, P., *Methods of Numerical Integration*, Second Edition, Academic Press, New York, New York, 1984.
70. IOAKIMIDIS, N. I., *A Superconvergence Result for the Numerical Determination of Stress Intensity Factors*, International Journal for Numerical Methods in Engineering, Vol. 21, pp. 1391-1401, 1985.
71. IOAKIMIDIS, N. I., *A Remark on the Application of Closed and Semiclosed Quadrature Rules to the Direct Solution of Singular Integral Equations*, Journal of Computational Physics, Vol. 42, pp. 396-402, 1981.
72. BAKER, C. T. H., *The Numerical Treatment of Integral Equations*, Oxford University Press, Oxford, England, 1977.
73. COHEN, H., *Accurate Numerical Solution of Integral Equations Containing Poles*, Journal of Computational Physics, Vol. 26, pp. 257-276, 1978.
74. IOAKIMIDIS, N. I., *On Kalandiya's Method for the Numerical Solution of Singular Integral Equations*, International Journal of Computer Mathematics, Vol. 13, pp. 287-299, 1983.
75. MAJUMDAR, R. S., *The Numerical Solution of Singular Integral Equations*, Journal of Mathematical and Physical Sciences, Vol. 8, pp. 517-533, 1975.
76. BOSE, S. C., AND KUNDU, N. C., *A Note on the Numerical Solution of Singular Fredholm Linear Integral Equations*, Zeitschrift für Angewandte Mathematik und Mechanik, Vol. 64, pp. 372-374, 1984.
77. HASHMI, S. M., AND DELVES, L. M., *The Fast Galerkin Method for the Solution of Cauchy Singular Integral Equations*, Treatment of Integral Equations by Numerical Methods, Edited by C. T. H. Baker and G. F. Miller, Academic Press, London, England, 1982.

78. Moss, W., AND CHRISTENSEN, M. J., *Scattering and Heat Transfer by a Strip*, Journal of Integral Equations, Vol. 4, pp. 299–317, 1980.
79. GUSEINOV, E. A., *Justification of the Projection Method for an Integral Equation of the First Kind with a Logarithmic Singularity*, Akademia Nauk Azerbaidzhan SSR Doklady, Vol. 38, pp. 10–12, 1982.
80. MACASKILL, L. C., *Numerical Solution of Some Fluid Flow Problems by Boundary Integral Equation Techniques*, University of Adelaide, PhD Thesis, 1977.
81. STROUD, A., AND SECREST, D., *Gaussian Quadrature Formulas*, Prentice-Hall, Englewood Cliffs, New Jersey, 1966.
82. PETERS, A. S., *Abel's Equation and the Cauchy Integral Equation of the Second Kind*, Communications on Pure and Applied Mathematics, Vol. 21, pp. 51–65, 1968.
83. KRENK, S., *On Quadrature Formulas of Closed Type for Solution of Singular Integral Equations*, Journal of the Institute of Mathematics and Applications, Vol. 22, pp. 99–107, 1978.
84. CHAWLA, M. M., AND KUMAR, S., *Chebyshev Series Approximate Solutions of Fredholm Integral Equations with Cauchy Type Singular Kernels*, Journal of Mathematical and Physical Sciences, Vol. 12, pp. 149–158, 1978.
85. DUSKOV, P. N., AND GABDULHAEV, B. G., *Direct Methods of Solution of Singular Integral Equations of First Kind*, Izvestiya Vissihikh Uchebnykzavendii Matematika, Vol. 7, pp. 112–124, 1973.
86. DZISKARIANI, A. V., *The Solution of Singular Integral Equations by Approximate Projection Methods*, Zhurnal Vychislitel'noi Matematiki i Matematicheskoi Fiziki, Vol. 21, pp. 355–362, 1981.
87. KRASNOSELSKII, M. A., et al., *Approximate Solution of Operators Equations*, Walters–Noordhoff, Groningen, Holland, 1972.
88. MIEL, G., *Perturbed Projection Methods for Split Equations of the First Kind*, Technical Report No. 80, Department of Mathematics, Arizona State University, 1983.

Additional Bibliography

89. ABD-ELAL, L. F., *A Galerkin Algorithm for Solving Cauchy-Type Singular Equations*, Indian Journal of Pure and Applied Mathematics, Vol. 11, pp. 699–709, 1980.
90. BELOTSERKOVSKII, S. M., AND LIFANOV, I. K., *Numerical Methods in Singular Integral Equations, Elasticity Theory, and Aerodynamics*, Nauka, Moscow, 1985 (in Russian).
91. CHAKRABARTI, A., *Solution of a Singular Integral Equation*, Journal of Engineering Mathematics, Vol. 15, pp. 201–210, 1981.
92. DANILOVIC, V. P., *Application of Polynomial Splines to Approximate the Solution of Nonlinear Singular Integral Equations*, Soviet Mathematics, Vol. 25, pp. 15–19, 1981.

93. DANG, D. Q., AND NORRIE, D. H., *A Finite Element Method for the Solution of Singular Integral Equations*, Computers and Mathematics with Applications, Vol. 4, pp. 219-224, 1978.
94. DIETMAR, D., AND JUNGHANNS, P., *Direct Multiple Grid Methods for Cauchy-Type Singular Integral Equations*, Wissenschaftliche Zeitschrift, Technische Hochschule Karl Marx Stadt, Vol. 29, pp. 180-186, 1987.
95. DRISCOLL, M. A., AND SRIVASTAV, R. P., *The Numerical Solution of Singular Integral Equations via Rational Function Approximations*, Numerical Solution of Singular Integral Equations, Edited by A. Gerasoulis and R. Vichnevetsky, IMACS, pp. 108-112, 1984.
96. DZHISKARIANI, A. V., *On the Solution of Singular Integral Equations by Collocation Methods*, Zhurnal Vychislitel'noi Matematiki i Matematicheskoi Fiziki, Vol. 21, pp. 355-362, 1981.
97. ELSCHNER, J., *Galerkin Methods with Splines for Singular Integral Equations Over (0, 1)*, Numerische Mathematik, Vol. 43, pp. 265-281, 1984.
98. ELSCHNER, J., *On Spline Approximations for Singular Integral Equations on an Interval*, Preprint P-04/87, Karl Weirstrass Institute for Mathematics, DDR, Berlin, 1987.
99. FABRIKANT, V., HOA, S. V., AND SANKAR, J. S., *On the Approximate Solution of Singular Integral Equations*, Computer Methods in Applied Mechanics and Engineering, Vol. 29, pp. 19-33, 1981.
100. FROMME, J. A., AND GOLBERG, M. A., *Aerodynamic Interference Effects on Oscillating Airfoils with Controls in Ventilated Wind Tunnels*, AIAA Journal, Vol. 18, pp. 951-957, 1980.
101. FROMME, J. A., AND GOLBERG, M. A., *Convergence and Stability of a Collocation Method for the Generalized Airfoil Equation*, Applied Mathematics and Computation, Vol. 8, pp. 281-292, 1981.
102. GABDULHAEV, B. G., *Direct Methods for the Solution of the Equation of Wing Theory*, Izvestiya Vissihikh Uchebnykzavendii Matematika, Vol. 18, pp. 29-44, 1974.
103. GAUTESAN, A. K., *On the Solution to a Cauchy Principal Value Integral Equation which Arises in Fracture Mechanics*, SIAM Journal of Applied Mathematics, Voll. 47, pp. 109-116, 1987.
104. GERA, A. E., *Singular Integral Equations with a Cauchy Kernel*, Journal of Computers and Applied Mathematics, Vol. 14, pp. 311-318, 1986.
105. GERASOULIS, A., *On The Existence of Approximate Solutions for Singular Integral Equations Discretized by Gauss-Chebyshev Quadrature*, BIT, Vol. 21, pp. 377-380, 1981.
106. GERASOULIS, A., *The Use of Piecewise Quadratic Polynomials for the Solution of Singular Integral Equations*, Computers and Mathematics with Applications, Vol. 8, pp. 15-22, 1982.
107. GERASOULIS, A., *On the Solvability of Singular Integral Equations via Gauss-Jacobi Quadrature*, International Journal of Computer Mathematics, Vol. 12, pp. 59-75, 1983.
108. GERASOULIS, A., *Singular Integral Equations—The Convergence of the Gauss-Jacobi Quadrature Method*, International Journal of Computer Mathematics, Vol. 15, pp. 143-161, 1984.

109. GERASOULIS, A., AND SRIVASTAV, R. P., *A Method for the Numerical Solution of Singular Integral Equations with a Principal Value Integral*, International Journal of Engineering Science, Vol. 19, pp. 1293-1298, 1981.
110. GERASOULIS, A., AND SRIVASTAV, R. P., *The Stability of the Gauss-Chebyshev Method for Cauchy Singular Integral Equations*, Computer Mathematics and Applications, Vol. 14, pp. 81-90, 1987.
111. GOLBERG, M. A., *The Convergence of a Collocation Method for a Class of Cauchy Singular Integral Equations*, Journal of Mathematical Analysis and Applications, Vol. 100, pp. 500-512, 1984.
112. GOLBERG, M. A., *Galerkin's Method for a Class of Operator Equations with Non-Negative Index*, Journal of Mathematical Analysis and Applications, Vol. 91, pp. 394-409, 1983.
113. GOLBERG, M. A., *Discrete Projection Methods for Cauchy Singular Integral Equations with Constant Coefficients*, to appear.
114. GOLBERG, M. A., AND FROMME, J. A., *On the L_2 Convergence of Collocation for the Generalized Airfoil Equation*, Journal of Mathematical Analysis and Applications, Vol. 71, pp. 271-286, 1979.
115. HAFTMANN, R., *Quadrature Formulas of Gauss Type for Singular Integrals and Their Application to the Solution of Singular Integral Equations*, Wissenshaftlichte Informationen, Vol. 17, Technische Hochshule Karl Marx Stadt, 1980.
116. IOAKIMIDIS, N. I., *Application of Interpolation Formulas to the Numerical Solution of Singular Integral Equations*, Technical Report, University of Patras, 1979.
117. IOAKIMIDIS, N. I., *A Remark on the Application of Interpolatory Quadrature Rules to the Numerical Solution of Singular Integral Equations*, Technical Report, University of Patras, 1981.
118. IOAKIMIDIS, N. I., *A Modification of the Quadrature Method for the Direct Solution of Singular Integral Equations*, Technical Report, University of Patras, 1981.
119. IOAKIMIDIS, N. I., *Application of the Method of Singular Integral Equations to Elasticity Problems with Concentrated Loads*, Acta Mechanica, Vol. 40, pp. 159-168, 1981.
120. IOAKIMIDIS, N. I., *Application of the Gauss and Radau-Laguerre Quadrature Rules to the Numerical Solution of Cauchy Type Singular Integral Equations*, Computers and Structures, Vol. 14, pp. 63-70, 1981.
121. IOAKIMIDIS, N. I., *Three Iterative Methods for the Numerical Solution of Singular Integral Equations Appearing in Crack and Other Elasticity Problems*, Acta Mechanica, Vol. 39, pp. 117-125, 1981.
122. IOAKIMIDIS, N. I., *A Method for the Numerical Solution of Singular Integral Equations with Logarithmic Singularities*, International Journal of Computer Mathematics, Vol. 9, pp. 363-372, 1981.
123. IOAKIMIDIS, N. I., *An Iterative Algorithm for the Numerical Solution of Singular Integral Equations*, Journal of Computational Physics, Vol. 43, pp. 164-176, 1981.
124. IOAKIMIDIS, N. I., *The Successive Approximations Method for the Airfoil Equation*, Technical Report, University of Patras, 1981.

125. IOAKIMIDIS, N. I., *A Generalization of the Concept of Cauchy-Type Principle Value Integrals for Plane Elasticity Problems*, ZAMM, Vol. 62, pp. 697–699, 1982.
126. IOAKIMIDIS, N. I., *On the Validity of the Singular Integral Equations of Elasticity Problems at Points of Loading Discontinuities*, Technical Report, University of Patras, 1982.
127. IOAKIMIDIS, N. I., *A Natural Interpolation Formula for Cauchy-Type Singular Integral Equations With Generalized Kernels*, Journal of Computational Physics, Vol. 48, pp. 117–126, 1982.
128. IOAKIMIDIS, N. I., *A Modification of the Generalized Airfoil Equation and the Corresponding Numerical Methods*, Technical Report, University of Patras, 1982.
129. IOAKIMIDIS, N. I., *Bounds for the Dislocation Densities and the Stress Intensity Factors in Elastic Crack Problems*, International Journal of Fracture, Vol. 20, pp. 133–145, 1982.
130. IOAKIMIDIS, N. I., *Two Methods for the Numerical Solution of Bueckner's Singular Integral Equation for Plane Elasticity Crack Problems*, Computer Methods in Applied Mechanics and Engineering, Vol. 31, pp. 169–177, 1982.
131. IOAKIMIDIS, N. I., *On the Quadrature Methods for the Numerical Solution of Singular Integral Equations*, Journal of Computational and Applied Mathematics, Vol. 8, pp. 81–86, 1982.
132. IOAKIMIDIS, N. I., *Simple Bounds for the Stress Intensity Factors by the Method of Singular Integral Equations*, Engineering Fracture Mechanics, Vol. 18, pp. 1191–1198, 1983.
133. IOAKIMIDIS, N. I., *A Natural Interpolation Formula for the Numerical Solution of Singular Integral Equations With Hilbert Kernel*, BIT, Vol. 23, pp. 92–104, 1983.
134. IOAKIMIDIS, N. I., *On the Numerical Solution of Cauchy Type Integral Equations by the Collocation Method*, Applied Mathematics and Computation, Vol. 12, pp. 49–60, 1983.
135. IOAKIMIDIS, N. I., *Numerical Comparison of Two Regularization Methods for Singular Integral Equations*, Numerical Solution of Singular Integral Equations, Edited by A. Gerasoulis and R. Vichnevetsky, IMACS, pp. 51–53, 1984.
136. IOAKIMIDIS, N. I., *Modified Algorithm for the Numerical Solution of Singular Integral Equations with Index One*, International Journal of Computer Mathematics, Vol. 15, pp. 65–75, 1984.
137. IOAKIMIDIS, N. I., AND THEOCARIS, P. S., *The Gauss-Hermite Numerical Integration Method for the Solution of the Plane Elastic Problem of Semi-Infinite Periodic Cracks*, International Journal of Engineering Science, Vol. 15, pp. 277–280, 1977.
138. IOAKIMIDIS, N. I., AND THEOCARIS, P. S., *Numerical Solution of Cauchy Type Singular Integral Equations by Use of the Lobatto-Jacobi Integration Rule*, Aplikacy Matematiky, Vol. 23, pp. 439–453, 1978.
139. IOAKIMIDIS, N. I., AND THEOCARIS, P. S., *Numerical Determination of a Class of Generalized Stress Intensity Factors*, International Journal for Numerical Methods in Engineering, Vol. 14, pp. 949–959, 1979.

140. IOAKIMIDIS, N. I., AND THEOCARIS, P. S., *On the Numerical Solution of Singular Integro-differential Equations*, Quarterly of Applied Mathematics, Vol. 37, pp. 325-331, 1979.
141. IOAKIMIDIS, N. I., AND THEOCARIS, P. S., *On the Convergence of Two Direct Methods for the Solution of Cauchy Type Singular Integral Equations of the First Kind*, BIT, Vol. 20, pp. 83-87, 1980.
142. IOAKIMIDIS, N. I., AND THEOCARIS, P. S., *A Comparison Between the Direct and Classical Numerical Methods for the Solution of Cauchy Type Singular Integral Equations*, SIAM Journal on Numerical Analysis, Vol. 17, pp. 115-118, 1980.
143. JUNGHANNS, P., *Collocation and Quadrature Methods for the Approximate Solution of a Class of Singular Integral Equations With One Fixed Singularity*, Wissenschaftliche Zeitschrift, Technische Hochschule Karl Marx Stadt, Vol. 24, pp. 295-303, 1982.
144. JUNGHANNS, P., *Some Remarks on the Zero Distribution of Pairs of Polynomials Associated with Singular Integral Equations: A Convergence Theorem for the Quadrature Method*, Wissenschaftliche Zeitschrift, Technische Hochschule Karl Marx Stadt, Vol. 27, pp. 88-93, 1985.
145. JUNGHANNS, P., *Effective Solution of Systems of Algebraic Equations Occurring in the Approximate Solution of Singular Integral Equations by Means of the Method of Quadrature Formulae*, Wissenschaftliche Zeitschrift, Technische Hochschule, Karl Marx Stadt, Vol. 27, pp. 94-96, 1985.
146. KAS'JANOV, V. I., *Some Approximate Solutions of Linear Singular Equations*, Soviet Mathematics, Vol. 23, pp. 84-89, 1979.
147. KHAPAEV, M. M., JR., *Some Methods of Regularization and Numerical Solution of Integral Equations of the First Kind*, Izvestiya VUZ Matematika, Vol. 27, No. 7, pp. 81-85, 1983.
148. KRENK, S., *On the Use of the Interpolation Polynomial for Solutions of Singular Integral Equations*, Quarterly of Applied Mathematics, Vol. 32, pp. 479-484, 1975.
149. KRENK, S., *Numerical Quadrature of Periodic Singular Integral Equations*, Journal of the Institute of Mathematics and Applications, Vol. 22, pp. 99-107, 1978.
150. LIFANOV, I. K., AND POLONSKI, I. E., *Proof of the Method of Discrete Vortices for Solving Singular Integral Equations*, Journal of Applied Mathematics and Mechanics, Vol. 39, pp. 713-718, 1974.
151. MAJUMDAR, S. R., *Perturbation Analysis of a Certain Singular Integral Equation*, Journal of Mathematical and Physical Sciences, Vol. 9, pp. 517-533, 1975.
152. MIEL, G., *Rates of Convergence and Superconvergence of Galerkin's Method for the Generalized Airfoil Equation*, Numerical Solution of Singular Integral Equations, Edited by A. Gerasoulis and R. Vichnevetsky, IMACS, pp. 62-64, 1984.
153. MILLER, G. R., AND KEER, L. M., *A Numerical Technique for the Solution of Singular Integral Equations of the Second Kind*, Quarterly of Applied Mathematics, Vol. 42, pp. 455-565, 1985.

154. Moss, W. F., *The Two Dimensional Oscillating Airfoil: A New Implementation of Galerkin's Method*, SIAM Journal on Numerical Analysis, Vol. 20, pp. 391-399, 1983.
155. Moss, W. F., *A New Implementation of the Galerkin Method for a Class of Singular and Cauchy Singular Integral Equations*, Numerical Solution of Cauchy Singular Integral Equations, Edited by A. Gerasoulis and A. Vichnevetsky, IMACS, pp. 65-67, 1984.
156. Napolitano, M., *Fourth Order Accurate Direct and Inverse Cauchy Integral Solver*, Computers and Fluids, Vol. 8, pp. 435-441, 1980.
157. Nied, H. F., *Numerical Solution of a Coupled System of Singular Integral Equations with Fracture Mechanics Applications*, Numerical Solution of Integral Equations, Edited by A. Gerasoulis and R. Vichnevetsky, IMACS, pp. 73-79, 1984.
158. Niessner, H., *Significance of Kernel Singularities for the Numerical Solution of Fredholm Integral Equations*, Boundary Elements, Vol. 9, Edited by C. A. Brebbia and W. Wendland, Springer-Verlag, New York, New York, 1987.
159. Parihar, S. K., *Piecewise Cubic Interpolatory Polynomials and Approximate Solution of Singular Integral Equations*, Computer Mathematics and Applications, Vol. 12, pp. 1201-1215, 1986.
160. Prössdorf, S., *On Approximation Methods for the Solution of One-Dimensional Singular Integral Equations*, Applicable Analysis, Vol. 7, pp. 259-270, 1977.
161. Prössdorf, S., and Rathsfeld, A., *On Spline Galerkin Methods for Singular Integral Equations with Piecewise Continuous Coefficients*, Numerische Mathematik, Vol. 48, pp. 99-118, 1986.
162. Prössdorf, S., and Rathsfeld, A., *Quadrature Methods and Spline Approximations for Singular Integral Equations*, Boundary Elements, Vol. 9, Edited by C. A. Brebbia and W. Wendland, Springer-Verlag, New York, New York, 1987.
163. Prössdorf, S., and Schmidt, G., *Notwendige und Hinreichende Bedlingungen für die Konvergenz des Kollokationnerfahrens bei Singularen Integralgleichungen*, Mathematische Nachrichten, Vol. 89, pp. 203-215, 1974.
164. Rathsfeld, A., *Quadrarformeluerfahren fur Eindimensionale Singulare Integralgleichungen*, Seminar Analysis, Operator Equations and Numerical Analysis, Karl Weirstrass Institute for Mathematics, pp. 147-186, Berlin, Germany, 1986.
165. Razali, M. R., *The Treatment of Seepage Problems*, Numerical Solution of Singular Integral Equations, Edited by A. Gerasoulis and R. Vichnevetsky, IMACS, pp. 91-93, 1984.
166. Razali, M. R., and Thomas, K. S., *Singular Integral Equations and Mixed Boundary Value Problems for Harmonic Functions*, Treatment of Integral Equations by Numerical Methods, Edited by C. T. H. Baher and G. Miller, Academic Press, London-New York, 1982.
167. Sankar, T. S., Hoa, S. V., and Fabrikant, V. J., *Approximate Solution of Singular Integrodifferential Equations in Elastic Contact Problems*, International Journal for Numerical Methods in Engineering, Vol. 18, pp. 503-519, 1982.

168. SCHMIDT, G., *Spline Collocation for Singular Integrodifferential Equations over $(0, 1)$*, Numerische Mathematik, Vol. 50, pp. 337-352, 1987.
169. SHERMAN, D. I., *On Some Types of Singular Integral Equations Occurring in Applications*, Applied Mathematics and Mechanics, Vol. 43, pp. 559-570, 1979.
170. SOLVEIG, M., *On Singular Integral Equations of Kinked Cracks*, International Journal of Fracture, Vol. 30, pp. 57-65, 1986.
171. SRIVASTAV, R. P., *Numerical Solution of Singular Integral Equations Using Gauss-Type Formulae: I. Quadrature and Collocation on Chebyshev Nodes*, IMA Journal of Numerical Analysis, Vol. 3, pp. 305-318, 1983.
172. SRIVASTAV, R. P., *A Hybrid Technique—Gaussian Quadrature and Rational Function Approximation—For Numerical Solution of Cauchy Singular Integral Equations*, Numerical Solution of Singular Integral Equations, Edited by A. Gerasoulis and R. Vichnevetsky, IMACS, pp. 120-123, 1984.
173. SRIVASTAV, R. P., AND JEN, E., *On the Polynomials Interpolating Approximate Solutions of Singular Integral Equations*, Applicable Analysis, Vol. 14, pp. 275-285, 1983.
174. SRIVASTAV, R. P., AND JEN, E., *Numerical Solution of Singular Integral Equations Using Gauss-Type Formula: II. Lobatto-Chebyshev Quadrature and Collocation on Chebyshev Nodes*, IMA Journal on Numerical Analysis, Vol. 3, pp. 319-325, 1983.
175. STENGER, F., AND ELLIOTT, D., *Sinc Method of Solution of Singular Integral Equations*, Numerical Solution of Singular Integral Equations, Edited by A. Gerasoulis and R. Vichnevetsky, IMACS, pp. 27-35, 1984.
176. THEOCARIS, P. S., *On the Numerical Solution of Cauchy-Type Singular Integral Equations*, Serdica, Vol. 2, pp. 252-275, 1976.
177. THEOCARIS, P. S., *On the Numerical Solution of Cauchy Type Singular Integral Equations and the Determination of Stress Intensity Factors in the Case of Complex Singularities*, Journal of Applied Mathematics and Physics, Vol. 28, pp. 1086-1088, 1977.
178. THEOCARIS, P. S., *Numerical Solution of Singular Integral Equations: Methods*, Journal of the Engineering Mechanics Division, ASCE, Vol. 107, pp. 733-752, 1981.
179. THEOCARIS, P. S., CHRYSAKIS, A. C., AND IOAKIMIDIS, N. I., *Cauchy-Type Integrals and Imtegral Equations With Logarithmic Singularities*, Journal of Engineering Mathematics, Vol. 13, pp. 63-74, 1979.
180. THEOCARIS, P. S., AND IOAKIMIDIS, N. I., *Application of the Gauss, Radau, and Lobatto Integration Rules to the Solution of Singular Integral Equations*, Zeitschrift für Angewandte Mathematik und Mechanik, Vol. 58, pp. 520-522, 1978.
181. THEOCARIS, P. S., AND IOAKIMIDIS, N. I., *A Method of Numerical Solution of Cauchy-Type Singular Integral Equations with Generalized Kernels and Arbitrary Complex Singularities*, Journal of Computational Physics, Vol. 30, pp. 309-323, 1979.
182. THEOCARIS, P. S., AND IOAKIMIDIS, N. I., *On the Gauss-Jacobi Numerical Integration Method Applied to the Solution of Singular Integral Equations*, Bulletin of the Calcutta Mathematical Society, Vol. 71, pp. 29-43, 1979.

183. THEOCARIS, P. S., IOAKIMIDIS, N. I., AND CHRYSAKIS, A. C., *On the Application of Numerical Integration Rules to the Solution of Some Singular Integral Equations*, Computer Methods in Applied Mechanical Engineering, Vol. 24, pp. 7-11, 1980.
184. THEOCARIS, P. S., AND TSAMASPHYROS, G., *Numerical Solution of Systems of Singular Integral Equations with Variable Coefficients*, Applicable Analysis, Vol. 9, pp. 37-52, 1979.
185. THOMAS, K., *Collocation Points for Singular Integral Equations*, Numerical Solution of Singular Integral Equations, Edited by A. Gerasoulis and R. Vichnevetsky, IMACS, pp. 101-102, 1984.
186. TSAMASPHYROS, G., *Study of Factors Influencing the Solution of Singular Integral Equations*, Engineering Fracture Mechanics, Vol. 21, pp. 567-570, 1986.
187. TSAMASPHYROS, G., AND ANDROULIDAKIS, P., *Tanh Transformation for the Solution of Singular Integral Equations*, International Journal for Numerical Methods in Engineering, Vol. 24, pp. 543-556, 1987.
188. TSAMASPHYROS, G., AND STASSINAKIS, C. A., *A Numerical Solution of Singular Integral Equations without Using Special Collocation Points*, International Journal for Numerical Methods in Engineering, Vol. 19, pp. 421-430, 1983.
189. TSAMASPHYROS, G., AND THEOCARIS, P. S., *Equivalence of Direct and Indirect Methods for the Numerical Solution of Singular Integral Equations*, Computing, Vol. 24, pp. 325-331, 1980.
190. TSAMASPHYROS, G., AND THEOCARIS, P. S., *Are Special Collocation Points Necessary for the Numerical Solution of Singular Integral Equations?* International Journal of Fracture, Vol. 17, R 21-24, 1981.
191. TSAMASPHYROS, G., AND THEOCARIS, P. S., *A Recurrence Formula for the Direct Solution of Singular Integral Equations*, Computer Methods in Applied Mechanics and Engineering, Vol. 31, pp. 79-89, 1982.
192. WELSTEAD, S. T., *Orthogonal Polynomials Applied to the Solution of Singular Integral Equations*, Numerical Solution of Singular Integral Equations, Edited by A. Gerasoulis and R. Vichnevetsky, IMACS, 1984.
193. WOLFERSDORF, L. V., *On the Theory of Nonlinear Singular Integral Equations of Cauchy Type*, Mathematical Methods in Applied Science, Vol. 9, pp. 493-517, 1985.

6
Convergence Theorems for Singular Integral Equations

D. ELLIOTT

Abstract. After describing a theory for singular integral equations with Cauchy kernel on the arc $(-1, 1)$, a general convergence theory for approximate methods of solving such equations is proposed which includes estimates of rates of convergence. This theory is then applied to two well-known direct approximate methods.

1. Introduction

In this chapter we shall consider convergence theorems for approximate methods of solution of the singular integral equation, taken over the arc $(-1, 1)$, given by

$$a(t)\phi(t) + \frac{b(t)}{\pi} \fint_{-1}^{1} \frac{\phi(\tau)\, d\tau}{\tau - t} + \int_{-1}^{1} k(t, \tau)\phi(\tau)\, d\tau = f(t). \qquad (1)$$

The first integral appearing in Eq. (1) is a Cauchy principal-value integral which is defined by

$$\fint_{-1}^{1} \frac{\phi(\tau)\, d\tau}{\tau - t} = \lim_{\varepsilon \to 0+} \left(\int_{-1}^{t-\varepsilon} + \int_{t+\varepsilon}^{1} \right) \frac{\phi(\tau)\, d\tau}{\tau - t}, \qquad (2)$$

provided of course that the limit exists. In Eq. (1) we assume that the functions a, b, k, and f are given and it is required to determine the function ϕ. In another chapter, details concerning various algorithms for the solution of Eq. (1) when a and b are constants have been given. But for convergence theory we have no need to be quite as restrictive since many of the techniques carry over to this more general equation. The problem then becomes one

---

D. ELLIOTT • Department of Applied Mathematics, The University of Tasmania, Hobart, Tasmania, Australia 7001.

of implementation of the algorithm, a topic with which we shall not concern ourselves in this chapter. It is trite to say, but it is nevertheless important, that one needs a good understanding of the theory of the solution of Eq. (1) before we can consider the behavior of approximate methods. The standard reference for this theory is the classic text by Muskhelishvili (Ref. 1). However, other theories have been produced over the years and we might mention the theory in L_p space considered by Tricomi (Ref. 2) and Widom (Ref. 3). Levinson (Ref. 4) assumed that ϕ was analytic at interior points of $[-1, 1]$ but infinite at the endpoints such that $|\phi(t)| \le C(1 + t)^{-\alpha}$ near $t = -1$ and $|\phi(t)| \le C(1 - t)^{-\beta}$ near $t = 1$ where $0 < \alpha, \beta < 1$. In the next section we shall consider a modification of these theories that appears to be appropriate to the equations which arise in practice. The theory is slightly more general than that given by Muskhelishvili but simplifications do arise. Having established the theory in Section 2 we shall consider approximate methods and their convergence in the remaining sections. In Section 3 we briefly touch upon indirect methods of approximate solution. In Sections 4-5 we consider, in more detail, direct methods. The general theory is given in Section 4 and particular applications of the theory are given in Section 5.

Throughout this chapter we shall omit many proofs. Frequent reference will be made to two Technical Reports by the author, Refs. 5 and 6, where proofs of many of the theorems quoted here will be found.

2. A Theory for the Singular Integral Equation

Throughout we shall write

$$T\phi(t) = \frac{1}{\pi} \int_{-1}^{1} \frac{\phi(\tau) \, d\tau}{\tau - t}, \qquad -1 < t < 1, \tag{3}$$

and, referring to Eq. (1), we shall write

$$M\phi = a\phi + bT\phi, \tag{4}$$

this being known as the *dominant* part of Eq. (1) which, when $k \ne 0$, is known as the *complete* equation.

The starting point for all our subsequent analysis is the observation that if $1 < p < \infty$ then T is a bounded linear operator on $L_p(-1, 1)$, the space of p-integrable functions defined on $(-1, 1)$, into itself. This is an immediate consequence of M. Riesz's famous theorem (Ref. 7) concerning the Hilbert transform. If, as usual, for a given p we define q by

$$\frac{1}{p} + \frac{1}{q} = 1, \tag{5}$$

the following theorem, due to Khevedelidze (Ref. 8), will be of considerable importance.

Theorem 2.1. Let $w(\tau) = (1 - \tau)^\alpha (1 + \tau)^\beta$. Suppose $1 < p < \infty$ and $-1/p < \alpha, \beta < 1/q$. Then there exists a constant $C = C(p; \alpha, \beta)$ such that

$$\|wT(\phi/w)\|_p \leq C\|\phi\|_p, \tag{6}$$

for all $\phi \in L_p(-1, 1)$.

With these preliminary comments let us now introduce the space of functions which we shall denote by $H(p_1, p_2)$ and define as follows:

Definition 2.1. A function ϕ defined on the interval $(-1, 1)$ is said to be in the space $H(p_1, p_2)$, $1 \leq p_1, p_2 \leq \infty$, if it satisfies the following conditions:
  (i) ϕ is Hölder continuous on every closed subinterval of $(-1, 1)$,
  (ii) near $t = -1$, $|\phi(t)| \leq C(1 + t)^{-\gamma_1}$, $0 \leq \gamma_1 < 1/p_1$,
  (iii) near $t = +1$, $|\phi(t)| \leq C(1 - t)^{-\gamma_2}$, $0 \leq \gamma_2 < 1/p_2$,
where C is some constant.

We allow p_1, p_2 to take the value $+\infty$ if we require ϕ to be continuous and therefore bounded at an endpoint. Thus, for example, the space $H(\infty, 1)$ contains those functions which are Hölder continuous on every closed subinterval of $[-1, 1)$ but integrable at $+1$. Muskhelishvili (Ref. 1) developed his theory for the space $H(1, 1)$. A norm on the space $H(p_1, p_2)$ will be defined as follows:

Definition 2.2.
  (i) For $\phi \in H(p_1, p_2)$, $1 \leq p_1, p_2 < \infty$ a norm will be denoted and defined by

$$\|\phi\|_{(p_1, p_2)} = \left(\int_{-1}^{1} |(1 - \tau)^{1/p_2 - 1/p_1} \phi(\tau)|^{p_1} \, d\tau \right)^{1/p_1}. \tag{7a}$$

(ii) For $\phi \in H(p_1, \infty)$, $1 \le p_1 < \infty$

$$\|\phi\|_{(p_1, \infty)} = \left(\int_{-1}^{1} (1 - \tau)^{-1} |\phi(\tau)|^{p_1} \, d\tau \right)^{1/p_1}. \tag{7b}$$

(iii) For $\phi \in H(\infty, p_2)$, $1 \le p_2 < \infty$,

$$\|\phi\|_{(\infty, p_2)} = \left(\int_{-1}^{1} (1 + \tau)^{-1} |\phi(\tau)|^{p_2} \, d\tau \right)^{1/p_2}. \tag{7c}$$

(iv) For $\phi \in H(\infty, \infty)$,

$$\|\phi\|_{(\infty, \infty)} = \max_{\tau \in [-1, 1]} |\phi(\tau)|. \tag{7d}$$

For given values of p_1, p_2 the numbers q_1, q_2 will always be defined such that

$$1/p_1 + 1/q_1 = 1, \tag{8a}$$

$$1/p_2 + 1/q_2 = 1. \tag{8b}$$

In the remainder of this chapter, $H(p_1, p_2)$ will denote the normed space of functions satisfying both Definitions 2.1 and 2.2.

From these definitions we can prove the following theorem.

Theorem 2.2.

(a) Suppose $\phi \in H(p_1, p_2)$ and $\psi \in H(q_1, q_2)$ where $1 \le p_1, p_2 \le \infty$. Then

$$\left| \int_{-1}^{1} \phi(\tau) \psi(\tau) \, d\tau \right| \le \|\phi\|_{(p_1, p_2)} \|\psi\|_{(q_1, q_2)}. \tag{9}$$

(b) If $1 < p_1, p_2 < \infty$, then T is a bounded linear operator on $H(p_1, p_2)$ into itself.

(c) If $1 < p_1, p_2 < \infty$ and a and b are Hölder continuous on $[-1, 1]$, then M is a bounded linear operator on $H(p_1, p_2)$ into itself.

Proof. See Elliott (Ref. 5). □

In addition we need to introduce the Parseval theorem and the Poincaré-Bertrand formula as applied to the space $H(p_1, p_2)$. These two results prove to be invaluable analytic tools in the development of the theory.

Theorem 2.3 (Parseval's theorem). Suppose $1 \le p_1, p_2 \le \infty$. If $\phi \in H(p_1, p_2)$ and $\psi \in H(q_1, q_2)$ then

$$\int_{-1}^{1} \{\phi(\tau) T\psi(\tau) + \psi(\tau) T\phi(\tau)\} \, d\tau = 0. \tag{10}$$

Theorem 2.4 (Poincaré–Bertrand formula). Suppose $1 < p_1, p_2 < \infty$. If $\phi \in H(p_1, p_2)$ and $\psi \in H(q_1, q_2)$, then at all points of $(-1, 1)$

$$T(\phi T\psi + \psi T\phi) = T\phi \cdot T\psi - \phi \cdot \psi. \tag{11}$$

For the proofs of both these theorems, see Elliott (Ref. 5).

It is well known that we cannot discuss the solution of the equation $M\phi = f$ without considering the adjoint $M^\times$ say of M. We shall define the adjoint as follows:

Definition 2.3. For a given operator $M: H(p_1, p_2) \to H(p_1, p_2)$ the adjoint operator $M^\times$ is such that for all $\phi \in H(p_1, p_2)$ and $\psi \in H(q_1, q_2)$ we have

$$\int_{-1}^{1} M\phi(\tau)\psi(\tau)\, d\tau = \int_{-1}^{1} \phi(\tau) M^\times \psi(\tau)\, d\tau. \tag{12}$$

As an immediate consequence of this definition and bearing in mind Parseval's theorem we have

$$M^\times \psi = a\psi - T(b\psi). \tag{13}$$

With these preliminaries established we can now commence to solve Eq. (1), but this we must take in easy stages; we shall start by considering the homogeneous, dominant equation $M\phi = 0$. Before proceeding we shall make a further assumption that the equation is *regular*, that is, neither $a + ib$ nor $a - ib$ vanishes on $[-1, 1]$. If we define

$$r^2(t) = a^2(t) + b^2(t), \qquad t \in [-1, 1], \tag{14}$$

then we shall assume that r^2 is never zero on $[-1, 1]$. Related to these coefficients we need to define a further function θ which is continuous on $[-1, 1]$ and satisfies the equations

$$\exp[\pm \pi i \theta(t)] = [a(t) \mp ib(t)]/r(t), \qquad t \in [-1, 1], \tag{15}$$

where the upper and lower signs go together. The insistence on the continuity of θ implies that r, which is obtained from (14), must also be continuous on $[-1, 1]$ so that appropriate branches of the square-root function must be chosen. We shall not here give the details of solving $M\phi = 0$; suffice it to say that we proceed, as does Muskhelishvili, by introducing the Cauchy integral of θ in the complex plane with $[-1, 1]$ deleted and applying the

Sokhotski–Plemelj formulas. This reduces the problem to a homogeneous Riemann–Hilbert problem from which our desired solution can be found. To state it explicitly we first introduce integers n_1, n_2 as follows. If Re $\theta(-1)$ is an integer we choose

$$n_1 = \text{Re } \theta(-1); \qquad (16a)$$

otherwise we require

$$[\text{Re } \theta(-1) - 1/p_1] < n_1 < 1 + [\text{Re } \theta(-1) - 1/p_1]. \qquad (16b)$$

Similarly, if Re $\theta(1)$ is an integer we choose

$$n_2 = -\text{Re } \theta(1); \qquad (17a)$$

otherwise we require

$$-[\text{Re } \theta(1) + 1/p_2] < n_2 < 1 - [\text{Re } \theta(1) + 1/p_2]. \qquad (17b)$$

Having determined n_1, n_2 we define a third integer, the *index* κ, by

$$\kappa(p_1, p_2) = -(n_1 + n_2). \qquad (18)$$

The significance of this integer will be apparent later.

We now introduce the *fundamental function* Z which is given by

$$Z(t) = (1-t)^\alpha (1+t)^\beta \Omega(t), \qquad t \in (-1, 1), \qquad (19a)$$

where

$$\alpha = n_2 + \text{Re } \theta(1), \qquad \beta = n_1 - \text{Re } \theta(-1), \qquad (19b)$$

and

$$\Omega(t) = \exp\left\{\int_{-1}^{1} \left(\frac{\theta(\tau) - \theta(t)}{\tau - t}\right) d\tau + [\theta(t) - \text{Re } \theta(1)]\log(1-t)\right.$$
$$\left. - [\theta(t) - \text{Re } \theta(-1)]\log(1+t)\right\}, \qquad (19c)$$

for $t \in [-1, 1]$. We observe that

$$-1/p_2 < \alpha < 1/q_2 \qquad \text{and} \qquad -1/p_1 < \beta < 1/q_1. \qquad (20)$$

Furthermore, we have that Ω never vanishes on $[-1, 1]$ and is also bounded, so that there exist constants c, C such that

$$0 < c < |\Omega(t)| < C < \infty \qquad \text{on } [-1, 1]. \tag{21}$$

Also the fundamental function Z has been constructed in such a way that

$$Z \in H(p_1, p_2) \quad \text{and} \quad 1/Z \in H(q_1, q_2). \tag{22}$$

From (19a) we see that Z is defined only on $(-1, 1)$. We shall however require an extension of Z into the complex plane and this gives rise to the *canonical function* of M in $H(p_1, p_2)$. This is denoted and defined by

$$X(z) = (1 + z)^{n_1}(1 - z)^{n_2} \exp\left[\int_{-1}^{1} \frac{\theta(\tau)\, d\tau}{\tau - z}\right], \qquad z \in \mathbf{C}\setminus[-1, 1]. \tag{23}$$

The relation between X and Z on $(-1, 1)$ is given by

$$[X^+(t)X^-(t)]^{1/2} = Z(t), \tag{24a}$$

and

$$Z(t) = [a(t) \pm ib(t)]X^{\pm}(t)/r(t), \tag{24b}$$

where again the upper and lower signs go together.

We are now in a position to state the solution of the homogeneous equation $M\phi = 0$. When $\kappa \leq 0$ it transpires that the only solution in $H(p_1, p_2)$ is the trivial solution $\phi = 0$. However, when $\kappa > 0$ we shall find κ linearly independent solutions given by

$$\phi_k(t) = b(t)t^k Z(t)/r(t), \qquad k = 0(1)(\kappa - 1). \tag{25}$$

This analysis of the dominant equation has enabled us to introduce the functions Z, X and the index κ which are so important in the general solution of the complete Eq. (1). But before we proceed with this, first by considering the inhomogeneous dominant equation $M\phi = f$, let us consider briefly the homogeneous adjoint equation. We may proceed as before to consider the solution of the homogeneous equation $M^{\times}\psi = 0$ for $\psi \in H(q_1, q_2)$ where $M^{\times}$ is given by (13). We find that if $\kappa^{\times}$ denotes the index of $M^{\times}$ then

$$\kappa(p_1, p_2) + \kappa^{\times}(q_1, q_2) = 0. \tag{26}$$

Furthermore, when $\kappa^\times \leq 0$ the equation $M^\times \psi = 0$ possesses only the trivial solution $\psi = 0$. However, when $\kappa^\times \geq 1$ then $M^\times \psi = 0$ has $\kappa^\times$ nontrivial solutions in $H(q_1, q_2)$ given by

$$\psi_k(t) = t^k / [Z(t)r(t)], \qquad k = 0(1)(\kappa^\times - 1). \tag{27}$$

The fundamental function of $M^\times$ in $H(q_1, q_2)$ is given on $(-1, 1)$ by Z^{-1} and the canonical function in the deleted complex plane is defined by X^{-1}. Thus we see that for all values of $\kappa(p_1, p_2)$ we have

$$\kappa = \dim(\ker M) - \dim(\ker M^\times), \tag{28}$$

which justifies the use of the word "index" for the integer κ.

So much for the adjoint equation. We now require some further relationships between a, b, r, and Z and to this end we introduce the idea of the *principal part* of a function.

Definition 2.4. Suppose that a function f has a Laurent expansion about the point at infinity given by $f(z) = \sum_{j=-\infty}^{N} f_j z^j$ where $f_N \neq 0$. The principal part of f at any point $z_0 \in \mathbf{C}$ is denoted and defined by

$$\text{p.p.}(f; z_0) = \sum_{j=0}^{N} f_j z_0^j. \tag{29}$$

We can now establish the following result.

Theorem 2.5. Let P be any polynomial. Then for $t \in (-1, 1)$,

$$\frac{a(t)Z(t)P(t)}{r(t)} + \left(T\left(\frac{bZP}{r}\right) \right)(t) = \text{p.p.}(PX; t); \tag{30}$$

$$\frac{a(t)P(t)}{r(t)Z(t)} - \left(T\left(\frac{bP}{rZ}\right) \right)(t) = \text{p.p.}(PX^{-1}; t). \tag{31}$$

Proof. See Elliott (Ref. 5). □

Let us consider now the dominant equation $M\phi = f$ where we assume that $f \in H(p_1, p_2)$ is given and we require $\phi \in H(p_1, p_2)$. When $\kappa(p_1, p_2) \geq 0$, we have seen that the homogeneous adjoint equation $M^\times \psi = 0$ has only the trivial solution in $H(q_1, q_2)$ so that $M\phi = f$ will then possess a solution for every $f \in H(p_1, p_2)$. However, when $\kappa(p_1, p_2) < 0$ the equation $M^\times \psi = 0$ possesses nontrivial solutions in $H(q_1, q_2)$ [see (27)] so that we cannot solve $M\phi = f$ for every f in $H(p_1, p_2)$ but only for those f such that

$$\int_{-1}^{1} \frac{f(\tau)\tau^k}{r(\tau)Z(\tau)} d\tau = 0 \qquad \text{for } k = 0(1)(-\kappa - 1). \tag{32}$$

These equations are known as the *consistency conditions*.

To solve $M\phi = f$, if we multiply this equation by $1/rZ$, take the T transform of each side, use the Poincaré–Bertrand formula, and use (31) we find that

$$\phi = \hat{M}^I f + \phi^{(0)}, \tag{33}$$

where $\phi^{(0)}$ is any solution of $M\phi^{(0)} = 0$ [see (25)] and $\hat{M}^I$ is a *pseudoinverse* of M defined by

$$\hat{M}^I f = \frac{af}{r^2} - \frac{bZ}{r} T\left(\frac{f}{rZ}\right) = \frac{Z}{r}\left[\frac{af}{rZ} - bT\left(\frac{f}{rZ}\right)\right]. \tag{34}$$

Obviously this pseudoinverse is an important operator in our analysis and one of its most important properties is given by the next theorem.

Theorem 2.6. If $1 < p_1, p_2 < \infty$, then $\hat{M}^I$ is a bounded linear operator from $H(p_1, p_2)$ into itself.

The proof of this theorem requires Khevedelidze's theorem (Theorem 2.1) and details are given in Elliott (Ref. 5). We observe that the case of $p_1 = p_2 = 1$, which is the case considered by Muskhelishvili, is excluded from this theorem.

Returning to (34) it is not difficult to show that the homogeneous equation $\hat{M}^I \phi = 0$ only possesses nontrivial solutions when $\kappa(p_1, p_2) < 0$ and these are given by

$$\phi_k(t) = bt^k, \quad k = 0(1)(-\kappa - 1). \tag{35}$$

When $\kappa(p_1, p_2) \geq 0$ the only solution of $\hat{M}^I \phi = 0$ in $H(p_1, p_2)$ is $\phi = 0$. Again the Poincaré–Bertrand formula can be used to show the following theorem.

Theorem 2.7. If $\phi^{(0)}$ and $\hat{\phi}^{(0)}$ denote any elements in the null spaces of M and $\hat{M}^I$ respectively, then

$$\hat{M}^I M \phi = \phi + \phi^{(0)}, \tag{36a}$$

$$M\hat{M}^I \phi = \phi + \hat{\phi}^{(0)}. \tag{36b}$$

As an immediate corollary of this we have

$$\begin{cases} M\hat{M}^I M = M, & \text{(37a)} \\ \hat{M}^I M \hat{M}^I = \hat{M}^I, & \text{(37b)} \end{cases}$$

a pair of equations which justify the use of the phrase "pseudoinverse" to describe the operator $\hat{M}^I$.

Let us now consider the complete equation (1) which we shall write as

$$M\phi + K\phi = f, \tag{38}$$

where

$$(K\phi)(t) = \int_{-1}^{1} k(t, \tau)\phi(\tau)\, d\tau, \qquad t \in (-1, 1). \tag{39}$$

If we rewrite Eq. (38) as $M\phi = f - K\phi$ and assume, for the moment, that the right-hand side is known, then we have from (32) and (33) that provided

$$\int_{-1}^{1} \frac{\tau^k}{r(\tau)Z(\tau)} \left[f(\tau) - K\phi(\tau) \right] d\tau = 0, \qquad k = 0(1)(-\kappa - 1), \tag{40}$$

then

$$\phi = \hat{M}^I(f - K\phi) + \phi^{(0)},$$

which can be rewritten as

$$(I + \hat{M}^I K)\phi = \hat{M}^I f + \phi^{(0)}. \tag{41}$$

We shall assume, without further ado, that Eqs. (40) are always satisfied so let us consider (41). Suppose that $1 < p_1, p_2 < \infty$, then if we define

$$|||k|||_{(p_1, p_2)} = \left\{ \int_{-1}^{1} (1-t)^{p_1/p_2 - 1} \left(\int_{-1}^{1} |(1-\tau)^{1/q_2 - 1/q_1} k(t, \tau)|^{q_1} d\tau \right)^{p_1/q_1} dt \right\}^{1/p_1}, \tag{42}$$

we shall have that if $|||k|||_{(p_1, p_2)}$ is finite then K is a *compact* linear operator on $H(p_1, p_2)$ into itself. Assuming that this condition is satisfied, then since $\hat{M}^I$ is a bounded linear operator on $H(p_1, p_2)$ into itself, $\hat{M}^I K$ is compact. Consequently (41) is a Fredholm equation to which we can apply the well-known theory of such equations. This, as we shall see later, is a most important observation which will be of great assistance in our analysis of algorithms.

Let us return to the operators M and $\hat{M}^I$ and the relations given by Eqs. (36). Since the null spaces of both M and $\hat{M}^I$ are finite dimensional we can rewrite (36) in operator form as

$$\begin{cases} \hat{M}^I M = I - K_1, \\ M\hat{M}^I = I - K_2, \end{cases} \tag{43}$$

respectively, where K_1 and K_2 are compact operators on $H(p_1, p_2)$ into itself. As a consequence we identify M and $\hat{M}^I$ as *Noether operators* on $H(p_1, p_2)$ into itself; see Schechter (Ref. 9, Chapter 5, Theorem 2.1) (note that Schechter calls such operators "Fredholm operators" but many Russian authors call them Noether operators and the latter seems to this author to be the more appropriate). Having identified these operators the following theorem is of considerable importance.

Theorem 2.8. Suppose $1 < (p_1, p_2) < \infty$. Let M_1, M_2 be Noether operators with indices $\kappa(M_1)$ and $\kappa(M_2)$ respectively. Let K be a compact operator from $H(p_1, p_2)$ into itself. Then

$$\kappa(M_1 M_2) = \kappa(M_1) + \kappa(M_2), \tag{44a}$$

$$\kappa(M_1 + K) = \kappa(M_1). \tag{44b}$$

Proof. See Schechter (Ref. 9, Chapter 5, Theorems 3.1 and 3.2). □

Thus we see that adding a compact operator to a Noether operator does not affect its index, so that the complete equation has the same index as the dominant equation. Furthermore, since the index of $I - K_1$ is zero it follows from (43) and (44a) that

$$\kappa(M) = -\kappa(\hat{M}^I), \tag{45}$$

so that on recalling (26) we see that $\hat{M}^I$ and $M^\times$ have the same index.

Finally, in this summary of the theory of singular integral equations on the arc $(-1, 1)$ we shall consider the two systems of orthogonal polynomials associated with the operator M. First let us assume that a and b are *real*. Then, in particular, Z and r will be real and, without loss of generality, we can choose them to be positive. If Z does take the value zero then this can only occur at the endpoints ± 1. We can now define two weight functions w_1 and w_2 by

$$w_1 = Z/r, \tag{46a}$$

$$w_2 = 1/Zr. \tag{46b}$$

Since $Z \in H(p_1, p_2)$ and $1/Z \in H(q_1, q_2)$ it follows that both w_1 and w_2 are integrable on $[-1, 1]$. Suppose now that $\{t_n\}$ denotes a sequence of orthogonal polynomials such that

$$\int_{-1}^{1} w_1(\tau) t_j(\tau) t_k(\tau) \, d\tau = h_j \delta_{j,k}, \qquad j, k = 0, 1, 2, \ldots. \tag{47}$$

Next, on recalling the definition of the canonical function of M in $H(p_1, p_2)$ [see (23)], let us define a second sequence of polynomials $\{u_n\}$ by

$$u_n(z) = \text{p.p.}\{Xt_{n+\kappa}; z\}, \tag{48a}$$

for all $z \in \mathbf{C}$ and $n \geq \max(0, -\kappa)$.

It is readily shown that equally well we can write, for $n \geq \max(0, \kappa)$ and $z \in \mathbf{C}$,

$$t_n(z) = \text{p.p.}\{X^{-1}u_{n-\kappa}; z\}. \tag{48b}$$

In order to obtain an alternative expression for u_n valid on $(-1, 1)$ we need to make an assumption.

Assumption A. The coefficient b of M is a polynomial of degree μ say with all its zeros on $[-1, 1]$.

This assumption is not, in practice, as restrictive as it first appears, and leads to the following important representation for u_n.

Theorem 2.9. Given Assumption A, for $n \geq \max(0, \mu - \kappa)$ and $t \in (-1, 1)$,

$$u_n(t) = M(Zt_{n+\kappa}/r) = \frac{a(t)Z(t)}{r(t)} t_{n+\kappa}(t) + b(t) T\left(\frac{Zt_{n+\kappa}}{r}\right)(t). \tag{49}$$

From this representation we find that

$$\int_{-1}^{1} w_2(\tau) u_j(\tau) u_k(\tau) \, d\tau = h_{j+\kappa} \delta_{j,k}, \tag{50}$$

for $j, k \geq \max(0, \mu - \kappa)$, thus establishing the orthogonality of the sequence of polynomials $\{u_n\}$ with respect to the weight function w_2. We can now derive a result for t_n in terms of $u_{n-\kappa}$ similar to Theorem 2.9.

Theorem 2.10. Given Assumption A, for $n \geq \max(0, \kappa + \mu)$ and $t \in (-1, 1)$,

$$t_n(t) = \frac{r(t)}{Z(t)} \hat{M}^I\left(\frac{u_{n-\kappa}}{rZ}\right)(t) = \frac{a(t)u_{n-\kappa}(t)}{r(t)Z(t)} - b(t) T\left(\frac{u_{n-\kappa}}{rZ}\right)(t). \tag{51}$$

Two further results connecting the polynomials t_n and u_n are given in the following theorem.

Theorem 2.11. Given Assumption A,
(a) if the polynomials t_n satisfy the recurrence relation

$$t_{n+1}(z) - (A_n z + B_n)t_n(z) + C_n t_{n-1}(z) = 0, \qquad (52)$$

for $n \geq 1$, then for $n \geq \max(1, \mu - \kappa + 1)$,

$$u_{n+1}(z) - (A_{n+\kappa} z + B_{n+\kappa})u_n(z) + C_{n+\kappa} u_{n-1}(z) = 0; \qquad (53)$$

(b) for $n \geq \max(\kappa, \mu)$,

$$t_n(z)u_{n-\kappa+1}(z) - t_{n+1}(z)u_{n-\kappa}(z) = \frac{\alpha_{n+1,n+1} h_n}{\pi \alpha_{n,n}} b(z), \qquad (54)$$

where $t_n(z) = \alpha_{n,n} z^n + \cdots$.

For a more detailed discussion on these orthogonal polynomials when a and b are real, the reader is referred to Elliott (Ref. 10) or Welstead (Ref. 11).

Finally, in this section on the theory of singular integral equations let us consider the case where a and b are *complex*. Then Z and r will in general be complex so that the functions w_1 and w_2 as defined by (46) will be complex. Let us now define

$$w_1 = |Z/r|, \qquad (55a)$$

$$w_2 = 1/|Zr|, \qquad (55b)$$

then these will be integrable weight functions which can be used to generate orthogonal polynomials on $(-1, 1)$. However, we would like these polynomials to be related through a *real* singular integral operator $\tilde{M}$ say, where $\tilde{M}\phi = \tilde{a}\phi + \tilde{b}T\phi$, $\tilde{a}$ and $\tilde{b}$ being real. Then the polynomials generated by w_1 and w_2 will satisfy equations such as (49) and (51). It can be shown that if we choose

$$\tilde{a}(t) = [(|r^2(t)| + |a(t)|^2 - |b(t)|^2)/2]^{1/2}, \qquad (56)$$

and

$$\tilde{b}(t) = \begin{cases} |b(t)|, & \text{if } \tilde{a}(t) = 0, \\ \text{sgn}\{\text{Re}[a(t)b(t)]\}\{[|r^2(t)| - |a(t)|^2 + |b(t)|^2]/2\}^{1/2}, & \text{if } \tilde{a}(t) \neq 0, \end{cases} \qquad (57)$$

then we obtain the appropriate real singular integral equation operator connecting the orthogonal polynomial generated by w_1 and w_2 as defined in (55). For further details see Elliott (Ref. 5).

So much for an overview of the theory of the singular integral equation (1). The theory as outlined is in one respect slightly more general than that given by Muskhelishvili (Ref. 1), but it leads to a simpler discussion of the index and it also shows that for $1 < p_1, p_2 < \infty$ both M and $\hat{M}^I$ are bounded linear operators on $H(p_1, p_2)$ into itself. We have also attempted, once we have introduced the fundamental and canonical functions, to obtain the solution of the dominant equation $M\phi = f$ by using the Poincaré-Bertrand formula which is the nearest thing one has to a convolution theorem for the finite Hilbert transform. This contrasts with Muskhelishvili's approach where the corresponding inhomogeneous Riemann-Hilbert problem was solved directly. It would be satisfying to develop a method which depends entirely upon transform theory without the use of function-theoretic techniques. Peters (Ref. 12) has established such a method for the special case when $a = 0$ and $b = 1$. The development of such a method for arbitrary a and b will have to wait for another day.

We are now in a position to consider discrete methods for the approximate solution of Eq. (1) and to construct a general theory for convergence.

3. The Approximate Solution of Singular Integral Equations

Much of the theory of singular integral equations has been known for the past fifty or sixty years due to the work of mathematicians such as Carleman, Muskhelishvili, Noether, Privalov, and Tricomi to mention just a few. Inevitably the study of the approximate solution of these equations has lagged behind the theory. The book by Ivanov (Ref. 13) gives an account of such methods until about the mid-1960s. Here attention is primarily focused on equations over closed contours and, in particular, the unit circle.

The advent of computers has led inevitably to a surge in interest in approximate methods and the past fifteen years have seen considerable advances in the development of approximate methods for solving Eq. (1). The effect of both integrable singularities at the endpoints of the arc $(-1, 1)$ and a nonzero index have added a certain piquancy to the development of appropriate algorithms. Methods for approximate solution of Eq. (1) fall readily into two categories, the so-called *direct* and *indirect* methods. In direct methods we consider Eq. (1) without any further transformation and look for appropriate discretizations of the operators M and K. For indirect methods we first reduce Eq. (1) to the equivalent Fredholm equation (41), which is then solved approximately. The past two decades have seen considerable development of approximate methods for the solution of Fredholm equations and it is certainly a sensible tactic to reduce an unsolved problem to one that has been well worked over. However, the main disadvantage of the indirect method is that the operator $\hat{M}^I K$ is of a rather complicated structure and this brings us back to consider direct methods in order to avoid this complication.

In Section 3.1 we shall take a brief look at indirect methods. This is a useful exercise, since the analysis of convergence of indirect methods introduces techniques which we shall exploit further when we look at direct methods. Our primary concern, in this second part of the chapter, is with the analysis of both convergence and rates of convergence of approximate methods. Such analyses for singular integral equations are a fairly recent development and at this stage may be considered as being far from complete. Sections 4 and 5 are devoted to direct methods. In Section 4 we consider the general theory. In Section 5 we apply this theory to a couple of the most widely used algorithms. Again we shall, in places, skip proofs although we do propose to provide somewhat more detail than we have done in Section 2; for further information the reader is referred to Elliott (Ref. 6). Throughout the remainder of this chapter we shall assume that a, b, k, and f are *real* functions. The extension of these methods to complex a and b does not yet seem to have been fully addressed in the literature.

3.1. An Indirect Method. Let us recall from Section 2 that the equation $M\phi + K\phi = f$ is equivalent to

$$(I + \hat{M}^I K)\phi = \hat{M}^I f + \phi^{(0)}, \tag{58}$$

where $\phi^{(0)}$ is the general solution of $M\phi = 0$; see (25). Equation (58) is obtained under the assumption that $f - K\phi$ satisfies the consistency conditions of (40) which are nontrivial when $\kappa \leq -1$. Suppose we have assumed that $\phi \in H(p_1, p_2)$ where p_1 and p_2 are chosen from $(1, \infty)$. In this case we know that $\hat{M}^I$ is a bounded linear operator on $H(p_1, p_2)$ into itself, so that if K is compact then $\hat{M}^I K$ will also be a compact linear operator on $H(p_1, p_2)$ into itself. Of all the methods for the approximate solution of (58) when $\hat{M}^I K$ is compact, the most elegant mathematically is the Galerkin method so let us consider it. We shall assume that $p_1 = p_2 = 2$ and we shall write

$$J = \hat{M}^I K, \tag{59a}$$

$$g = \hat{M}^I f + \phi^{(0)}, \tag{59b}$$

so that we are considering the equation

$$(I + J)\phi = g, \tag{60}$$

where J is compact. Let $\mathcal{H}$ denote a real, separable Hilbert space with inner product denoted by $\langle\ ,\ \rangle$ and norm $\|\cdot\|_\mathcal{H}$ and we shall suppose that $g \in \mathcal{H}$ and that $J: \mathcal{H} \to \mathcal{H}$. In the Galerkin method for finding an approximation to $\phi \in \mathcal{H}$ we shall suppose that $\{t_1, t_2, \ldots, t_n\}$ is a set of n linearly independent functions in $\mathcal{H}$ which span a subspace H_n say in $\mathcal{H}$. For a given $n \in \mathbf{N}$ we look for an approximate solution of (60) in the form

$$\phi_n = \sum_{j=1}^{n} \xi_j t_j. \tag{61}$$

Defining a residual, $(\text{res})_n$ by

$$(\text{res})_n = \phi_n + J\phi_n - g, \tag{62}$$

we can obtain n equations for the n unknowns ξ_j by requiring that

$$\langle t_i, (\text{res})_n \rangle = 0, \qquad i = 1(1)n. \tag{63}$$

This gives rise to the Galerkin equations which we shall write as

$$(G_n + J_n)v_n = w_n, \tag{64}$$

where $v_n = (\xi_1, \xi_2, \ldots, \xi_n)^\text{T}$, $G_n = (\langle t_i, t_j \rangle)$ is the Gram matrix of the basis functions, $J_n = (\langle t_i, Jt_j \rangle)$, and $w_n = (\langle t_1, g \rangle, \ldots, \langle t_n, g \rangle)^\text{T}$. The question now arises as to whether Eqs. (64) possess a unique solution and, if they do, how this solution compares with that of the original equation (60). There are many approaches to this problem, but the one preferred by the author is that based on the use of restriction and prolongation operators. Since we shall be using these extensively in the subsequent sections, it is of some interest to see how they can be used in this particular case. We shall give an outline; for further details the reader is referred to the papers by Noble (Ref. 14) and Spence and Thomas (Ref. 15).

Let X_n denote the Euclidean n-dimensional space and recall that we have already defined $H_n = \text{span}\{t_1, t_2, \ldots, t_n\}$. If $v_n = (\xi_1, \xi_2, \ldots, \xi_n)^\text{T}$ is an element of X_n, we define a *prolongation operator* $p_n: X_n \to H_n$ by

$$p_n v_n = \sum_{i=1}^{n} \xi_i t_i. \tag{65}$$

A *restriction operator* $r_n : \mathcal{H} \to X_n$ is defined by

$$r_n g = G_n^I(\langle t_1, g \rangle, \ldots, \langle t_n, g \rangle)^T, \tag{66}$$

for any $g \in \mathcal{H}$ where G_n^I denotes the inverse of the Gram matrix. From these definitions we have

$$r_n p_n = I_n, \tag{67}$$

where I_n is the unit matrix on X_n. Furthermore, if we write

$$P_n = p_n r_n, \tag{68}$$

then we shall also require that $P_n g \to g$ as $n \to \infty$ for every $g \in \mathcal{H}$. We also have that both P_n and $I - P_n$ are self-adjoint operators and P_n is the orthogonal projection from $\mathcal{H}$ into H_n. We shall define the norm in X_n by

$$\|v_n\|_{X_n} = \|p_n v_n\|_{\mathcal{H}} = \|G_n^{1/2} v_n\|_2, \tag{69}$$

where $\|\cdot\|_2$ denotes the usual 2-norm on the Euclidean space X_n. From these definitions we can show that

$$\|p_n\| = \|r_n\| = 1, \tag{70}$$

and, for all $g \in \mathcal{H}$, we shall have

$$\|r_n g\|_{X_n} \leq \|g\|_{\mathcal{H}}. \tag{71}$$

Let us return now to Eq. (64). In terms of the above definitions we see that it can be rewritten as

$$(I_n + r_n J p_n) v_n = r_n g, \tag{72}$$

which is of a form similar to Eq. (60) and allows a comparison as the next theorem shows.

Theorem 3.1. Suppose $(I + J)^I$ exists. If

$$\delta_n = \|(I + J)^I\|_{\mathcal{H}} \cdot \|(I - P_n)J\|_{\mathcal{H}} < 1, \tag{73}$$

then

$$\|(I_n + r_n J p_n)^I\|_{X_n} \leq \|(I + J)^I\|_{\mathcal{H}} / (1 - \delta_n). \tag{74}$$

Proof. See Thomas (Ref. 16, Theorem 3.2) or Noble (Ref. 14). □

This theorem is based on a well-known result due originally to Banach. Since J is compact we know that $\|(I - P_n)J\|_{\mathcal{H}} \to 0$ as $n \to \infty$, so we see that for n large enough, inequality (73) will be satisfied. Consequently for n large enough we see that Eq. (72) will possess a unique solution since from (74) we have that $(I_n + r_n J p_n)^I$ exists. The question then arises as to how the solutions ϕ and v_n compare and this is given in the next theorem.

Theorem 3.2. Let

$$c(n) = \|(I + J)^I\|_{\mathcal{H}} \|(I - P_n)J^*\|_{\mathcal{H}} / (1 - \delta_n), \tag{75}$$

where J^* denotes the Hilbert-adjoint of J. Then if n is large enough so that $\delta_n < 1$ we have

$$\|r_n \phi - v_n\|_{X_n} \leq c(n) \|(I - P_n)\phi\|_{\mathcal{H}}, \tag{76}$$

$$\|\phi - p_n v_n\|_{\mathcal{H}} \leq [1 + c(n)] \|(I - P_n)\phi\|_{\mathcal{H}}. \tag{77}$$

Proof. From Eqs. (60) and (72) we have

$$r_n \phi - v_n = -r_n J \phi + r_n J p_n v_n = -r_n J(1 - p_n r_n)\phi - r_n J p_n (r_n \phi - v_n).$$

Consequently we have

$$(I_n + r_n J p_n)(r_n \phi - v_n) = -r_n J(I - P_n)\phi.$$

For n large enough we can solve these equations, and using (74) gives

$$\|r_n \phi - v_n\|_{X_n} \leq \|(I + J)^I\|_{\mathcal{H}} \cdot \|r_n J(I - P_n)\phi\|_{X_n} / (1 - \delta_n).$$

However, we have

$$\|r_n J(I - P_n)\phi\|_{X_n} \leq \|J(I - P_n)\phi\|_{\mathcal{H}} \quad \text{by (71)}$$

$$= \|J(I - P_n)^2 \phi\|_{\mathcal{H}} \quad \text{since } P_n^2 = P_n$$

$$\leq \|J(I - P_n)\|_{\mathcal{H}} \cdot \|(I - P_n)\phi\|_{\mathcal{H}}$$

$$= \|(I - P_n)J^*\|_{\mathcal{H}} \cdot \|(I - P_n)\phi\|_{\mathcal{H}}.$$

Inequality (76) now follows at once. To obtain (77) we observe that we can write

$$\phi - p_n v_n = (I - P_n)\phi + p_n(r_n\phi - v_n),$$

from which the required result follows at once on using (70) and (76). □

A couple of comments on this result will not go amiss. We note that (76) gives the *discretization* error; it compares the vectors v_n with $r_n\phi$. On the other hand (77) gives a measure of the *global* error between ϕ and the Galerkin solution which we shall denote by $\phi_n^G = p_n v_n$. We should also note that since J is compact so is J^*, in which case $\|(I - P_n)J^*\|_{\mathcal{H}}$ will tend to zero as $n \to \infty$ since we have assumed pointwise convergence of P_n in $\mathcal{H}$. Consequently we have demonstrated the discrete convergence of v_n to ϕ and the rate of convergence will be determined by the two factors on the right-hand side of (76). The rate of global convergence of v_n to ϕ, as given by (77), will be determined by the pointwise convergence rate of $P_n\phi$ to ϕ.

It is worth mentioning here a variant of the Galerkin method as described above, the so-called *iterated* Galerkin method [see Sloan (Ref. 17)]. Having obtained $\phi_n^G = p_n v_n$ we define a further approximation ϕ_n^* say by

$$\phi_n^* = g - J\phi_n^G. \tag{78}$$

It is readily shown that $P_n\phi_n^* = \phi_n^G$ so that ϕ_n^* satisfies the equation

$$(I + JP_n)\phi_n^* = g. \tag{79}$$

Since $\|JP_n - J\|_{\mathcal{H}} \to 0$ as $n \to \infty$ (recall that J is compact) it follows that for large enough n, $(I + JP_n)^I$ exists and in fact is uniformly bounded. Consequently we find that there exists a constant $c > 0$ such that

$$\|\phi_n^* - \phi\|_{\mathcal{H}} \le c\|J - JP_n\|_{\mathcal{H}}\|\phi - \phi_n^G\|_{\mathcal{H}}, \tag{80}$$

which shows that the iterated solution ϕ_n^* converges globally to ϕ more quickly than does the Galerkin solution ϕ_n^G. It is for this reason that this first iteration of the Galerkin solution is attractive.

So much for an outline of the Galerkin method as an indirect method of solving the singular integral equation on $(-1, 1)$. However, much detail remains to be resolved. The choice of a basis H_n has not been discussed. Once H_n has been chosen we need to evaluate the inner products, and in general these can only be evaluated approximately by the use of a quadrature rule. This gives rise to additional errors and some discussion on the analysis of this "discrete" Galerkin method has been given by Spence and Thomas

(Ref. 15). Returning to the spaces H_n, one possible choice is to use spline functions. But recall that we are approximating to a function ϕ which is likely to be unbounded at the endpoints of the interval $(-1, 1)$. To get a uniform approximation to ϕ on $(-1, 1)$ will require the use of splines on a nonuniform grid. This problem was first discussed in this context by Thomas (Ref. 18) and his results have subsequently been refined by Elschner (Ref. 19).

An indirect method based on collocation has been described by Dow and Elliott (Ref. 20), although they do not solve directly for ϕ but for a closely related function ψ [as given in Eq. (81)]. More recently, Stenger and Elliott (Ref. 21) have considered an indirect Galerkin method based on the sinc-function, the inner products being evaluated approximately by an appropriate quadrature rule. The use of the sinc-function gives rise to errors which decay exponentially and promises to be a fruitful area for further development.

4. Direct Methods and Analysis of Their Convergence

In this and the next section we shall consider direct methods where we do not first perform the process of regularization. In all these methods it appears that we do not find approximations to ϕ but to a closely related function ψ. An inspection of the results of Section 2 suggests that we might write

$$\phi = (Z/r)\psi, \tag{81}$$

so that our complete equation can now be written as

$$A\psi + \mathcal{K}\psi = f, \tag{82}$$

where

$$A\psi = M(Z\psi/r) = \frac{aZ\psi}{r} + bT\left(\frac{Z\psi}{r}\right), \tag{83}$$

and

$$\mathcal{K}\psi = K(Z\psi/r) = \int_{-1}^{1} [k(t, \tau)Z(\tau)\psi(\tau)/r(\tau)] \, d\tau. \tag{84}$$

The fundamental function Z is of course in $H(p_1, p_2)$ and $1/Z \in H(q_1, q_2)$ where we have prescribed $p_1, p_2 \in [1, \infty)$. With A defined in this way it follows that if we define [see Eq. (34)]

$$\hat{A}^I f = (af)/(Zr) - bT(f/rZ) = (r/Z)\hat{M}^I f, \tag{85}$$

then we shall have, for all values of the index κ,

$$\begin{cases} \hat{A}^I A \psi = \psi + \psi^{(0)}, & \text{(86a)} \\ A\hat{A}^I f = f + f^{(0)}. & \text{(86b)} \end{cases}$$

Here $\psi^{(0)}$ is an arbitrary element out of ker(A) and $f^{(0)}$ is an arbitrary element out of ker($\hat{A}^I$). From Eqs. (25) and (35) respectively we see that

$$\begin{cases} \ker(A) = \text{span}\{b, bt, \ldots, bt^{\kappa-1}\}, & \text{(87a)} \\ \ker(\hat{A}^I) = \text{span}\{b, bt, \ldots, bt^{-\kappa-1}\}, & \text{(87b)} \end{cases}$$

where ker(A) = $\{0\}$ if $\kappa \leq 0$ and ker($\hat{A}^I$) = $\{0\}$ if $\kappa \geq 0$. Only when $\kappa = 0$ does A have an inverse so that then $\hat{A}^I = A^I$. At this stage of the development we shall not specify the spaces on which A, $\hat{A}^I$, and $\mathcal{K}$ operate. We shall assume only that X and Y are normed spaces and that

$$A : X \to Y, \quad \mathcal{K} : X \to Y, \quad \text{and} \quad \hat{A}^I : \text{ran}(A) \to X. \tag{88}$$

The choice of different methods (see Sections 5.1–5.3) will determine the spaces X and Y.

Let us return to Eq. (82). In the case when $\kappa \leq 0$ the solution ψ will be unique, but only provided that $(-\kappa)$ consistency conditions are satisfied when $\kappa < 0$. We may write these as

$$s_j(f - \mathcal{K}\psi) = 0, \quad \text{for } j = 1(1)(-\kappa). \tag{89a}$$

On the other hand, when $\kappa > 0$ then Eq. (82) is solvable for every $f \in H(p_1, p_2)$ but we shall need to specify a further κ conditions for uniqueness. We shall write these as

$$s_j \psi = 0, \quad \text{for } j = 1(1)\kappa. \tag{89b}$$

Finally we note that from the theory of Section 2, Eq. (82) is equivalent to the equation

$$(I + \hat{A}^I \mathcal{K})\psi = \hat{A}^I f + \psi^{(0)}, \tag{90}$$

provided that $f - \mathcal{K}\psi \in \text{ran}(A)$. The equivalence of (82) to (90) will be of great assistance to us in our analysis of approximate methods to which we now turn.

The bottom line in all approximate methods is that we must replace Eq. (82) by a system of linear algebraic equations whose behavior mimics, as closely as possible, that of (82). One of the first questions raised is that of coping with the index and here a simple observation is very pertinent. The index of any $m \times n$ matrix is given by

$$\kappa = n - m, \tag{91}$$

for if A denotes any such matrix of rank $k \leq \min(m, n)$ then $\dim(\ker A) = n - k$ while $\dim(\ker A^T) = m - k$, the transpose A^T of A being the adjoint of A. Recalling (28), Eq. (91) follows at once.

Let us recall now the theory of the solution of linear algebraic equations of the form

$$A_n x_n = y_m, \tag{92}$$

where A_n is an $m \times n$ matrix of rank $= \min(m, n)$ (we choose equality for reasons which will become apparent below), y_m is an $m \times 1$ vector, and we require the $n \times 1$ vector, x_n. This theory can be conveniently given in terms of the *singular value decomposition* of the matrix A_n [see, for example, Noble (Ref. 22, Section 10.8)]. First, when $\kappa = 0$ we have, since the rank of A_n is n, that

$$x_n = A_n^I y_m. \tag{93}$$

Suppose now that $\kappa \neq 0$; then there exist unitary matrices U_n, V_m say of order n, m respectively such that

$$V_m^H A_n U_n = \Sigma_n, \quad \text{say}, \tag{94}$$

where Σ_n is an $m \times n$ matrix of the form

$$\Sigma_n = \begin{cases} (\Lambda_m, 0), & \kappa > 0, \\ \begin{pmatrix} \Lambda_n \\ 0 \end{pmatrix}, & \kappa < 0. \end{cases} \tag{95}$$

Here the matrix Λ_k with $k = \min(m, n)$ is a diagonal matrix given by

$$\Lambda_k = \text{diag}(\mu_1, \mu_2, \ldots, \mu_k), \tag{96}$$

where we may write $\mu_1 \geq \mu_2 \geq \cdots \geq \mu_k > 0$. The positive numbers μ_i^2, $i = 1(1)k$, are the nonzero eigenvalues of $A_n^H A_n$ (or $A_n A_n^H$). The numbers μ_i, $i = 1(1)k$ are the *singular values* of A_n. From (94) we have

$$A_n = V_m \Sigma_n U_n^H, \tag{97}$$

so that we may rewrite Eq. (92) as

$$V_m^H A_n (U_n U_n^H) x_n = V_m^H y_m,$$

from which it follows on using (94) and (95) that

$$\begin{cases} (\Lambda_m, 0)(U_n^H x_n) = V_m^H y_m, & \kappa > 0, \quad (98a) \\ \begin{pmatrix} \Lambda_n \\ 0 \end{pmatrix} (U_n^H x_n) = V_m^H y_m, & \kappa < 0. \quad (98b) \end{cases}$$

When $\kappa < 0$ we see from Eq. (98b) that if we write $V_m = (v_1, v_2, \ldots, v_m)$ then

$$0 = v_i^H y_m, \quad i = (n+1)(1)(n-\kappa). \tag{99}$$

In order that a solution exists in this case, the vector y_m cannot be arbitrary but must satisfy these $(-\kappa)$ *consistency* conditions. When $\kappa \geq 0$ no such conditions are needed.

We can solve Eq. (92) by introducing the $n \times m$ matrix $\hat{A}_m^I$ defined by

$$\hat{A}_m^I = U_n \hat{\Sigma}_m^I V_m^H, \tag{100}$$

where

$$\hat{\Sigma}_m^I = \begin{cases} \begin{pmatrix} \Lambda_m^I \\ 0 \end{pmatrix}, & \kappa > 0, \\ (\Lambda_n^I, 0), & \kappa < 0. \end{cases} \tag{101}$$

If we recall that $\mu_1, \mu_2, \ldots, \mu_k$ are positive, then

$$\Lambda_k^I = \mathrm{diag}(\mu_1^{-1}, \mu_2^{-1}, \ldots, \mu_k^{-1}), \tag{102}$$

where $k = \min(m, n)$. Suppose that $\kappa > 0$; then from Eqs. (97) and (100) we have that

$$A_n \hat{A}_m^I = I_m, \tag{103a}$$

so that $\hat{A}_m^I$ is a right inverse of A_n. On the other hand, the same two equations give

$$\hat{A}_m^I A_n x_n = x_n + U_n \begin{pmatrix} 0 & 0 \\ 0 & I_\kappa \end{pmatrix} U_n^H x_n,$$

where I_κ denotes the unit matrix of order κ. Since it is readily shown that this last vector is in $\ker(A_n)$ we have

$$\hat{A}_m^I A_n x_n = x_n + x_n^{(0)}, \tag{104a}$$

where $x_n^{(0)}$ is any arbitrary vector out of $\ker(A_n)$. We can argue in an analogous way when $\kappa < 0$. We now find

$$\hat{A}_m^I A_n = I_n, \tag{103b}$$

so that $\hat{A}_m^I$ is a left inverse of A_n. Again

$$A_n \hat{A}_m^I y_m = y_m + y_m^{(0)}, \tag{104b}$$

where $y_m^{(0)}$ is an arbitrary vector out of $\ker(\hat{A}_m^I)$. No matter what value of κ is chosen we always have

$$\hat{A}_m^I A_n \hat{A}_m^I = \hat{A}_m^I \tag{105a}$$

and

$$A_n \hat{A}_m^I A_n = A_n, \tag{105b}$$

which should be compared with Eqs. (37) for the operator M. Thus, for all values of κ, we find that provided $y_m \in \operatorname{ran}(A_n)$, the solution of Eq. (92) is given by

$$x_n = \hat{A}_m^I y_m + x_n^{(0)}. \tag{106}$$

With this analysis we can now consider the solution of m linear algebraic equations in n unknowns given by

$$(A_n + K_n) x_n = y_m, \tag{107}$$

where A_n is of index κ. On assuming that $y_m - K_n x_n \in \operatorname{ran}(A_n)$ then we have

$$x_n = \hat{A}_m^I (y_m - K_n x_n) + x_n^{(0)},$$

so that x_n also satisfies the n equations in n unknowns given by

$$(I_n + \hat{A}_m^I K_n) x_n = \hat{A}_m^I y_m + x_n^{(0)}, \tag{108}$$

and we may consider comparing solutions of this equation with those of Eq. (90). As in the preceding section we shall do this by matching up the spaces X, Y with Euclidean spaces X_n, Y_m respectively by the use of restriction and prolongation operators.

Since A and $\mathcal{H}$ are considered as mappings between normed spaces X and Y, we shall consider A_n as a mapping between Euclidean spaces X_n and Y_m of dimension n and m respectively, where throughout we shall always have $n - m = \kappa$. Thus we shall require two sets of restriction and prolongation operators relating X_n with X and Y_m with Y. These are defined as follows, where we note that throughout the chapter c will denote a generic constant which will take different values at different places.

Definition 4.1. (a) To each $n \in \mathbf{N}$, a restriction operator r_n maps X into X_n and a prolongation operator p_n maps X_n into X subject to the following conditions:
(i) $\sup_n \|r_n\| \leq r < \infty$, $\|p_n\| \leq p(n)$;
(ii) $\lim_{n \to \infty} \|r_n \psi\| = \|\psi\|$ for all $\psi \in X$;
(iii) $r_n p_n = I_n$;
(iv) $\|p_n r_n \psi - \psi\| \leq c n^{-\delta_1}$, $\delta_1 > 0$, for all $\psi \in \mathrm{dom}(A)$.

(b) To each $m \in \mathbf{N}$, a restriction operator s_m maps Y into Y_m and a prolongation operator q_m maps Y_m into Y subject to the following conditions:
(i) $\sup_m \|s_m\| \leq s < \infty$, $\|q_m\| \leq q(m)$;
(ii) $\lim_{m \to \infty} \|s_m f\| = \|f\|$ $\forall f \in Y$;
(iii) $s_m q_m = I_m$;
(iv) $\|q_m s_m f - f\| \leq c n^{-\delta_2}$, $\delta_2 > 0$ $\forall f \in \mathrm{ran}(A)$.

Some comments are in order on these assumptions. Note that although we require the restriction operators to be uniformly bounded we do not require this of the prolongation operators. The functions $p(n)$ and $q(m)$ may diverge as $n \to \infty$ although one would expect that the rate of divergence will not be too great. Although we look upon A as defined on X we shall at times choose $\mathrm{dom}(A) \subset X$ and for such x we expect Definition 4.1(a)(iv) to be satisfied. However, one might not have $\|p_n r_n \psi - \psi\| \to 0$ as $n \to \infty$ for all $\psi \in X$. A similar comment applies to Definition 4.1(b)(iv). Note that our use of the norm symbol $\|\cdot\|$ will mean different things in different spaces. However, the context will always make clear which norm is being used and, when we consider specific choices of spaces, we shall always give a precise definition of the norm. As in Section 3, we shall define projection operators P_n and Q_m where

$$P_n = p_n r_n, \tag{109a}$$

$$Q_m = q_m s_m. \tag{109b}$$

P_n maps X into itself and it is immediately obvious that $P_n^2 = P_n$. Similarly Q_m maps Y into itself and $Q_m^2 = Q_m$. Conditions (iv) give pointwise convergence of $P_n \psi$ to ψ and $Q_m f$ to f on appropriate domains.

When we consider convergence we need to define two different sorts of convergence.

Definition 4.2. A sequence $\{x_n\}$ with $x_n \in X_n$, $n \in \mathbf{N}$, is said to converge discretely to $\psi \in X$ if $\lim_{n \to \infty} \|r_n \psi - x_n\| = 0$ and is said to converge globally to ψ if $\lim_{n \to \infty} \|\psi - p_n x_n\| = 0$.

The extension of this definition to the spaces Y_m and Y is immediately obvious.

We must now relate operators acting on the finite-dimensional and infinite-dimensional spaces. To do this the idea of *consistency* is extremely important.

Definition 4.3. (a) Suppose operators $B: X \to Y$ and $\{B_n\}$, where $B_n: X_n \to Y_m$ (with $\min(m, n) \geq 1$), are given. The sequence of operators $\{B_n\}$ is said to be consistent (of order $\gamma > 0$) with B if

$$\|\delta_n^B \psi\| \leq cn^{-\gamma}, \quad \text{for all } \psi \in \text{dom}(B), \tag{110}$$

where

$$\delta_n^B \psi = s_m B \psi - B_n r_n \psi. \tag{111}$$

(b) Suppose operators $C: Y \to X$ and $\{C_m\}$, where $C_m: Y_m \to X_n$ (with $\min\{m, n\} > 1$), are given. The sequence of operators $\{C_m\}$ is said to be consistent (of order $\gamma > 0$) with C if

$$\|\delta_n^C f\| < cn^{-\gamma}, \quad \text{for all } f \in \text{dom}(C), \tag{112}$$

where

$$\delta_n^C f = r_n C f - C_m s_m f. \tag{113}$$

As we shall see later the matrix $\hat{A}_m^I$ is of fundamental importance in our convergence analysis and the behavior of its norm as a function of n is crucial. This leads to the idea of *stability*.

Definition 4.4. The sequence of operators $\{A_n\}$ is said to be weakly stable if

$$\|\hat{A}_m^I\| \leq \hat{a}(n), \qquad \text{for all } m, n \in \mathbf{N}, \tag{114}$$

where there exists $r > 0$ such that $\lim_{n \to \infty} n^{-r} \hat{a}(n) = 0$ and $\hat{A}_m^I$ are matrices satisfying Eqs. (105).

If $\{\hat{a}(n)\}$ is uniformly bounded then we shall say that the sequence of operators is *stable*.

We shall now see how these ideas come together by considering an error analysis for the dominant equation $A\psi = f$ where we assume that $f \in \text{ran}(A)$. The discretized system of equations is given by $A_n x_n = y_m$ where we assume that $y_m \in \text{ran}(A_n)$. Thus we need to compare the solutions

$$\begin{cases} \psi = \hat{A}^I f + \psi^{(0)}, & (115a) \\ x_n = \hat{A}_m^I y_m + x_n^{(0)}, & (115b) \end{cases}$$

and this is done in the next theorem.

Theorem 4.1. Suppose that
  (i) $\{A_n\}$ is consistent of order γ_1 with A;
  (ii) $\{A_n\}$ is weakly stable;
  (iii) $y_m = s_m f$ and $y_m \in \text{ran}(A_n)$ for all $m, n \in \mathbf{N}$.
Then
  (a) to each $\psi^{(0)} \in \ker(A)$ there exists $x_n^{(0)} \in \ker(A_n)$ such that

$$\|r_n \psi^{(0)} - x_n^{(0)}\| \leq c\hat{a}(n) n^{-\gamma_1};$$

  (b) $\|\delta_n^{\hat{A}^I} f\| \leq c\hat{a}(n) n^{-\gamma_1}$;
  (c) to each solution of $A\psi = f$ there is a solution of $A_m x_n = s_m f$ such that

$$\|r_n \psi - x_n\| \leq c\hat{a}(n) n^{-\gamma_1}, \tag{116a}$$

$$\|\psi - p_n x_n\| \leq c \max\{p(n)\hat{a}(n) n^{-\gamma_1}, n^{-\delta_1}\}. \tag{116b}$$

Proof. (a) From Definition 4.3(a), since $\psi^{(0)} \in \text{dom}(A)$ then $\|\delta_n^A \psi^{(0)}\| \leq c n^{-\gamma_1}$. However from Eq. (111) we have that $\delta_n^A \psi^{(0)} = -A_n r_n \psi^{(0)}$. Consequently from (106) we can solve this equation to give

$$r_n \psi^{(0)} = -\hat{A}_m^I \delta_n^A \psi^{(0)} + x_n^{(0)},$$

where $x_n^{(0)}$ is an arbitrary element from $\ker(A_n)$. Hence

$$\|r_n\psi^{(0)} - x_n^{(0)}\| = \|-\hat{A}_m^I \delta_n^A \psi^{(0)}\| \le c\hat{a}(n)n^{-\gamma_1},$$

as required (we observe that this result is trivially true when $\kappa \le 0$, since then $\psi^{(0)} = 0$ and $x_n^{(0)} = 0$).

(b) Since $f \in \mathrm{ran}(A)$, on writing $f = A\psi$ and using Eq. (113) we have

$$\delta_n^{\hat{A}^I} f = r_n \hat{A}^I A\psi - \hat{A}_m^I s_m A\psi = r_n(\psi + \psi^{(0)}) - \hat{A}_m^I \{\delta_n^A \psi + A_n r_n \psi\},$$

on using Eqs. (86) and (111),

$$= -\hat{A}_m^I \delta_n^A \psi + r_n \psi^{(0)} - x_n^{(0)},$$

on using Eq. (106). Hence

$$\|\delta_n^{\hat{A}^I} f\| \le \|\hat{A}_m^I\| \cdot \|\delta_n^A \psi\| + \|r_n \psi^{(0)} - x_n^{(0)}\|,$$

from which the result follows at once on using (a).

(c) From Eq. (115a) we have

$$r_n\psi = r_n \hat{A}^I f + r_n \psi^{(0)},$$

so that on subtracting Eq. (115b) and using Eq. (113) applied to the operator $\hat{A}^I$ we have

$$r_n\psi - x_n = \delta_n^{\hat{A}^I} f + (r_n\psi^{(0)} - x_n^{(0)}).$$

Equation (116a) follows at once on taking norms and using (a) and (b). To prove (116b) we note that

$$\psi - p_n x_n = p_n(r_n\psi - x_n) + (\psi - p_n r_n \psi), \tag{117}$$

so that the result follows on taking norms, using the result just obtained and recalling Definition 4.1(a). □

This theorem gives the rates of convergence for both discrete and global convergence of the dominant equation. It exhibits in particular the dependence of the convergence on both $p(n)$ and $\hat{a}(n)$, so that if these diverge as $n \to \infty$ they should not diverge too quickly. In one case (see Section 5.1) we see that both these quantities are uniformly bounded so that convergence is then guaranteed by the presence of the $n^{-\gamma_1}$ and $n^{-\delta_1}$ terms; in the second case (see Section 5.2) $p(n)$ and $\hat{a}(n)$ diverge only as $\log n$ which does not affect the rate of convergence very much. We shall now turn our attention to the complete equation and its approximate solution.

Let us recall that we are starting with the singular integral equation $A\psi + \mathcal{K}\psi = f$ [see (82)], which we know is equivalent to [see (90)]

$$(I + \hat{A}^I\mathcal{K})\psi = \hat{A}^I f + \psi^{(0)} = g, \quad \text{say.} \qquad (118)$$

Our original equation is discretized to give a set of m linear algebraic equations in n unknowns which we write as $(A_n + K_n)x_n = y_m$ [see (107)] and we know that this set of equations is equivalent to

$$(I_n + \hat{A}_m^I K_n)x_n = \hat{A}_m^I y_m + x_n^{(0)} = g_n, \quad \text{say} \qquad (119)$$

[see (108)]. We shall be assuming throughout that $(I + \hat{A}^I\mathcal{K})^I$ exists or, in other words, that -1 is not an eigenvalue of the (compact) operator $\hat{A}^I\mathcal{K}$. We would like to show that, for large enough n, the matrix $(I_n + \hat{A}_m^I K_n)$ occurring in (119) possesses an inverse and we should then like to obtain bounds for $\|r_n\psi - x_n\|$ and $\|\psi - p_n x_n\|$. We shall write

$$J = \hat{A}^I\mathcal{K}, \qquad (120a)$$

$$J_n = \hat{A}_m^I K_n, \qquad (120b)$$

and since J is a compact operator on X into X we can, as we did in Section 3, quote results from the theory of the approximate solution of such equations. A convenient starting point is the next theorem, a particular form of which was given as Theorem 3.1.

Theorem 4.2. Suppose $(I + J)^I$ exists. Define

$$R_n = r_n(I + J)^I(Jp_n - p_n J_n). \qquad (121)$$

If there exists n_0 such that $\Delta_n = \|R_n\| < 1$ for all $n > n_0$, then $(I_n + J_n)^I$ exists for all $n > n_0$ and

$$\|(I_n + J_n)^I\| \le cp(n)/(1 - \Delta_n). \qquad (122)$$

Proof. See Thomas (Ref. 16, Theorem 3.2). □

If we define the consistency error of J by

$$\delta_n^J \psi = r_n J\psi - J_n r_n \psi \qquad (123)$$

for all $\psi \in \text{dom}(J)$, then we have the following result.

Theorem 4.3. Suppose that $(I_n + J_n)^I$ exists, then

$$\|r_n\psi - x_n\| \le \|(I_n + J_n)^I\|\{\|r_n g - g_n\| + \|\delta_n^J \psi\|\}. \qquad (124)$$

Proof. Operating on (118) with r_n and subtracting (119) gives

$$r_n \psi - x_n + r_n J\psi - J_n x_n = r_n g - g_n.$$

From Eq. (123) we may rewrite this as

$$(I_n + J_n)(r_n \psi - x_n) = (r_n g - g_n) - \delta_n^J \psi, \qquad (125)$$

from which (124) follows immediately on taking norms. □

So much for the results on Fredholm equations. What we must do now is to relate these results to the operators and functions appearing in our singular integral equation. We shall always assume that the operator $\mathcal{K}$ is defined on dom(A) and furthermore that ran$(\mathcal{K}) \subseteq \text{ran}(A) = \text{dom}(\hat{A}^I)$. The following theorem is a first step in this process.

Theorem 4.4. Suppose that
(i) $\{A_n\}$ is consistent of order γ_1 with A,
(ii) $\{K_n\}$ is consistent of order γ_2 with $\mathcal{K}$,
(iii) $\|s_m f - y_m\| \le c n^{-\gamma_3}$, $\gamma_3 > 0$.

Then

$$\|\delta_n^J \psi\| \le c\hat{a}(n) n^{-\min(\gamma_1, \gamma_2)}, \qquad \text{for all } \psi \in \text{dom}(A), \qquad (126)$$

and

$$\|r_n g - g_n\| \le c\hat{a}(n) n^{-\min(\gamma_1, \gamma_3)}. \qquad (127)$$

Proof. Consider first (126). From Eqs. (120) and (123) we have

$$\delta_n^J \psi = r_n \hat{A}^I \mathcal{K} \psi - \hat{A}_m^I K_n r_n \psi = r_n \hat{A}^I (\mathcal{K}\psi) - \hat{A}_m^I s_m(\mathcal{K}\psi) + \hat{A}_m^I \delta_n^\mathcal{K} \psi,$$

on using (111) for the operator $\mathcal{K}$. Consequently

$$\delta_n^J \psi = \delta_n^{\hat{A}^I}(\mathcal{K}\psi) + \hat{A}_m^I \delta_n^\mathcal{K} \psi. \qquad (128)$$

From Theorem 4.1(b) and Definition 4.4 we find

$$\|\delta_n^J \psi\| \le c\hat{a}(n) n^{-\gamma_1} + c\hat{a}(n) n^{-\gamma_2},$$

and (126) follows at once.

To establish (127), we have from (118) and (119)

$$r_n g - g_n = (r_n \psi^{(0)} - x_n^{(0)}) + (r_n \hat{A}^I f - \hat{A}_m^I y_m)$$

$$= (r_n \psi^{(0)} - x_n^{(0)}) + \delta_n^{\hat{A}^I} f + \hat{A}_m^I (s_m f - y_m),$$

on using (113) for the operator $\hat{A}^I$. The result follows on taking norms and using Theorem 4.1(b). □

From (124) we see that we must now establish an upper bound for $\|(I_n + J_n)^I\|$. From Theorem 4.2 we see that this inverse exists only for those n such that $\Delta_n = \|R_n\| < 1$ and we require $\lim_{n \to \infty} \Delta_n = 0$. To express the operator R_n in terms of the operators of our singular integral equation we first define, for the operator $\mathcal{H}$,

$$\Theta_n^{\mathcal{H}} = s_m \mathcal{H} p_n - K_n. \tag{129}$$

Then, on recalling (120), we have from (121) that

$$R_n = r_n(I + \hat{A}^I K)^I \{ p_n \hat{A}_m^I \Theta_n^{\mathcal{H}} + (1 - p_n r_n) \hat{A}^I \mathcal{H} p_n + p_n \delta_n^{\hat{A}^I} \mathcal{H} p_n \}. \tag{130}$$

Consequently, since we have assumed that $(I + \hat{A}^I K)^I$ exists and is bounded, r_n is stable and, recalling that $\hat{A}^I \mathcal{H}$ is compact (being the product of a bounded and a compact operator), we have

$$\Delta_n = \|R_n\| \leq cp(n) \max\{\hat{a}(n) \|\Theta_n^{\mathcal{H}}\|, n^{-\delta_1}, p(n)\hat{a}(n)n^{-\gamma_1}\}, \tag{131}$$

where for the last term we recall Theorem 4.1(b). Given that $\Delta_n \to 0$ as $n \to \infty$, we then have from (122) that

$$\|(I_n + J_n)^I\| \leq cp(n), \quad \text{for } n \text{ large enough.} \tag{132}$$

We shall summarize the discussion thus far by stating a theorem which will essentially be a check list for the particular methods we shall be considering in the next section.

Theorem 4.5. Suppose that
 (i) $\{A_n\}$ is consistent of order γ_1 with A,
 (ii) $\{A_n\}$ is weakly stable with $\|\hat{A}_m^I\| \leq \hat{a}(n)$,
 (iii) $\{K_n\}$ is consistent of order γ_2 with $\mathcal{H}$,
 (iv) $\|s_m f - y_m\| \leq cn^{-\gamma_3}$,
 (v) $\lim_{n \to \infty} p(n)\hat{a}(n)\|\Theta_n^{\mathcal{H}}\| = \lim_{n \to \infty} p(n)n^{-\delta_1} =$
    $= \lim_{n \to \infty} p^2(n)\hat{a}(n)n^{-\gamma_1} = 0.$
Then there exists $n_0 \in \mathbb{N}$ such that for all $n \geq n_0$

$$\|r_n \psi - x_n\| \leq cp(n)\hat{a}(n)n^{-\min(\gamma_1, \gamma_2, \gamma_3)}, \tag{133}$$

where c is independent of n.

Proof. This follows from the above discussion. □

Inequality (133) determines both the discrete convergence of x_n to ψ and also gives an upper bound for the rate of convergence. It is readily observed that if p_n is stable, so that the sequence $\{p(n)\}$ is uniformly bounded, then there is a considerable simplification of the above analysis. A similar comment applies if $\{\hat{a}(n)\}$ is uniformly bounded. We might note that we can obtain an upper bound for the global convergence of x_n to ψ by writing $\psi - p_n x_n$ as in Eq. (117) and using (133). This gives

$$\|\psi - p_n x_n\| \leq c \max\{n^{-\delta_1}, p^2(n)\hat{a}(n)n^{-\min(\gamma_1, \gamma_2, \gamma_3)}\}. \tag{134}$$

We conclude this section by noting important simplifications which occur when some further assumptions are made. These will enable us to identify the matrices A_n and $\hat{A}'_m$ in terms of A and $\hat{A}'$, respectively. Let us recall that $P_n = p_n r_n$ is a projection operator on X and that similarly $Q_m = q_m s_m$ is a projection operator on Y. For the operators A and $\hat{A}'$ we make the further assumptions that for large enough n

$$P_n\{\ker(A)\} = \ker(A), \tag{135a}$$

$$Q_m\{\ker(\hat{A}')\} = \ker(\hat{A}'), \tag{135b}$$

and

$$Q_m A p_n = A p_n, \tag{136a}$$

$$P_n \hat{A}' q_m = \hat{A}' q_m. \tag{136b}$$

We can readily prove the following theorem.

Theorem 4.6. Given that (135) and (136) are satisfied, if we choose

$$A_n = s_m A p_n, \tag{137}$$

then

$$r_n(\ker(A)) = \ker(A_n), \tag{138a}$$

$$s_m(\ker(\hat{A}')) = \ker(\hat{A}'_m), \tag{138b}$$

and

$$\hat{A}'_m = r_n \hat{A}' q_m. \tag{139}$$

Convergence Theorems for Singular Integral Equations

Proof. First we observe that if $\psi^{(0)} \in \ker(A)$ then

$$A_n r_n \psi^{(0)} = s_m A P_n \psi^{(0)} = s_m A \psi^{(0)} = 0,$$

where we have used Eq. (135a). Since if $\kappa > 0$ this is true for each of κ linearly independent functions in $\ker(A)$, (138a) is established. Assuming that $\hat{A}_m^I$ is as defined by (139), (138b) follows for $\kappa < 0$. Finally let us look at (139). We have

$$\hat{A}_m^I A_n x_n = r_n \hat{A}^I Q_m A p_n x_n$$

$$= r_n (\hat{A}^I A) p_n x_n \qquad \text{by (136a)}$$

$$= r_n \{ p_n x_n + \psi^{(0)} \} \qquad \text{by (86a)}$$

$$= x_n + x_n^{(0)}.$$

Thus (104a) is established. By considering $A_n \hat{A}_m^I y_m$ we recover, by similar arguments, Eq. (104b) and this establishes the theorem. □

Bearing in mind (137), which gives A_n in terms of A, suppose we similarly defined

$$K_n = s_m \mathcal{K} p_n. \qquad (140)$$

With this choice of K_n we see from (129) that $\Theta_n^{\mathcal{K}} = 0$ and this simplifies the definition of R_n as given by (130). Making use of (136) we see that we can write

$$R_n = r_n (I + \hat{A}^I K)^I \hat{A}^I (I - Q_m) \mathcal{K} p_n. \qquad (141)$$

Since the first three operators appearing on the right of (141) are stable, we have simply that

$$\Delta_n = \|R_n\| \leq cp(n) \cdot \|(I - Q_m)\mathcal{K}\|. \qquad (142)$$

Since we assume $\mathcal{K}$ to be compact, we know that $\|(I - Q_m)\mathcal{K}\| \to 0$ as $n \to \infty$ and, provided it goes to zero more quickly than $p(n)$ goes to $+\infty$, then $\Delta_n \to 0$ as $n \to \infty$. This simplifies somewhat condition (v) of Theorem 4.5 in this particular case.

5. Some Examples of Direct Methods

In the last section we set up a theoretical framework for the convergence of approximate solutions of singular integral equations based on the direct discretization of the operators A and $\mathcal{H}$. We shall now consider some particular cases. These are all based on the properties of the two sets of orthogonal polynomials which we introduced at the end of Section 2. First we shall consider a Galerkin-Petrov method based on weighted L_2 spaces, the weights being those from which the orthogonal polynomials are derived. Assuming that all the inner products may be evaluated exactly, we obtain an elegant convergence analysis whose results are given in Theorem 5.3. Next we shall discuss a collocation method whose collocation points are chosen as the zeros of the polynomial u_m. Here the mathematics becomes much more difficult and the application is restricted to equations where f and $k(\cdot, \tau)$ are Hölder continuous on $[-1, 1]$. Much of the mathematical detail is omitted here. Finally, in Section 5.3, we shall see how these two methods come together when we use appropriate Gauss quadrature rules to evaluate the inner products in the Galerkin-Petrov method.

5.1. The Galerkin–Petrov Method.

Recall from Section 2 that when a and b are real, both the fundamental function Z and r are positive so that we can introduce two positive weight functions w_1 and w_2 where $w_1 \in H(p_1, p_2)$ and $w_2 \in H(q_1, q_2)$. Following Section 2 we introduce two sets of polynomials $\{t_n\}$ and $\{u_n\}$ which, in this section, we assume to be *orthonormal* with respect to the weight functions w_1 and w_2, respectively. Then if we assume that b is a polynomial of degree μ with all its zeros on $[-1, 1]$ (this was Assumption A), we shall have

$$u_n = At_{n+\kappa}, \quad \text{for } n \geq \max(0, \mu - \kappa), \tag{143}$$

and

$$t_n = \hat{A}^I u_{n-\kappa}, \quad \text{for } n \geq \max(0, \kappa + \mu); \tag{144}$$

see Eqs. (49) and (51), respectively.

Let $\mathcal{H}_i$, $i = 1, 2$ denote two real, separable Hilbert spaces with inner products denoted and defined by

$$\langle \psi_1, \psi_2 \rangle_i = \int_{-1}^{1} w_i(\tau) \psi_1(\tau) \psi_2(\tau) \, d\tau; \quad i = 1, 2. \tag{145}$$

For $\psi \in \mathcal{H}_i$ its norm will be given by $\|\psi\|_{\mathcal{H}_i} = \langle \psi, \psi \rangle_i^{1/2}$. The operator A will now be considered as an operator on $\mathcal{H}_1$ into $\mathcal{H}_2$ and similarly $\hat{A}^I$ will be an operator on $\mathcal{H}_2$ into $\mathcal{H}_1$. Let us consider first the operator A. The Hilbert-adjoint of A, $A^*: \mathcal{H}_2 \to \mathcal{H}_1$ is such that

$$\langle A\psi_1, \psi_2 \rangle_2 = \langle \psi_1, A^*\psi_2 \rangle_1 \tag{146}$$

for every $\psi_1 \in \mathcal{H}_1$ and $\psi_2 \in \mathcal{H}_2$. From the definition of inner product we have immediately that

$$A^*\psi = M^\times(\psi/Zr), \tag{147}$$

on recalling Parseval's theorem and Eq. (13). The norm of A will be defined by

$$\|A\| = \sup_{\psi \in \mathcal{H}_1} \|A\psi\|_{\mathcal{H}_2} / \|\psi\|_{\mathcal{H}_1}. \tag{148}$$

It is not difficult to show that A is a bounded linear operator from $\mathcal{H}_1$ into $\mathcal{H}_2$. First, we have that

$$\|A\psi\|_{\mathcal{H}_2}^2 = \langle A\psi, A\psi \rangle_2 = \langle \psi, A^*A\psi \rangle_1. \tag{149}$$

Next, from the Poincaré-Bertrand formula [Eq. (11)] and Eq. (31), where we are assuming that b is a polynomial, we find, after some algebra, that for $t \in (-1, 1)$

$$(A^*A\psi)(t) = \psi(t) - \frac{1}{\pi} \int_{-1}^{1} w_1(\tau) \left(\frac{\text{p.p.}(bX^{-1}; \tau) - \text{p.p.}(bX^{-1}; t)}{\tau - t} \right) \psi(\tau)\, d\tau. \tag{150}$$

Finally, on applying Hölder's inequality to this equation we find that there exists a constant $c > 0$ such that

$$\|A\psi\|_{\mathcal{H}_2}^2 \leq c\|\psi\|_{\mathcal{H}_1}^2. \tag{151}$$

The boundedness of A follows at once. Similarly we can show that $\hat{A}^I$ is a bounded linear operator on $\mathcal{H}_2$ into $\mathcal{H}_1$, although we shall omit the details.

With these preliminaries we can now describe the Galerkin-Petrov method. We assume an approximate solution $\psi_n \in \mathcal{H}_1$ of the form

$$\psi_n = \sum_{j=1}^{n} \xi_j t_{j-1}, \tag{152}$$

which gives rise to a residual, $(\text{res})_n$, where

$$(\text{res})_n = (A + \mathcal{K})\psi_n - f. \tag{153}$$

The n unknown constants in (152) are obtained by requiring that

$$\langle u_{i-1}, (\text{res})_n \rangle_2 = 0, \quad \text{for } i = 1(1)m. \tag{154}$$

Equation (154) gives rise to our required system of m linear algebraic equations in n unknowns which we write in the usual way as

$$(A_n + K_n)x_n = y_m. \tag{155}$$

In this case

$$A_n = (\langle u_{i-1}, At_{j-1} \rangle_2), \tag{156a}$$

$$K_n = (\langle u_{i-1}, \mathcal{K}t_{j-1} \rangle_2), \tag{156b}$$

for $i = 1(1)m$, $j = 1(1)n$, and $x_n = (\xi_1, \xi_2, \ldots, \xi_n)^T$ with

$$y_m = (\langle u_{i-1}, f \rangle_2), \quad i = 1(1)m. \tag{157}$$

From Eq. (143) we see that for $i = 1(1)m$ and $j \geq \max(1 + \mu, 1 + \kappa)$,

$$\langle u_{i-1}, At_{j-1} \rangle_2 = \delta_{i-1, j-1-\kappa}, \tag{158}$$

so that much of the matrix A_n has a very simple structure.

At this point it is convenient to identify appropriate restriction and prolongation operators. We have

$r_n: \mathcal{H}_1 \to X_n$ such that $r_n x = (\langle t_0, x \rangle_1, \ldots, \langle t_{n-1}, x \rangle_1)^T$;

$p_n: X_n \to \mathcal{H}_1$ such that $p_n x_n = \sum_{i=1}^{n} \xi_i t_{i-1}$ where $x_n = (\xi_1, \xi_2, \ldots, \xi_n)^T$.

For $x_n \in X_n$ we define

$$\|x_n\|_{X_n} = \|p_n x_n\|_{\mathcal{H}_1} = \left(\sum_{i=1}^{n} \xi_i^2 \right)^{1/2}, \tag{159}$$

so that we have chosen the 2-norm in X_n. Then it follows that

$$\|p_n\| = \sup_{x_n \in X_n} \|p_n x_n\|_{\mathcal{H}_1} / \|x_n\|_{X_n} = 1, \tag{160}$$

Convergence Theorems for Singular Integral Equations

so that in this case the prolongation operators are uniformly bounded. Similarly $\|r_n\| = 1$. Between the spaces $\mathcal{H}_2$ and Y_m we define

$$s_m : \mathcal{H}_2 \to Y_m \text{ such that } s_m y = (\langle u_0, y \rangle_2, \ldots, \langle u_{m-1}, y \rangle_2)^T;$$

$$q_m : Y_m \to \mathcal{H}_2 \text{ with } q_m y_m = \sum_{j=1}^{m} \eta_j u_{j-1} \text{ where } y_m = (\eta_1, \eta_2, \ldots, \eta_n)^T.$$

Again with the use of the 2-norm in Y_m we find that $\|s_m\| = \|q_m\| = 1, m \in \mathbb{N}$. In terms of these operators we see at once that

$$A_n = s_m A p_n, \tag{161a}$$

$$K_n = s_m \mathcal{K} p_n. \tag{161b}$$

Thus we appear to be constructing an example similar to that discussed in Theorem 4.6. Let $T_n = \text{span}\{t_0, t_1, \ldots, t_{n-1}\}$; then $P_n = p_n r_n$ is the orthogonal projection of $\mathcal{H}_1$ onto T_n. Similarly, if we write $U_m = \text{span}\{u_0, u_1, \ldots, u_{m-1}\}$ then $Q_m = q_m s_m$ is the orthogonal projection of $\mathcal{H}_2$ onto U_m. In this case we have even more useful relationships than those given by Eqs. (136) as the following theorem shows.

Theorem 5.1. For $n \geq \max(\mu, \kappa)$

$$AP_n = Q_m A, \tag{162a}$$

$$P_n \hat{A}^I = \hat{A}^I Q_m. \tag{162b}$$

Proof. We shall consider just the first of these two equations. For any $\psi \in \mathcal{H}_1$ we have $\psi = \sum_{j=0}^{\infty} \langle t_j, \psi \rangle_1 t_j$, so that $P_n \psi = \sum_{j=0}^{n-1} \langle t_j, \psi \rangle_1 t_j$ and we have that, for $n \geq \max(\mu, \kappa)$,

$$A(I - P_n)\psi = \sum_{j=n}^{\infty} \langle t_j, \psi \rangle_1 A t_j = \sum_{j=n}^{\infty} \langle t_j, \psi \rangle_1 u_{j-\kappa} \quad \text{by (143)}.$$

Now recall that

$$t_j = \text{p.p.}(u_{j-k} X^{-1}) \quad \text{by Eq. (48(b))}$$

$$= M^\times(u_{j-\kappa}/Zr) \quad \text{by Eqs. (13) and (31)}$$

$$= A^* u_{j-\kappa} \quad \text{by Eq. (147)}.$$

Consequently we can write

$$A(I - P_n)\psi = \sum_{j=n}^{\infty} \langle A^* u_{j-\kappa}, \psi \rangle_1 u_{j-\kappa} = \sum_{j=n}^{\infty} \langle u_{j-\kappa}, A\psi \rangle_2 u_{j-\kappa}$$

$$= \sum_{j=m}^{\infty} \langle u_j, A\psi \rangle_2 u_j = (I - Q_m)A\psi.$$

We can proceed similarly for the second of (162) and the theorem is proved. □

It remains to show that Eqs. (135) are satisfied. From Eqs. (87) we see that ker(A) is either {0} when $\kappa \leq 0$ or a polynomial of degree $(\mu + \kappa - 1)$ when $\kappa \geq 1$. Similarly ker($\hat{A}'$) is either {0} when $\kappa \geq 0$ or a polynomial of degree $(\mu - \kappa - 1)$ whenever $\kappa \leq -1$. Consequently Eqs. (135) will be satisfied in all cases whenever $n \geq \max\{\mu - 1, \mu + \kappa - 1\}$. Thus all the conditions of Theorem 4.6 are satisfied in this case so that

$$\hat{A}_m^I = r_n \hat{A}^I q_m. \tag{163}$$

We shall now gather together some useful results before applying the convergence theorem (Theorem 4.5) to this method.

Lemma 5.2. For the Galerkin-Petrov method
(i) $\delta_n^A \psi = 0$, for all $\psi \in \mathcal{H}_1$,
(ii) $\Theta_n^{\mathcal{H}} x_n = 0$, for all $x_n \in X_n$,
(iii) $\hat{a}(n) \leq c$, for some constant c.

Proof. (i) From Eq. (111) with A_n given by (161a) we have

$$\delta_n^A \psi = s_m A\psi - s_m A P_n \psi \quad \text{since } P_n = p_n r_n$$

$$= s_m A\psi - s_m Q_m A\psi \quad \text{by (162a)}$$

$$= 0 \quad \text{since } s_m Q_m = (s_m q_m) s_m = s_m.$$

Thus we can write $\gamma_1 = \infty$ in this case.

(ii) This follows at once from the definition of $\Theta_n^{\mathcal{H}}$ as given by Eq. (129) and the definition of K_n in this case; see Eq. (161b).

(iii) Consider any $y_m \in \text{ran}(A_n)$; then

$$\|\hat{A}_m^I y_m\|_{X_n} = \|r_n \hat{A}^I q_m y_m\|_{X_n} \quad \text{by (163)}$$

$$= \|p_n r_n \hat{A}^I q_m y_m\|_{\mathcal{H}_1} \quad \text{by (159)}$$

$$= \|(\hat{A}^I Q_m) q_m y_m\|_{\mathcal{H}_1} \quad \text{by (162b)}$$

$$= \|\hat{A}^I q_m y_m\|_{\mathcal{H}_1} \quad \text{since } Q_m = q_m s_m \text{ and } s_m q_m = I_m$$

$$\leq \|\hat{A}^I\| \cdot \|y_m\|_{Y_m}.$$

Since $\hat{A}^I$ is a bounded linear operator from $\mathcal{H}_2$ into $\mathcal{H}_1$, it follows at once that $\|\hat{A}_m^I\| \leq \|\hat{A}^I\|$ so that $\hat{a}(n) \leq c$ for some constant c. □

In the light of this analysis let us restate the convergence theorem for this method.

Theorem 5.3. Let $\mathcal{K}$ be a compact operator on $\mathcal{H}_1$ into $\mathcal{H}_2$. Suppose $f \in \text{ran}(A)$ and let A_n, K_n be given by Eqs. (161). Then for all $n \geq \max(1 + \mu, 1 + \kappa)$ the equations $(A_n + K_n)x_n = s_m f$ possess a solution such that

$$\|r_n \psi - x_n\| \leq c \cdot n^{-\gamma_2}, \tag{164}$$

$$\|\psi - p_n x_n\| \leq c \cdot \max\{n^{-\delta_1}, n^{-\gamma_2}\}. \tag{165}$$

Proof. Since we have chosen $y_m = s_m f$, then we can also choose $\gamma_3 = \infty$. We have already commented that $\gamma_1 = \infty$ so that the discrete convergence will depend only on the rate at which $\|\delta_n^{\mathcal{K}} \psi\|$ tends to zero. Global convergence as given by (165) is obtained at once by noting that

$$\psi - p_n x_n = p_n(r_n \psi - x_n) + (\psi - p_n r_n \psi),$$

taking norms and using (164). □

We might comment that

$$\|\delta_n^{\mathcal{K}} \psi\| = \|s_m \mathcal{K}(I - P_n)\psi\| \leq \|\mathcal{K}(I - P_n)\psi\| \leq c n^{-\delta_1},$$

since $\mathcal{K}$, being compact, is certainly bounded. The value of δ_1 will depend upon properties of the function ψ which, in turn, depend upon those of f and $k(\cdot, \tau)$. We shall pursue this topic in a little more detail in the next section.

So much for the analysis of this method. When it comes to implementing the algorithm there are many problems to be considered. We must, for example, know Z explicitly or at least be able to compute it for a given

$t \in (-1, 1)$. We must be able to evaluate the inner products in both $\mathcal{H}_1$ and $\mathcal{H}_2$ and this will inevitably require the use of some appropriate quadrature rules. In Section 5.3 we shall consider these evaluations using Gauss-type quadrature rules based on the zeros of the polynomials t_n and u_m. But these zeros, together with the associated Christoffel numbers, must also be computed and this is not always straightforward. The polynomials t_n and u_m are in fact generalized Jacobi polynomials and, for a review of their properties, the reader is referred to Nevai (Ref. 23). However, in the special case when both a and b are constants, the polynomials t_n and u_m turn out to be Jacobi polynomials and much can then be made of the known analytic properties of these polynomials.

Finally, mention should be made of the use of the iterated Galerkin method. This will be similar to that already discussed in Section 3 and for further details the reader is referred to Elliott (Ref. 24).

5.2. A Collocation Method. For this method we must put further restrictions on both f and k. In particular we shall assume that both f and $k(\cdot, \tau)$ are Hölder continuous on $[-1, 1]$ (a function f is said to be Hölder continuous on $[-1, 1]$ with Hölder index ν, $0 < \nu \leq 1$ if there exists a constant $A \geq 0$ such that

$$|f(t_1) - f(t_2)| \leq A|t_1 - t_2|^\nu, \tag{166}$$

for all $t_1, t_2 \in [-1, 1]$). Later we shall be even more restrictive on the functions f and k.

In order to discretize (82) we shall first introduce two quadrature rules. These will be based on the zeros of the polynomial t_n which, from Section 2, were orthogonal on $[-1, 1]$ with respect to the weight function w_1. Since most of the results of Section 2 are based on Assumption A we shall, throughout Section 5.2, assume that b is a polynomial.

For any $n \in \mathbf{N}$, let $\tau_{j,n}, j = 1(1)n$, where

$$-1 < \tau_{n,n} < \tau_{n-1,n} < \cdots < \tau_{1,n} < 1 \tag{167}$$

denote the n simple zeros of the polynomial t_n [see Eq. (47)]. Let $\mu_{j,n}$, $j = 1(1)n$, where

$$\mu_{j,n} = \pi T(w_1 t_n; \tau_{j,n}) / t'_n(\tau_{j,n}) \tag{168}$$

denote the corresponding Cotes numbers where, as before, $w_1 = Z/r$. The Gauss quadrature rule based on the zeros of t_n is given by

$$\int_{-1}^{1} w_1(\tau)\psi(\tau) \, d\tau = \sum_{j=1}^{n} \mu_{j,n}\psi(\tau_{j,n}) + R_n^G(\psi), \tag{169}$$

Convergence Theorems for Singular Integral Equations

where $R_n^G(\psi) = 0$ whenever $\psi \in \mathbf{P}_{2n-1}$. For Cauchy principal-value integrals there exists a comparable quadrature rule [due initially to Hunter (Ref. 25), but see also Rabinowitz (Ref. 26)] which can be written as

$$\fint_{-1}^{1} \frac{w_1(\tau)\psi(\tau)}{\tau - t} d\tau = \sum_{j=1}^{n} \mu_{j,n} \frac{\psi(\tau_{j,n})}{\tau_{j,n} - t} + \frac{\pi \psi(t)}{t_n(t)} T(w_1 t_n; t) + R_n^H(\psi; t), \quad (170)$$

where $t \in (-1, 1)$ and we assume $t \neq \tau_{j,n}$, $j = 1(1)n$. The remainder is such that $R_n^H(\psi; t) = 0$ whenever $\psi \in \mathbf{P}_{2n}$. The right-hand side of (170) needs modification if $t = \tau_{j,n}$ for some j, but we shall not consider this modification here. With these two quadrature rules we can now consider the discretization of Eq. (82). Substituting (169) and (170) into (82)–(84) and recalling (49), which defines the polynomial u_m in terms of t_n, we find

$$\frac{u_m(t)}{t_n(t)} \psi(t) + \sum_{j=1}^{n} \frac{\mu_{j,n} b(t) \psi(\tau_{j,n})}{\pi(\tau_{j,n} - t)} + \sum_{j=1}^{n} \mu_{j,n} k(t, \tau_{j,n}) \psi(\tau_{j,n})$$

$$= f(t) - \frac{b(t)}{\pi} R_n^H(\psi; t) - R_n^G(k(t, \cdot)\psi), \quad (171)$$

where $t \neq \tau_{j,n}$. We shall obtain m equations for the n unknowns $\psi(\tau_{j,n})$ by choosing $t = t_{i,m}$, $i = 1(1)m$, where

$$-1 < t_{m,m} < t_{m-1,m} < \cdots < t_{1,m} < 1 \quad (172)$$

are the m simple zeros of the polynomial u_m [recall from Eq. (50) that the polynomials u_m are orthogonal on $(-1, 1)$ with respect to the weight function $w_2 = 1/(Zr)$]. If we also neglect the remainders R_n^H and R_n^G arising in (171), we obtain a system of m linear algebraic equations in n unknowns given by

$$(A_n + K_n) x_n = y_m, \quad (173)$$

where, if we write $A_n = (a_{i,j}^{(n)})$ and $K_n = (k_{i,j}^{(n)})$, we find, provided $\tau_{j,n} \neq t_{i,m}$, that

$$a_{i,j}^{(n)} = \frac{\mu_{j,n} b(t_{i,m})}{\pi(\tau_{j,n} - t_{i,m})}, \quad (174a)$$

$$k_{i,j}^{(n)} = \mu_{j,n} k(t_{i,m}, \tau_{j,n}), \quad (174b)$$

for $i = 1(1)m$ and $j = 1(1)n$. If it should happen that $\tau_{J,n} = t_{I,m}$ then Elliott (Ref. 27) has shown that

$$a_{I,j}^{(n)} = \begin{cases} 0, & j \neq J \\ Z(t_{I,m}) \operatorname{sgn} a(t_{I,m}), & j = J. \end{cases} \quad (175)$$

Finally, in Eq. (173) we have

$$y_m = (f(t_{1,m}), f(t_{2,m}), \ldots, f(t_{m,m}))^T, \tag{176}$$

and we want to solve for $x_n = (\xi_1, \xi_2, \ldots, \xi_n)^T$, where ξ_j will hopefully be a good approximation to $\psi(\tau_{j,n})$.

We now have the required discretization for the method of classical collocation and in order to discuss convergence we need to choose appropriate spaces and the restriction and prolongation operators between these spaces. Let us choose $X = C[-1, 1]$, the space of all continuous functions on $[-1, 1]$ equipped with the uniform norm. Then $\text{dom}(A)$ will be a subspace of $C[-1, 1]$ which we shall identify later. Since $f \in H_\nu[-1, 1]$, $0 < \nu < 1$, we shall choose Y to be the space of all such functions equipped with the Hölder norm. That is, if $g \in H_\nu[-1, 1]$ then

$$\|g\|_Y = \max\{\|g\|_\infty, \sup |g(t_1) - g(t_2)|/|t_1 - t_2|^\nu\}, \tag{177}$$

where the supremum is taken over all $t_1, t_2 \in [-1, 1]$ provided $t_1 \neq t_2$. The operator $\hat{A}^I$ will now be considered as mapping Y into X and we shall define the norm of $\hat{A}^I$ by

$$\|\hat{A}^I\| = \sup_{g \in H_\nu[-1,1]} \|\hat{A}^I g\|_\infty / \|g\|_Y. \tag{178}$$

Before proceeding any further we need to show that $\hat{A}^I$ is a bounded linear operator on Y into X. To do this we need the following result.

Theorem 5.4. Suppose $\phi \in H_\nu[-1, 1]$, $0 < \nu < 1$ and $\phi(\pm 1) = 0$. Then $T\phi \in H_\nu[-1, 1]$.

Proof. See Muskhelishvili (Ref. 1, Section 19). □

Before proceeding we need to make assumptions concerning the values of α and β and the weight w_2.

Assumption B. We shall impose the following conditions:
(i) $-1/p_2 < \alpha < 0$, $-1/p_1 < \beta < 0$;
(ii) $w_2 \in H_\nu[-1, 1]$, $0 < \nu < 1$.

In practice, Assumption B(i) is not too restrictive even though in (20) we allowed $-1/p_2 < \alpha < 1/q_2$ and $-1/p_1 < \beta < 1/q_1$. As a consequence of this we have, since $w_2 = 1/(Zr) = (1 - \tau)^{-\alpha}(1 + \tau)^{-\beta}\Omega^{-1}(\tau)/r(\tau)$, that $w_2(\pm 1) = 0$ and in any case we assumed w_2 was Hölder continuous on every closed subinterval $[c, d]$ say of $(-1, 1)$. This leads to the following result.

Theorem 5.5. Suppose $g \in H_\nu [-1, 1]$, $0 < \nu < 1$ and w_2 satisfies Assumption B. If

$$G(t) = \int_{-1}^{1} w_2(\tau)\left(\frac{g(\tau) - g(t)}{\tau - t}\right) d\tau, \qquad t \in [-1, 1], \qquad (179)$$

then $G \in H_\nu[-1, 1]$.

Proof. We can write

$$G(t) = T(w_2 g)(t) - g(t)(Tw_2)(t).$$

Since both w_2 and g are in $H_\nu[-1, 1]$, so is $w_2 g$. Furthermore, $w_2(\pm 1) = 0$ so that we can apply Theorem 5.4 to each term in this equation and the result follows at once. □

We now have the result that we require.

Theorem 5.6. Under Assumptions A and B, $\hat{A}^I$ maps $H_\nu[-1, 1]$ into itself when $0 < \nu < 1$ and is a bounded linear operator on $H_\nu[-1, 1]$ into $C[-1, 1]$.

Proof. The linearity of $\hat{A}^I$ is obvious from (85). From (31) with $P \equiv 1$ we have

$$aw_2 = \text{p.p.}(X^{-1}) + T(bw_2),$$

so that substituting this into (85) gives

$$(\hat{A}^I g)(t) = g(t)Q(t) - \frac{b(t)}{\pi}\int_{-1}^{1} w_2(\tau)\left(\frac{g(\tau) - g(t)}{\tau - t}\right) d\tau, \qquad (180a)$$

where

$$Q(t) = \text{p.p.}(X^{-1}; t) + \frac{1}{\pi}\int_{-1}^{1} w_2(\tau)\left(\frac{b(\tau) - b(t)}{\tau - t}\right) d\tau. \qquad (180b)$$

Now Q is a polynomial so that $gQ \in H_\nu[-1, 1]$. That the second term in (180a) is in $H_\nu[-1, 1]$ follows from Theorem 5.5. Hence $\hat{A}^I$ maps $H_\nu[-1, 1]$ into itself and therefore into $C[-1, 1]$. From (180a)

$$|(\hat{A}^I g)(t)| \le c\left\{|g(t)| + \int_{-1}^{1} w_2(\tau)\left|\frac{g(\tau) - g(t)}{\tau - t}\right| d\tau\right\}, \qquad (181)$$

where c is a constant independent of both g and t. Recalling (177) we have that

$$\|\hat{A}^I g\|_\infty \le c \cdot \|g\|_Y, \tag{182}$$

where c is independent of g. This establishes boundedness. □

Having established that $\hat{A}^I$ is a bounded linear operator on Y into X let us now consider restriction and prolongation operators between the Euclidean spaces X_n, Y_m and the spaces X, Y respectively. We define

$$r_n : X \to X_n \text{ such that } r_n\psi = (\psi(\tau_{1,n}), \psi(\tau_{2,n}), \ldots, \psi(\tau_{n,n}))^T,$$

$$p_n : X_n \to X \text{ such that } p_n x_n = (L^X_{n-1} x_n)(t), \, x_n = (\xi_1, \xi_2, \ldots, \xi_n)^T,$$

where

$$(L^X_{n-1} x_n)(t) = \sum_{j=1}^n \frac{t_n(t)\xi_j}{t'_n(\tau_{j,n})(t - \tau_{j,n})}, \tag{183}$$

this being the Lagrangian interpolation polynomial of degree $(n-1)$ based on the zeros of t_n. Obviously $r_n p_n = I_n$. The projection operator $P_n = p_n r_n$ is such that for all $\psi \in C[-1, 1]$, $\|\psi - P_n\psi\|_\infty$ does not tend to 0 as $n \to \infty$. We define

$$\|r_n\| = \sup_{\psi \in X} \frac{\|r_n\psi\|_\infty}{\|\psi\|_\infty}, \qquad \|p_n\| = \sup_{x_n \in X_n} \frac{\|p_n x_n\|_\infty}{\|x_n\|_\infty}, \tag{184}$$

where we have chosen the infinity norm on the Euclidean space X_n. Then $\|r_n\| = 1$ but $\|p_n\|$ will not be bounded as $n \to \infty$.

Again, between the spaces Y and Y_m we define

$$s_m : Y \to Y_m \text{ such that } s_m g = (g(t_{1,m}), g(t_{2,m}), \ldots, g(t_{m,m}))^T,$$

$$q_m : Y_m \to Y \text{ such that } q_m y_m = (L^Y_{m-1} y_m)(t), \, y_m = (\eta_1, \eta_2, \ldots, \eta_m)^T,$$

where L^Y_{m-1} denotes the Lagrangian interpolation polynomial of degree $(m-1)$ based on the zeros of the polynomial u_m and such that $(L^Y_{m-1} y_m)(t_{i,m}) = \eta_i$, $i = 1(1)m$. Again we have $s_m q_m = I_m$ and the projection operator $Q_m = q_m s_m$ is such that $Q_m g \in \mathbf{P}_{m-1}$ for every $g \in Y$. We define

$$\|s_m\| = \sup_{g \in Y} \frac{\|s_m g\|_\infty}{\|g\|_Y}, \qquad \|q_m\| = \sup_{y_m \in Y_m} \frac{\|q_m y_m\|_Y}{\|y_m\|_\infty}, \tag{185}$$

where the infinity norm has been used on Y_m.

Having defined these operators let us reconsider the discretization of the operators A and $\mathcal{K}$ as given by Eqs. (173) and (174). We see that

$$A_n = s_m A p_n, \tag{186a}$$

$$K_n = s_m \mathcal{K} p_n, \tag{186b}$$

so that again we have a situation similar to that given at the end of Section 4. Now $p_n x_n \in \mathbf{P}_{n-1}$ and since both $AP_{n-1} \in \mathbf{P}_{m-1}$ and $Q_m \mathbf{P}_{m-1} \in \mathbf{P}_{m-1}$ it follows that Eq. (136a) is satisfied. The same is true for Eq. (136b). Again, let us note that since $\ker(A)$ is either $\{0\}$ when $\kappa \leq 0$ or a polynomial of degree $(\mu + \kappa - 1)$ whenever $\kappa \geq 1$, then $P_n\{\ker(A)\} = \ker(A)$ whenever $n \geq \mu + \kappa - 1$. Consequently Eq. (135a) is satisfied and, by a similar argument, so is (135b). Hence by Theorem 4.6 we have

$$\hat{A}_m^I = r_n \hat{A}^I q_m, \tag{187}$$

and $r_n(\ker(A)) = \ker(A_n)$, $s_m(\ker(\hat{A}^I)) = \ker(\hat{A}_m^I)$.

We need to display the elements of $\hat{A}_m^I$ explicitly. In order to do this we need to introduce another quadrature rule, comparable to (170), based on the zeros $t_{i,m}$, $i = 1(1)m$, of the polynomial u_m. If $t \neq t_{j,m}$ and $\psi \in \mathbf{P}_{2m}$ then

$$\fint_{-1}^{1} \frac{w_2(\tau)\psi(\tau)}{\tau - t}\, d\tau = \sum_{j=1}^{m} \nu_{j,m} \frac{\psi(t_{j,m})}{t_{j,m} - t} + \frac{\pi \psi(t)}{u_m(t)} T(w_2 u_m; t). \tag{188}$$

Substituting this into (85) for $\hat{A}^I$ and recalling that $q_m y_m \in \mathbf{P}_{m-1}$, we find, after a little algebra, that if $\hat{A}_m^I = (\hat{a}_{i,j}^{(m)})$ then

$$\hat{a}_{i,j}^{(m)} = \frac{b(\tau_{i,n})\nu_{j,m}}{\pi(\tau_{i,n} - t_{j,m})}, \quad i = 1(1)n, \quad j = 1(1)m, \tag{189}$$

provided that $\tau_{i,n} \neq t_{j,m}$.

We shall now give an estimate for the upper bound of $\|\hat{A}_m^I\|_\infty$.

Theorem 5.7. Under Assumptions A and B

$$\|\hat{A}_m^I\|_\infty \leq c_1 + c_2 \log n, \tag{190}$$

where c_1 and c_2 are constants independent of n.

Proof. See Elliott (Ref. 6, Appendix B). □

In order to prove convergence we could evaluate all the quantities arising in Theorem 4.5, but we can obtain a better result in this case if we proceed a little differently. From Eq. (125) we have

$$(I_n + J_n)(r_n\psi - x_n) = (r_n g - g_n) - \delta_n^J \psi, \tag{191}$$

where $J = \hat{A}^I \mathcal{H}$, $J_n = \hat{A}_m^I K_n$, $g = \hat{A}^I f + \psi^{(0)}$, and $g_n = \hat{A}_m^I y_m + x_n^{(0)}$. Now we see that

$$r_n g - g_n = (r_n \hat{A}^I f - \hat{A}_m^I s_m f) + (r_n \psi^{(0)} - x_n^{(0)}),$$

where we have written

$$y_m = s_m f. \tag{192}$$

Now we can choose $r_n \psi^{(0)} - x_n^{(0)} = 0$ so that, on recalling (113), we have simply that

$$r_n g - g_n = \delta_n^{\hat{A}^I} f. \tag{193}$$

Again we have seen from (128) that

$$\delta_n^J \psi = \delta_n^{\hat{A}^I}(\mathcal{H}\psi) + \hat{A}_m^I \delta_n^{\mathcal{H}} \psi,$$

so that we can rewrite (191) as

$$(I_n + \hat{A}_m^I K_n)(r_n \psi - x_n) = \delta_n^{\hat{A}^I}(f - \mathcal{H}\psi) - \hat{A}_m^I \delta_n^{\mathcal{H}} \psi.$$

Now, as we have seen, for n large enough $(I_n + \hat{A}_m^I K_n)^I$ exists so that

$$r_n\psi - x_n = (I_n + \hat{A}_m^I K_n)^I \{\delta_n^{\hat{A}^I}(f - \mathcal{K}\psi) - \hat{A}_m^I \delta_n^{\mathcal{K}}\psi\}, \qquad (194)$$

and we shall take this as the starting point for our error analysis. We shall obtain an upper bound for $\|r_n\psi - x_n\|_\infty$ by putting a bound on each term on the right-hand side of (194). But before doing this we shall make one further assumption.

Assumption C. For some nonnegative integer r and $0 < \nu < 1$ we suppose
(i) $f^{(r)} \in H_\nu[-1, 1]$;
(ii) both $\partial^r k/\partial t^r$ and $\partial^r k/\partial \tau^r$ are in $H_\nu[-1, 1]$.

The case $r = 0$ corresponds to the function itself. With this we now have the following theorem:

Theorem 5.8. Under Assumptions A, B and C
(i) all solutions of $(A + \mathcal{K})\psi = f$ are such that $\psi^{(r)} \in H_\nu[-1, 1]$;
(ii) $\|\delta_n^{\mathcal{K}}\|_\infty \leq cn^{-(r+\nu)}$;
(iii) $\|\delta_n^{\hat{A}^I}(f - \mathcal{K}\psi)\|_\infty \leq n^{-(r+\nu)}(c_1 + c_2 \log n)$.

Proof. See Elliott (Ref. 6, Appendix C). □
From this result our convergence theorem follows immediately.

Theorem 5.9. Under Assumptions A, B, and C, if ψ is any solution of $(A + \mathcal{K})\psi = f$ then there is a solution x_n of $(A_n + K_n)x_n = s_m f$ such that

$$\|r_n\psi - x_n\|_\infty \leq p(n)n^{-(r+\nu)}\{c_1 + c_2 \log n\}, \qquad (195)$$

where c_1 and c_2 are constants independent of n.

Proof. This is obtained immediately by combining (132) with (194) and the results of Theorem 5.8. □

The question remains as to an estimate for $p(n)$. Now $p(n)$ is an upper bound for the norm of the prolongation operator p_n which is Lagrange interpolation based on the zeros of the polynomial t_n, which in turn are orthogonal with respect to $w_1 = (1 - \tau)^\alpha (1 + \tau)^\beta \Omega(\tau)/r(\tau)$. Nevai (Ref. 28) has shown that if $\gamma = \max(\alpha, \beta)$ then

$$p(n) \leq \begin{cases} cn^{\gamma+1/2}, & \text{if } \gamma > -\tfrac{1}{2}, \\ c \log n, & \text{if } \gamma \leq -\tfrac{1}{2}. \end{cases} \qquad (196)$$

This can be inserted into (195). Finally, for global error we have that since

$$\psi - p_n x_n = p_n(r_n\psi - x_n) + (\psi - p_n r_n \psi),$$

then

$$\|\psi - p_n x_n\|_\infty \leq p^2(n) \cdot n^{-(r+\nu)}(c_1 + c_2 \log n). \tag{197}$$

This completes the analysis of the rate of convergence for classical collocation under some assumptions which, in practice, turn out not to be too restrictive. The rate of convergence may be looked upon as being satisfactory within the restrictions given.

5.3. A Discrete Galerkin Method. In Section 5.1 we discussed a Galerkin-Petrov method for the approximate solution of $(A + \mathcal{K})\psi = f$ and in Section 5.2 we discussed the so-called classical collocation method. These two methods come together when we consider a particular discretization of the Galerkin-Petrov method. Let us recall from Section 5.1 that we looked for an approximate solution, which here we shall write as ψ_n^G, of the form

$$\psi_n^G = \sum_{j=1}^n \xi_j t_{j-1}. \tag{198}$$

On substituting this approximation into the complete equation we obtained a residual $(\text{res})_n$ given by

$$(\text{res})_n = (A + \mathcal{K})\psi_n^G - f. \tag{199}$$

The unknowns ξ_j, $j = 1(1)n$ of (198) are obtained by requiring that

$$\langle u_{i-1}, (\text{res})_n \rangle_2 = 0 \quad \text{for } i = 1(1)m. \tag{200}$$

This gave rise to $m \times n$ matrices A_n, K_n whose elements involved inner products. These were assumed to be evaluated exactly. In general, of course, this will not occur and use will have to be made of some appropriate quadrature rules. Let us suppose that for inner products $\langle \ , \ \rangle_1$ we use the Gauss quadrature rule given by (169) while for $\langle \ , \ \rangle_2$ we use

$$\int_{-1}^1 w_2(\tau)\psi(\tau) \, d\tau = \sum_{j=1}^m \nu_{j,m}\psi(t_{j,m}) + \text{remainder}, \tag{201}$$

where the remainder is zero whenever $\psi \in \mathbf{P}_{2m-1}$ [cf. (188)]. From (199) and (200) we have

$$\langle u_{i-1}, (A + \mathcal{K})\psi_n^G \rangle_2 = \langle u_{i-1}, f \rangle_2, \qquad i = 1(1)m,$$

which is of course equivalent to

$$\langle (A^* + \mathcal{K}^*)u_{i-1}, \psi_n^G \rangle_1 = \langle u_{i-1}, f \rangle_2, \qquad i = 1(1)m, \qquad (202)$$

where A^* is the Hilbert-adjoint of A [see (147)] and $\mathcal{K}^*$, the Hilbert-adjoint of $\mathcal{K}$, is defined by

$$(\mathcal{K}^*\psi)(t) = \int_{-1}^{1} w_2(\tau)k(\tau, t)\psi(\tau)\, d\tau. \qquad (203)$$

From the quadrature rule given by (188) we have

$$(A^*u_{i-1})(\tau_{j,n}) = \sum_{k=1}^{m} \frac{\nu_{k,m}b(t_{k,m})u_{i-1}(t_{k,m})}{\pi(\tau_{j,n} - t_{k,m})}, \qquad \text{exactly,} \qquad (204)$$

while from (201) we have

$$(\mathcal{K}^*u_{i-1})(\tau_{j,n}) = \sum_{k=1}^{m} \nu_{k,m}k(t_{k,m}, \tau_{j,n})u_{i-1}(t_{k,m}) + \text{remainder} \qquad (205)$$

and

$$\langle u_{i-1}, f \rangle_2 = \sum_{k=1}^{m} \nu_{k,m}u_{i-1}(t_{k,m})f(t_{k,m}) + \text{remainder}. \qquad (206)$$

On neglecting the remainder term in (205) and (206), substituting the resulting expressions and (204) into (202), and evaluating the inner product on the left using the quadrature rule (169) we obtain a system of linear algebraic equations given by

$$\sum_{j=1}^{n} \mu_{j,n} \left\{ \sum_{k=1}^{m} \nu_{k,m} \left[\frac{u_{i-1}(t_{k,m})b(t_{k,m})}{\pi(\tau_{j,n} - t_{k,m})} + k(t_{k,m}, \tau_{j,n})u_{i-1}(t_{k,m}) \right] \right\} \xi_j$$

$$= \sum_{k=1}^{m} \nu_{k,m}u_{i-1}(t_{k,m})f(t_{k,m}), \qquad \text{for } i = 1(1)m. \qquad (207)$$

On solving for ξ_j, $j = 1(1)n$, arising in (207) we obtain an approximate solution $\tilde{\psi}_n^G$ say to ψ where

$$\tilde{\psi}_n^G = \sum_{j=1}^{n} \xi_j t_{j-1}. \tag{208}$$

On interchanging the summations in (207) we obtain for $i = 1(1)m$ the equations

$$\sum_{k=1}^{m} \nu_{k,m} u_{i-1}(t_{k,m}) \left\{ \sum_{j=1}^{n} \left[\frac{b(t_{k,m})\mu_{j,n}}{\pi(\tau_{j,n} - t_{k,m})} + \mu_{j,n} k(t_{k,m}, \tau_{j,n}) \right] \xi_j - f(t_{k,m}) \right\} = 0. \tag{209}$$

If we introduce the $m \times m$ matrices

$$N_m = \mathrm{diag}(\nu_{k,m}),$$

$$U_m = (u_{i-1}(t_k)),$$

then, on recalling Eqs. (173) and (174), we may write (209) in matrix form as

$$U_m N_m \{(A_n + K_n)x_n - y_m\} = 0. \tag{210}$$

Since both the matrices N_m and U_m are nonsingular we have that (210) is equivalent to

$$(A_n + K_n)x_n = y_m, \tag{211}$$

which was the system of equations for classical collocation. Thus each method gives rise to the same vector x_n from which we obtain $\tilde{\psi}_n^G$ through (208) or ψ_n^C, say, through (183) which gives $\psi_n^C = L_{n-1}^X x_n$. From the theory of orthogonal polynomials we have that

$$\tilde{\psi}_n^G = \psi_n^C \quad \text{on } [-1, 1]; \tag{212}$$

consequently the results obtained in the previous section for $\|\psi - \psi_n^C\|_\infty$ now hold for the uniform norm of $\psi - \tilde{\psi}_n^G$.

Returning to Eq. (199) we may obtain m equations for the ξ_j arising in (198) by requiring that $(\mathrm{res})_n$ be zero at m distinct points of $(-1, 1)$. If we choose these zeros to be those of u_m then, using Gauss quadrature again, we recover the linear algebraic equations of classical collocation. A variant of this method is to choose the m points not to be the zeros of u_m but the zeros of a polynomial not directly related to the singular integral equation. Thus Cuminato (Ref. 29) has chosen this polynomial always to be T_m, the

Chebyshev polynomial of the first kind of degree m, and has given a convergence analysis in the case when a and b are constants. We shall not pursue this matter any further here.

One might also note that if instead of concerning ourselves with errors in the uniform norm we had considered $\psi - \tilde{\psi}_n^G$ in the space $\mathcal{H}_1$, then a much simpler analysis would have ensued. Again, we shall not pursue this matter here but refer the interested reader to the papers of Golberg (Ref. 30), Fromme and Golberg (Ref. 31), and Miel (Ref. 32).

6. Conclusions

This attempt at a unified theory of convergence is far from complete but does reflect the considerable progress that has been made in this topic in the past decade. Much remains to be done in the application of this theory to different algorithms. Many topics, such as the analysis of algorithms involving splines on the one hand and the Whittaker cardinal function on the other, have only been touched upon in this chapter. The use of splines with direct methods has been ignored and for some progress in this direction see, for example, Jen and Srivastav (Ref. 33), Miller and Keer (Ref. 34), and Gerasoulis (Ref. 35). Most of the work in the past decade has been influenced by global approximation techniques based on the orthogonal polynomials which arise so naturally with these equations. For an alternative approach to that taken in this chapter the reader is referred to the paper by Junghanns and Silbermann (Ref. 36).

We have made no mention here of methods for singular integral equations in more than one dimension, but it is hoped that a thorough understanding of the methods for one-dimensional equations will lead to a better understanding for methods in higher dimension. For the theory of such equations, the book by Mikhlin (Ref. 37) remains the standard text.

Acknowledgment

This research was supported by a grant from the Australian Research Grants Scheme (A.R.G.S.).

References

1. MUSKHELISHVILI, N. I., *Singular Integral Equations*, Noordhoff, Gröningen, Holland, 1953.
2. TRICOMI, F. G., *Integral Equations*, Interscience, New York, New York, 1957.

3. WIDOM, H., *Singular Integral Equations in L_p*, Transactions of the American Mathematical Society, Vol. 97, pp. 131-160, 1960.
4. LEVINSON, N., *Simplified Treatment of Integrals of Cauchy Type, the Hilbert Problem and Singular Integral Equations. Appendix: Poincaré-Bertrand Formula*, SIAM Review, Vol. 7, pp. 474-502, 1965.
5. ELLIOTT, D., *Singular Integral Equations on the Arc $(-1, 1)$: Theory and Approximate Solution, Part 1: Theory*, Mathematics Department, University of Tasmania, Technical Report No. 218, 84 pp., 1987.
6. ELLIOTT, D., *Singular Integral Equations on the Arc $(-1, 1)$: Theory and Approximate Solution, Part 2: Approximate Methods*, Mathematics Department, University of Tasmania, Technical Report No. 223, 78 pp., 1987.
7. RIESZ, M., *Sur les Fonctions Conjuguées*, Mathematische Zeitschrift, Vol. 27, pp. 218-244, 1927.
8. KHEVEDELIDZE, B. V., *Linear Discontinuous Boundary Problems in the Theory of Functions, Singular Integral Equations and Some of Their Applications*, Akademiya Nauk Gruzinskoi SSR, Vol. 23, pp. 3-158, 1956 (in Russian).
9. SCHECHTER, M., *Principles of Functional Analysis*, Academic Press, New York and London, 1971.
10. ELLIOTT, D., *Orthogonal Polynomials Associated with Singular Integral Equations having a Cauchy Kernel*, SIAM Journal on Numerical Analysis, Vol. 13, pp. 1041-1052, 1982.
11. WELSTEAD, S. T., *Singular Integral Operators in a Weighted L^2-Space*, Integral Equations and Operator Theory, Vol. 8, pp. 402-426, 1985.
12. PETERS, A. S., *A Note on the Integral Equation of the First Kind with a Cauchy Kernel*, Communications on Pure and Applied Mathematics, Vol. 16, pp. 57-61, 1963.
13. IVANOV, V. V., *The Theory of Approximate Methods and their Application to the Numerical Solution of Singular Integral Equations*, Noordhoff, Leyden, Holland, 1976.
14. NOBLE, B., *Error Analysis of Collocation Methods for Solving Fredholm Integral Equations*, Topics in Numerical Analysis, Edited by J. J. H. Miller, Academic Press, New York, New York, 1973.
15. SPENCE, A., AND THOMAS, K. S., *On Superconvergence Properties of Galerkin's Method for Compact Operator Equations*, IMA Journal of Numerical Analysis, Vol. 3, pp. 253-271, 1983.
16. THOMAS, K. S., *On the Approximate Solution of Operator Equations*, Numerische Mathematik, Vol. 23, pp. 231-239, 1975.
17. SLOAN, I. H., *Four Variants of the Galerkin Method for Integral Equations of the Second Kind*, IMA Journal of Numerical Analysis, Vol. 4, pp. 9-17, 1984.
18. THOMAS, K. S., *Galerkin Methods for Singular Integral Equations*, Mathematics of Computation, Vol. 36, pp. 193-205, 1981.
19. ELSCHNER, J., *Galerkin Methods with Splines for Singular Integral Equations over $(0, 1)$*, Numerische Mathematik, Vol. 43, pp. 265-281, 1984.
20. DOW, M. L., AND ELLIOTT, D., *The Numerical Solution of Singular Integral Equations over $(-1, 1)$*, SIAM Journal on Numerical Analysis, Vol. 16, pp. 115-134, 1979.

21. STENGER, F., AND ELLIOTT, D., *Sinc Method of Solution of Singular Integral Equations*, Numerical Solution of Singular Integral Equations, Edited by A. Gerasoulis and R. Vichnevetsky, IMACS, 1984.
22. NOBLE, B., *Applied Linear Algebra*, Prentice-Hall, Englewood Cliffs, New Jersey, 1969.
23. NEVAI, P., *Mean Convergence of Lagrange Interpolation III*, Transactions of the American Mathematical Society, Vol. 282, pp. 669-698, 1984.
24. ELLIOTT, D., *A Galerkin-Petrov Method for Singular Integral Equations*, Journal of the Australian Mathematical Society, Series B, Vol. 25, pp. 261-275, 1983.
25. HUNTER, D. B., *Some Gauss Type Formulas for the Evaluation of Cauchy Principal Value Integrals*, Numerische Mathematik, Vol. 19, pp. 419-424, 1972.
26. RABINOWITZ, P., *On the Convergence and Divergence of Hunter's Method for Cauchy Principal Value Integrals*, Numerical Solution of Singular Integral Equations, Edited by A. Gerasoulis and R. Vichnevetsky, IMACS, 1984.
27. ELLIOTT, D., *The Classical Collocation Method for Singular Integral Equations*, SIAM Journal on Numerical Analysis, Vol. 19, pp. 816-832, 1982.
28. NEVAI, P., *Orthogonal Polynomials*, Memoirs of the American Mathematical Society, Vol. 18, No. 213, American Mathematical Society, Providence, Rhode Island, 1979.
29. CUMINATO, J. A., *On the Uniform Convergence of a Collocation Method for a Class of Singular Integral Equations*, BIT, Vol. 27, pp. 190-202, 1987.
30. GOLBERG, M. A., *The Convergence of a Collocation Method for a Class of Cauchy Singular Integral Equations*, Journal of Mathematical Analysis and Applications, Vol. 100, pp. 500-512, 1984.
31. FROMME, J. A., AND GOLBERG, M. A., *Convergence and Stability of a Collocation Method for the Generalized Airfoil Equation*, Applied Mathematics and Computation, Vol. 8, pp. 281-292, 1981.
32. MIEL, G., *On the Galerkin and Collocation Methods for a Cauchy Singular Integral Equation*, SIAM Journal on Numerical Analysis, Vol. 23, pp. 135-143, 1986.
33. JEN, E., AND SRIVASTAV, R. P., *Cubic Splines and Approximate Solution of Singular Integral Equations*, Mathematics of Computation, Vol. 37, pp. 417-423, 1981.
34. MILLER, G. R., AND KEER, L. M., *A Numerical Technique for the Solution of Singular Integral Equations of the Second Kind*, Quarterly of Applied Mathematics, Vol. 42, pp. 455-465, 1985.
35. GERASOULIS, A., *Piecewise-polynomial Quadratures for Cauchy Singular Integrals*, SIAM Journal on Numerical Analysis, Vol. 23, pp. 891-902, 1986.
36. JUNGHANNS, P., AND SILBERMANN, B., *Zur Theorie der Näherungsverfahren für Singuläre Integralgleichungen auf Intervallen*, Mathematische Nachrichten, Vol. 103, pp. 199-244, 1981.
37. MIKHLIN, S. G., *Multidimensional Singular Integrals and Integral Equations*, Pergamon Press, Oxford, England, 1965.

7

Planing Surfaces

E. O. Tuck

Abstract. The planing surface problem is reviewed. Simplifications leading to the generalized airfoil equation can be solved by using known numerical techniques. The more complete two-dimensional case leads to novel integral equations whose numerical solution presents challenging computational problems.

1. Introduction

Planing is a state of steady motion of a boat in which the wetted draft is small compared to the wetted length and beam, and hydrodynamics, not hydrostatics, provides the support for its weight. Most hulls are capable of planing if driven sufficiently fast, the lift generated by hydrodynamic forces increasing like the square of speed, and causing the draft reduction necessary to maintain planing.

Planing is just one facet of the great unsolved problem of ship hydrodynamics. The basic ship-hydrodynamic problem is to solve Laplace's equation in three space dimensions, subject to a Neumann boundary condition on the given ship hull surface, free-stream and radiation conditions at infinity, and a pair of wickedly nonlinear boundary conditions on the (unknown) free surface of the water. No useful numerical results have ever been obtained for this problem.

Under some circumstances, the free-surface boundary conditions can be linearized and applied on the undisturbed plane surface $y = 0$. The resulting "Neumann–Kelvin" boundary-value problem, although linear, is still too difficult to solve for routine application to ship design. Doctors and Beck (Ref. 1) have recently come to quite pessimistic conclusions about

---

E. O. Tuck • Applied Mathematics Department, University of Adelaide, Adelaide, SA, Australia 5001.

this matter. There is also some doubt (Ref. 2; see also other articles in that conference proceedings) about the legitimacy of linearization for a general ship.

However, no such doubt exists for planing surfaces as defined above, linearization being justified on account of the smallness of draft. At the same time, the Neumann boundary condition on the hull can also be simplified, and applied on the projection of the hull onto the plane $y = 0$. Thus the linearized planing problem is a boundary-value problem defined in the whole lower-half space $y < 0$, with boundary conditions only on the plane $y = 0$ and at infinity. Such half-space problems have a much greater chance of being solvable, and indeed the planing problem is now in principle (and, to a large extent, in practice) solved.

In order to indicate how one might proceed to solve it, it is convenient first to discuss another ship-hydrodynamic problem for which linearization is justifiable, namely, that of a *hovercraft*. From the theoretical point of view, the action of the fans of a hovercraft can be modeled by a prescribed distribution of pressure $p(x, z)$ in the air over a portion of the undisturbed free-surface plane $y = 0$. This pressure (excess over atmospheric) will cause the free surface to be displaced into a surface with equation $y = \eta(x, z)$. Linearization is justified if the magnitude of $p(x, z)$ is sufficiently small, and then so is the magnitude of the free-surface displacement $\eta(x, z)$.

We can now observe that *every* planing surface is hydrodynamically equivalent to *some* hovercraft. That is, $y = \eta(x, z)$ may now be taken as the equation defining a planing surface, and our real task is to find $p(x, z)$, given $\eta(x, z)$. The direct problem of finding $\eta(x, z)$, given $p(x, z)$, has long been solved, and can be expressed as an integral over that portion W of the plane $y = 0$ where $p(x, z)$ is nonzero. The inverse problem, of finding $p(x, z)$, given $\eta(x, z)$, therefore reduces to an integral equation on the domain W.

The two-dimensional special case, on which most work has been done, is discussed in Section 2, and three-dimensional and other generalizations are given in Section 3. In the following, the pressure p has been made nondimensional with respect to ρU^2, where ρ is water density and U the speed of the boat. We also use a parameter $\gamma = g/U^2$, where g is the acceleration of gravity.

2. The Planing Equation

In the case of two-dimensional flow, when there is no dependence on the lateral coordinate z, the problem reduces to a standard Cauchy-singular integral equation in one variable. That is, suppose a pressure distribution

Planing Surfaces

$p(x)$ acts on the free surface, for $-l < x < l$. Then (Ref. 3) this causes the free surface to be displaced to $y = \eta(x)$, where

$$\eta(x) = \int_{-l}^{l} p(\xi) K(x - \xi) \, d\xi. \tag{1}$$

The kernel function $K(x)$ in (1) is expressable in terms of sine and cosine integrals, namely,

$$K(x) = -\frac{1}{\pi} g(\gamma x) - \begin{cases} 2 \sin \gamma x, & x > 0, \\ 0, & x < 0, \end{cases} \tag{2}$$

where

$$g(x) = \cos x \int_{|x|}^{\infty} \frac{\cos t}{t} \, dt + \sin|x| \int_{|x|}^{\infty} \frac{\sin t}{t} \, dt. \tag{3}$$

The even auxiliary function $g(x)$ can be computed very easily, e.g., by the rational approximations given in Ref. 4.

If the shape of the submerged hull is prescribed for $-l < x < l$, then $\eta(x)$ is known, and (1) is an integral equation to determine the unknown pressure $p(x)$. However, this integral equation is not yet of Cauchy-singular type, since it has a logarithmic singularity, with

$$K(x) \to -\frac{1}{\pi} \log|x|, \quad \text{as } x \to 0. \tag{4}$$

A Cauchy-singular equation arises if we differentiate once, giving

$$\eta'(x) = \int_{-l}^{l} p(\xi) K'(x - \xi) \, d\xi. \tag{5}$$

The integral in (5) has a Cauchy principal-value interpretation.

Equation (5) may be called the "planing equation," by analogy with the "airfoil equation"

$$\eta'(x) = -\frac{1}{\pi} \int_{-l}^{l} \frac{p(\xi)}{x - \xi} \, d\xi \tag{6}$$

(Ref. 5). In each case, we have a singular integral equation of the first kind on a finite interval, with a Cauchy-singular kernel. In each case, the unknown function $p(x)$ can be identified with the pressure on, and the given function $\eta'(x)$ with the slope of, a body in contact with a fluid.

Indeed, this analogy becomes an identity in the limit as $\gamma \to 0$, since in that limit (4) determines $K(x)$ and (5) reduces exactly to (6). This limit corresponds to planing at very high speed, and in such a case the flow beneath a planing surface is identical to that past the lower surface of a thin airfoil of the same shape.

The analogy with the airfoil equation enables us to assert that Eq. (5) does not possess a unique solution. Specifically, there exists a one-parameter family of solutions, of the form

$$p(x) = p_1(x) + Cp_0(x), \qquad (7)$$

where $p = p_1(x)$ is any solution with the given $\eta'(x)$, $p = p_0(x)$ is a solution of the equivalent *homogeneous* integral equation [i.e., that with $\eta'(x) \equiv 0$], and C is an arbitrary constant. In the case of the airfoil equation, $p_1(x)$ can be obtained by a quadrature, and

$$p_0(x) = (l^2 - x^2)^{-1/2}.$$

However, in the case of the planing equation, both $p_1(x)$ and $p_0(x)$ must be determined computationally.

The particular member of the family (7) that is of interest in the present application is the one that satisfies a "smooth-detachment" condition at the trailing edge $x = l$. This simply demands that the pressure $p(x)$ vanish there, i.e.,

$$p(l) = 0. \qquad (8)$$

Note that since p is defined to be the excess over atmospheric pressure, Eq. (8) really states that the actual pressure has returned to the ambient atmospheric value at the trailing edge.

Exactly the same requirement holds in the case of airfoil problems, where p means the jump in pressure between lower and upper surfaces, and (8) is called the "Kutta condition." We retain the name Kutta condition for it in the planing case. It is not hard to show that (8) guarantees smoothness of detachment from the trailing edge, in the sense that when (8) holds, the free surface at $x = l_+$ has the same displacement and slope as the hull at $x = l_-$.

Early work on the planing equation (5) was done by Sretenskii (Ref. 6) and Maruo (Ref. 7). These authors [and others such as Squire (Ref. 8), and Cumberbatch (Ref. 9)] borrowed heavily from techniques that had proved successful for the airfoil equation, particularly involving truncation of a Fourier-like series. In such methods, the Kutta condition (8) is built into the numerical solution of the integral equation, as a constraint on the form of the series, which also has a special first term to model the inverse-square-root pressure singularity at the leading edge.

An alternative procedure used by the present author (Ref. 10) and others (Refs. 11–12) involves actual computation of the separate constituents $p_1(x)$ and $p_0(x)$ of the general solution (7). The procedure for doing this is revealing, in that it throws light on an aspect of the problem not fully appreciated by the early investigators.

First we note that (5) is not only the planing equation for an input hull $\eta(x)$, but also corresponds to every hull that is obtained from it by a vertical shift. That is, if we integrate (5) back again, we do not necessarily get (6), but in general we get

$$\eta(x) + C = \int_{-l}^{l} p(\xi) K(x - \xi) \, d\xi, \tag{9}$$

where C is an arbitrary constant that can be identified with that in (7).

The numerical procedure is now quite straightforward. Equations such as (1) with a logarithmic kernel possess (almost always, the exceptions being not of practical importance here) a unique solution for any input $\eta(x)$, and this solution can be obtained numerically by methods discussed thoroughly in Ref. 13. Given an algorithm for solving (1), we use it twice, first with $\eta(x)$ taken as given, calling the output $p_1(x)$, and second with $\eta(x)$ replaced by 1 (for all x), calling the output $p_0(x)$. We now implement the Kutta condition (8) by setting

$$C = -\lim_{x \to l} p_1(x)/p_0(x), \tag{10}$$

noting that the numerical equivalent of this limit has to be performed carefully, since both $p_1(x)$ and $p_0(x)$ have inverse-square-root singularities at $x = l$.

The above numerical method works well and efficiently. Using it with minor modifications, Oertel (Ref. 11) obtained 2-3 figure accuracy with a 50-point grid. He provided plots of $p(x)$ for various surfaces $\eta(x)$, over a range of values of γl. Similar, but somewhat less accurate, pressure plots were obtained by earlier investigators.

It is notable that this procedure uniquely determines not only the pressure (and hence important practical outputs like the net lift force and center of pressure), but also the constant C. That is, there is *no solution* for a given general $\eta(x)$; instead, there is a solution for a modified input $\eta(x) + C$, where C is uniquely determined! At first sight, this is very disturbing; can we not solve the planing problem for a given hull? In fact, *we can not*, for a given hull of a given length $2l$, at a given speed (determining γ)! Instead, the flow insists on changing the given hull, by shifting it upward through a certain displacement C that it determines. In general, this shift C is of the same (small) order of magnitude as the input $\eta(x)$.

This y-shift also occurs (in principle) in the case of the airfoil equation, but is not of practical significance in the aerodynamic context; the piece of sky at $y = y_1$ is dynamically indistinguishable from that at $y = y_2$. This is far from the case for planing surfaces, since their vertical location relative to the free surface at $y = 0$ is crucial to the dynamics of their motion. This is manifest in the dramatic variations that occur in the angle of attack and wetted length of a high-speed planing boat as its speed is varied. In fact, it is the wetted length ($2l$) that provides the key to resolution of the paradox discussed above.

Although we fix the range of integration $(-l, l)$ in the planing equation (5), this is only for computational convenience. The problem would fit the application better if we were to allow this range to vary, i.e., to join the pressure $p(x)$ in the list of unknowns. We must pay a price for our perverse preference for a fixed wetted length, and that price is an inability fully to specify the shape function $\eta(x)$. Actually, the price is not too high, at least in two dimensions, since inverse methods can be used (Refs. 8, 11) in which results are computed with l fixed, and subsequently replotted as if l were an output parameter.

Another interesting issue for the planing equation concerns "premature detachment." Most planing surfaces have a "transom," i.e., are sharply cut off at their trailing edge. Therefore, when the above inverse method determines the wetted length $2l$, it is really determining the location of the *leading* edge, relative to the (fixed) trailing edge. Well-designed planing surfaces have $p(x) > 0$ for $-l < x < l$; after all, they are intended to provide positive lift forces. In fact, if $p(x) < 0$ near the trailing edge, it is almost certain that the flow will not remain attached to the hull up to the transom, and will detach at some point forward of it.

Premature detachment is undesirable in practice (and also in the mathematics, since the solution has been computed assuming it did not occur!). It cannot occur for flat or concave-down hulls and, in practice, planing-boat designers (see, e.g., Ref. 14) avoid positive (concave-up) curvature, at least near the transom. Nevertheless, the mathematical problem of computing flows with premature detachment on hulls with such positive curvature is a very interesting one (see Refs. 11, 15). One simple but laborious approach is to compute a family of solutions, with the supposed trailing edge moved progressively forward until negative pressures no longer occur. The point at which this first happens is the required premature-detachment point; clearly it will be such that not only the pressure, but also the pressure *gradient* vanishes. More direct methods could clearly be developed, but haven't been.

3. Generalizations

Within the context of two-dimensional flow, various other physical phenomena can be included to generalize the planing equation (5). For example, Doctors (Ref. 16) allows the planing surface to *deform* under the action of the pressure $p(x)$. This gives a coupled elastic-hydrodynamic problem, in which the shape of the planing surface determines the flow, which in turn determines the shape of the surface.

In the same spirit, the present author (Ref. 17) has considered flows caused by an airfoil in air moving very close to a water surface. Again, there is a coupling between the pressure $p(x)$ and water surface deformation $\eta(x)$, neither of which is an input quantity. The planing equation (5) is one branch of this coupling, and the aerodynamics of the air flow in the gap between airfoil and free surface provides the other. The work reported in Ref. 17 was of a preliminary nature, and used a hydrostatic ($\gamma \to \infty$) approximation to (5). The full problem at finite γ was solved by Grundy (Ref. 12), a special feature being that the resulting nonlinear integral equation possesses nonunique solutions for some γ.

Another interesting extension involves the inclusion of surface tension. In fact, it is surprisingly easy to modify the kernel $K(x)$ to take this into account, sine and cosine integrals being still the only necessary special functions. However, there are some major qualitative differences and, in particular, the integral equation is no longer of Cauchy-singular type. This has profound consequences for both the numerical algorithms and the physical interpretation of the results, and a full discussion is provided in Refs. 18-19.

It is also appropriate at this point to mention fully nonlinear planing theories, in which gravity is usually neglected, the free surface then becoming one of constant flow speed. The hodograph technique enables exact solution of some such two-dimensional problems (Ref. 20). In practice, such a zero-gravity situation is approximated for high speed, i.e., as $\gamma l \to 0$, a limit in which the lift becomes large, and hence the draft small, so that linearization could have again been justified. Nevertheless, there is considerable theoretical interest in this case, and one important class of studies is that in which one attempts to put gravity back into the problem, at least as a perturbation for small γl; see Ref. 21.

The above generalizations retain the two-dimensional flow assumption, and hence from the mathematical and numerical point of view the advantage of reducing to an integral equation for a function of one variable only. The most important generalization in practice is however to three-dimensional flow, and hence to an integral equation involving a double integral, with an unknown function $p(x, z)$ of two variables.

Specifically, the three-dimensional equivalent of Eq. (1) is

$$\eta(x, z) = \int\int_W p(\xi, \zeta) K(x - \xi, z - \zeta) \, d\xi \, d\zeta, \tag{11}$$

where W is the projection onto $y = 0$ of the hull surface, and (Ref. 3)

$$K(x, z) = \int_0^{\pi/2} d\theta \frac{\sec^2 \theta}{\pi^2} \left[\int_0^\infty \frac{\cos(kx \cos \theta) \cos(kz \sin \theta)}{k - \gamma \sec^2 \theta} k \, dk \right.$$

$$\left. - \gamma\pi \sec^2 \theta \sin(\gamma x \sec^2 \theta) \cos(\gamma z \sec^2 \theta \sin \theta) \right]. \tag{12}$$

The k-integral in (12) is a Cauchy principal value, and should be interpreted as the limit $y \uparrow 0$ of an integral containing an extra factor $\exp(ky)$, where $y < 0$ is the depth coordinate.

There are two obvious increases in the difficulty of the problem. In the first place, the integral equation involves an extra variable, and is bound to be harder to solve, irrespective of the kernel K. But, of perhaps greater significance, the new kernel $K(x, z)$ is very much harder to compute with efficiency (Ref. 22) than the simple combination of sine and cosine integrals in Section 2. Mainly for this reason, few attempts have been made to solve the full three-dimensional problem (see Ref. 11 for an attempt and a discussion of the difficulties).

It should be noted that in the limit $\gamma \to 0$, we recover the "lifting surface" integral equation, which is the aerodynamic generalization to three dimensions of the airfoil equation (6). Since even the lifting-surface problem presents considerable numerical difficulties (see Refs. 23–24), it is hardly surprising that little progress has been made with the very much harder case of nonzero γ.

There is a third respect in which (11) is more difficult to work with than (1). This is the fact that instead of the single wetted-length parameter $2l$, we now have an unknown region W of the plane $y = 0$ to determine in an inverse manner. If we prescribe W in advance, the flow will not accept any input $\eta(x, z)$, but will *distort* such an input to the form $\eta(x, z) + C(z)$, where $C(z)$ is uniquely determined. Such a distortion of the desired input hull (in constrast to the simple vertical shift when C is independent of z) is unlikely to be acceptable in practice, so that there appears little choice but to allow some freedom for the program to determine W.

Some progress has been made by geometrical simplification. The most obvious case to examine is that of "nearly-two-dimensional" flow, in which one seeks three-dimensional corrections to the results of Section 2. The aerodynamic analog is with Prandtl's lifting-line theory (Ref. 25) for wings of high aspect ratio, and Shen (Ref. 26) has exploited this analogy to give results for planing surfaces of high aspect ratio.

At the other extreme, one can consider planing surfaces of low aspect ratio, i.e., those whose width is much less than their length. This case has been studied in Refs. 15, 27–30. The simplification is quite worthwhile, and numerical results seem to be obtainable for arbitrary hulls $\eta(x, z)$. In view of the fact that the hulls of most actual planing boats (at least as seen outside the water!) are slender, and therefore fall in the low-aspect-ratio category, this would seem to be one of the more promising theoretical approaches.

However, it is worth noting yet again that planing is a high-speed phenomenon, and that the wetted length varies with flow conditions. Specifically, for a fixed weight and centre of mass, as we increase speed, the wetted length tends to decrease dramatically, until at very high speed the boat is "flying," with only its extreme rear end in the water. From the hydrodynamical point of view, such a configuration is no longer necessarily of low aspect ratio, even if the hull is slender. It is this variability in the domain of the problem, associated with dramatic variations in the wetted area, that makes the planing-surface problem difficult and mathematically challenging.

References

1. DOCTORS, L. J., AND BECK, R. F., *Numerical Aspects of the Neumann-Kelvin Problem*, Journal of Ship Research, Vol. 31, pp. 1-13, 1987.
2. TUCK, E. O., *Invited Discussion*, 2nd International Conference on Numerical Ship Hydrodynamics, Edited by J. V. Wehausen and N. Salvesen, University of California, Berkeley, California, pp. 104-106, 1977.
3. WEHAUSEN, J. V., AND LAITONE, E. V., *Surface Waves*, Handbuch der Physik, Vol. 9, Edited by S. Flügge, Springer-Verlag, Berlin, Germany, pp. 446-815, 1960.
4. ABRAMOWITZ, M., AND STEGUN, I. A., *Handbook of Mathematical Functions*, Dover, New York, New York, 1964.
5. TRICOMI, F. G., *Integral Equations*, Interscience, New York, New York, 1957.
6. SRETENSKII, L. N., *On the Motion of a Glider on Deep Water*, Izvestiya Akademii Nauk SSSR, OTDEL Mat. Estest. Nauk, pp. 817-835, 1933.
7. MARUO, H., *Two-Dimensional Theory of the Hydroplane*, 1st Japan National Congress of Applied Mechanics, Science Council of Japan, Tokyo, Japan, pp. 409-415, 1952.
8. SQUIRE, H. B., *The Motion of a Simple Wedge Along the Water Surface*, Proceedings of the Royal Society of London, Series A, Vol. 243, pp. 48-64, 1957.
9. CUMBERBATCH, E., *Two-Dimensional Planing at High Froude Number*, Journal of Fluid Mechanics, Vol. 4, pp. 466-478, 1958.
10. TUCK, E. O., *The Effect of Span-Wise Oscillations on the Thrust-Generating Performance of a Flapping Thin Wing*, Symposium on Swimming and Flying in Nature, Vol. 2, Edited by T. Y.-T. Wu, C. J. Brokaw, and C. Brennen, Plenum, New York, New York, pp. 953-974, 1975.
11. OERTEL, R. P., *The Steady Motion of a Flat Ship, Including an Investigation of Local Flow Near the Bow*, University of Adelaide, PhD Thesis, 1975.

12. GRUNDY, I. H., *Airfoils Moving in Air Close to a Dynamic Water Surface*, Journal of the Australian Mathematical Society, Series B, Vol. 27, pp. 327–345, 1986.
13. TUCK, E. O., *Application and Numerical Solution of Cauchy-Singular Integral Equations*, The Application and Numerical Solution of Integral Equations, Edited by R. S. Anderssen, F. R. de Hoog, and M. A. Lukas, Sijthoff and Noordhoff, The Netherlands, pp. 21–49, 1980.
14. SAVITSKY, D., *Hydrodynamic Design of Planing Hulls*, Journal of Marine Technology, Vol. 1, pp. 71–95, 1964.
15. CASLING, E. M., *Planing of a Low-Aspect-Ratio Flat Ship at Infinite Froude Number*, Journal of Engineering Mathematics, Vol. 12, pp. 43–58, 1978.
16. DOCTORS, L. J., *Theory of Compliant Planing Surfaces*, 2nd International Conference on Numerical Ship Hydrodynamics, Edited by J. V. Wehausen and N. Salvesen, University of California, Berkeley, California, pp. 185–197, 1977.
17. TUCK, E. O., *A Simple One-Dimensional Theory for Air-Supported Vehicles Over Water*, Journal of Ship Research, Vol. 28, pp. 290–292, 1984.
18. TUCK, E. O., *Linearised Planing-Surface Theory with Surface Tension. Part 1: Smooth Detachment*, Journal of the Australian Mathematical Society, Series B, Vol. 23, pp. 241–258, 1982.
19. TUCK, E. O., *Linearised Planing-Surface Theory with Surface Tension. Part 2: Detachment with Discontinuous Slope*, Journal of the Australian Mathematical Society, Series B, Vol. 23, pp. 259–277, 1982.
20. GREEN, A. E., *The Gliding of a Plate on a Stream of Finite Depth*, Proceedings of the Cambridge Philosophical Society, Vol. 31, pp. 589–603, 1936.
21. WU, T. Y.-T., *A Singular Perturbation Theory for Nonlinear Free Surface Flow Problems*, International Shipbuilding Progress, Vol. 14, pp. 88–97, 1967.
22. NEWMAN, J. N., *Evaluation of the Wave-Resistance Green Function. Part 1: The Double Integral*, Journal of Ship Research, Vol. 31, pp. 79–90, 1987.
23. WANG, H. T., *Comprehensive Evaluation of Six Thin-Wing Lifting-Surface Computer Programs*, U.S. Department of the Navy, David Taylor Naval Ship Research and Development Center, Report No. 4333, 1974.
24. KERWIN, J. E., AND LEE, C. S., *Prediction of Steady and Unsteady Marine Propellor Performance by Numerical Lifting Surface Theory*, Transactions of the Society of Naval Architects and Marine Engineers, Vol. 86, pp. 218–253, 1978.
25. SCLAVOUNOS, P. D., *An Unsteady Lifting-Line Theory*, Journal of Engineering Mathematics, Vol. 21, pp. 201–226, 1987.
26. SHEN, Y. T., *Theory of High-Aspect-Ratio Planing Surfaces*, University of Michigan, Naval Architecture and Marine Engineering Report No. 102, 1970.
27. TULIN, M. P., *The Theory of Slender Surfaces Planing at High Speeds*, Schiffstechnik, Vol. 4, pp. 125–133, 1956.
28. MARUO, H., *High and Low Aspect Ratio Approximations of Planing Surfaces*, Schiffstechnik, Vol. 14, pp. 57–64, 1967.
29. TUCK, E. O., *Low-Aspect-Ratio Flat-Ship Theory*, Journal of Hydronautics, Vol. 9, pp. 3–12, 1975.
30. COLE, S., *A Simple Example From Flat Ship Theory*, Journal of Fluid Mechanics, Vol. 189, pp. 301–310, 1988.

8

Abel Integral Equations

R. S. ANDERSSEN AND F. R. DE HOOG

Abstract. Even though they have a rather specialized structure, Abel equations form an important class of integral equations in applications. This happens because completely independent problems lead to the solution of such equations. After an initial survey of Abel integral equations, this chapter focuses on the numerical solution of these equations when the only available data are observational. Computationally, this is quite challenging, because Abel equations are (weakly) improperly posed. The chapter ends with some general advice about choosing numerical methods.

1. Introduction

The integral equation derived by Abel (Ref. 1), which now takes his name, had the form

$$\int_0^x (x-y)^{-1/2} u(y) \, dy = s(x), \qquad 0 \le y \le x < \infty, \tag{1}$$

where the right-hand side $s(x)$ is a known function and Eq. (1) must be solved for $u(y)$. Its importance then related to how Abel used the relationship between the kinetic and potential energies of falling bodies instead of Newton's laws, to determine the path along which a particle must be constrained to fall, under constant vertical acceleration, in order that its time of fall be a prescribed function $s(x)$ of the distance fallen (see Lonseth, Ref. 2, Section 2). Abel not only formulated the integral equation (1) but also derived (Ref. 1 and Ref. 2, respectively) the following two inversion

---

R. S. ANDERSSEN AND F. R. DE HOOG • Division of Mathematics and Statistics, Institute of Information and Communications Technologies, CSIRO, GPO Box 1965, Canberra, Australia 2601.

formulas for Eq. (1):

$$u(y) = \int_0^y (y-x)^{-1/2} s'(x)\,dx, \qquad s'(x) = ds(x)/dx, \qquad (2)$$

$$u(y) = \frac{1}{\pi} \frac{d}{dy}\left\{\int_0^y (y-x)^{-1/2} s(x)\,dx\right\}. \qquad (3)$$

In the time since Abel formulated Eq. (1), many authors have generalized it [see, for example, Linz (Ref. 3)]. However, we shall consider only equivalent forms of Eq. (1) obtained by some simple transformations of the dependent and independent variables—that is, equations obtained from Eq. (1) by the transformations

$$x = g(r), \qquad y = g(t),$$

where g is a monotone function. Equations obtained in this way obviously have inversion formulas corresponding to (2)-(3) and particular examples of practical interest will be given in the sequel.

Because explicit inversion formulas exist for Eq. (1), it might be thought that obtaining a numerical approximation to u is straightforward, since a cursory glance at (2) suggests that a differentiation followed by an integration should be stable. The fallacy of this argument is the singularity in the integral and it turns out that the inversion is actually (weakly) unstable. We can see this by defining

$$(\mathbf{A}s) = \int_0^x (x-y)^{-1/2} s(y)\,dy, \qquad (4)$$

$$(\mathbf{D}s) = ds(x)/dx, \qquad (5)$$

from which we obtain, using (2)-(3) and then (1),

$$\mathbf{A}\mathbf{D}s = \mathbf{D}\mathbf{A}s = \pi u, \qquad (6)$$

and

$$\mathbf{A}^2 \mathbf{D}s = \mathbf{A}\mathbf{D}\mathbf{A}s = \pi s, \qquad (7)$$

when $s(0) = 0$. Thus, the operator $\mathbf{A}$ can be thought of as a half integration and its inverse, a half differentiation. In fact, the definition of a fractional derivative can be based on the Abel integral equation [Eq. (36) in the sequel].

It may therefore be expected that difficulties will arise when seeking the solution of Eq. (1), and generalizations of it, which are similar to those associated with obtaining estimates of derivatives. As is well known [see, for example, Anderssen and Bloomfield (Ref. 4)], difficulties associated with estimating derivatives can become quite severe when the data is in error. It turns out that similar difficulties arise in the numerical solution of Eq. (1). Consequently, the problems which Abel integral equations model are *improperly posed* in the sense that certain small perturbations in the data $s(x)$ can lead to arbitrarily large perturbations in the solution $u(y)$.

A further complication arises due to the fact that, in general, s and u are not both smooth. For example, if $u(x) = 1$ then $s(x) = 2\sqrt{x}$ while, on the other hand, if $u(x) = \sqrt{x}$ then $s(x) = \pi x/2$. This means that care needs to be exercised when constructing numerical schemes as some of the obvious candidates may have a slow rate of convergence.

We shall see in Section 2 that Abel integral equations arise in a large number of physically unrelated ways, and that the available information about $s(x)$ varies greatly, depending on the context of the application.

Since the analysis of such equations will vary greatly depending on the structure of $s(x)$, the subsequent examination is categorized in terms of the properties of $s(x)$ in the following ways:

(i) *Exact Analytic.* When $s(x)$ corresponds to a known function of x, then, formally in terms of (2) and (3), we also know $u(y)$. This knowledge is explicit when either (2) or (3) can be evaluated exactly. Consequently, exact solutions of Abel-type integral equations are known for a wide variety of situations and are tabulated as Riemann–Liouville fractional integrals (see Erdélyi, Ref. 5, pp. 185–201). The existence of such solutions plays an important role in the construction and analysis of numerical methods for Abel integral equations, for example, in the construction of pseudoanalytic methods, which will be discussed in some detail below, and the study of certain interrelationships in integral geometry (geometric probability) [Moran (Ref. 6) and Santaló (Ref. 7)].

(ii) *Discrete (Numerical).* When the inversion formulas (2)–(3) cannot be evaluated exactly to determine $u(y)$, it will be necessary to work with discretized (numerical) values of $s(x)$ and to resort to numerical (computational) procedures in order to determine an approximation to $u(y)$. Because of the inherent improperly posedness of Eq. (1) (mentioned above), the construction of stable numerical processes for the solution of (1) poses additional challenges. Even though some standard numerical methods yield good approximations to $u(y)$ when the discretized values of $s(x)$ are evaluated with sufficient accuracy (Ref. 8), it is much more difficult, from the point of view of the underlying numerical analysis, to characterize when such numerical methods will work successfully. But discretized (numerical) data are like very accurate observational data, and thus only a mild form of stabilization is necessary to control the inherent improperly posedness and thereby yield good approximations to $u(y)$ [Anderssen (Ref. 9) and Lukas (Ref. 10)].

(iii) *Nonexact Observational.* Here, the only available information about the exact data $s(x)$ is the discrete observational data

$$d_j = s(x_j) + \varepsilon_j, \qquad j = 1, 2, \ldots, n, \qquad a < x_1 < x_2 < \cdots < x_n < b, \qquad (8)$$

where the ε_j denote observational (measurement) errors. This typifies the circumstances of a wide variety of applications (as the discussion of Section 2 will indicate) including the situation where $s(x)$ is a probability distribution function and the only available information is a histogram approximation. Now, the problem is to recover as much information about $s(x)$ and $u(y)$ consistent with the information in the available data $d_j, j = 1, 2, \ldots, n$. Because of the underlying numerical differentiation which must be performed implicitly or explicitly on the data $d_j, j = 1, 2, \ldots, n$, in order to recover information about the solution $u(y)$, some form of stabilization must be used. There are different ways in which this can be done including regularization [Lukas (Ref. 10), Baev and Glasko (Ref. 11)], constrained optimization, linear programming inversion [Anderssen and Gustafson (Ref. 12)], and low-dimensional parameterization in the characterization of the approximations. An alternative strategy is to simply limit the information determined and used for inference purposes about the solution $u(y)$ to bounded linear functionals

$$L_\theta(u) = \int_a^b \theta(y)u(y)\, dy, \qquad \theta(y) \equiv \text{known}, \qquad (9)$$

defined on $u(y)$ such as the statistical moments of mean, variance, and kurtosis which are defined in terms of linear functionals characterized by $\theta(y) = y, y^2, y^3$.

Though much of the literature devoted to the numerical solution of Abel-type integral equation is only concerned with the discrete (numerical) situation, it is the nonexact observational situation which is the most important, from the point of view of applications. Consequently, attention is focused on the latter aspect in this chapter.

2. Abel Integral Equations in Applications

Because Abel integral equations arise naturally when formulating mathematical models from a variety of independent starting assumptions, such equations are central to the analysis of a wide variety of applications [Anderssen (Ref. 13) and Jakeman and Anderssen (Ref. 14)]. The following examples aim to give some impression of the scope of the computational considerations which arise with respect to the numerical solution of Abel integral equations. In particular, they indicate that such equations must be solved regularly when the only available information about $s(x)$ is observational data.

2.1. First-Kind Abel Integral Equations in Geometric Probability.
See Wicksell (Refs. 15-16), Santaló (Ref. 7), and Moran (Ref. 6):

$$s(y) = \frac{y}{m} \int_y^\infty \frac{u(x)}{(x^2 - y^2)^{1/2}} \, dx, \qquad 0 \le y \le x < \infty, \tag{10}$$

$$m = \int_0^\infty u(y) \, dy = \pi/2H, \qquad H = \int_0^\infty \frac{s(y)}{y} \, dy, \tag{11}$$

where
(i) $u(x)$ denotes the size distribution of spheres distributed by a Poisson process in three-dimensional space;
(ii) $s(y)$ denotes the size distribution of the circles obtained on random plane sections through the space;
(iii) m denotes the average radius of the spheres.

In such applications, one is interested in the analytic properties of $u(x)$ relative to those of $s(y)$, and thus the data $s(y)$ are analytic. Clearly, it is necessary to first evaluate H in order to remove the inherent nonlinearity in the formulation (10) and (11).

2.2. First-Kind Abel Integral Equations in Interferometry.
See Merzkirch (Ref. 17, Chapter 8, Section 3.A):

$$s(y) = \frac{2}{\lambda} \int_y^R \frac{ru(r)}{(r^2 - y^2)^{1/2}} \, dr, \qquad 0 \le y \le r \le R < \infty, \tag{12}$$

where
(i) $u(r)$ denotes the refractive index (or gas density) at a particular cross-section in an axial symmetric flow;
(ii) $s(y)$ denotes the fringe shift function seen on interferograms of the flow at different cross-sections obtained using an interferometer;
(iii) λ denotes the wavelength of the light used in the interferometer;
(iv) R denotes the radius of the cylinder containing the flow.

Since the fringe shifts can only be measured accurately relative to the peaks and troughs in the fringes, the available observation

$$d_i = s(y_i) + \varepsilon_i, \qquad i = 0, 1, 2, \ldots, n, \qquad \varepsilon_i = \text{random errors}, \qquad (13)$$

though quite accurate, are limited in number (typically, 25-50). In such applications, one is interested in estimating the change in density of the gas at different points along an axial symmetric flow past some obstacle. Thus, it is necessary to solve (12) at a sequence of cross-sections using the accurate but limited number of data available.

2.3. First-Kind Abel Integral Equations in Stereology. See Wicksell (Refs. 15-16), Moran (Ref. 6), and Jakeman and Anderssen (Ref. 14):

$$s(y) = \frac{y}{m} \int_y^a \frac{u(x)}{(x^2 - y^2)^{1/2}} dx, \qquad 0 \le y \le x \le a < \infty, \qquad (14)$$

$$m = \int_0^a u(y)\, dy = \pi/2H, \qquad H = \int_0^a \frac{s(y)}{y}\, dy, \qquad (15)$$

where
  (i) $u(x)$ denotes the size distribution of spherical particles (such as carbon particles in steel [Hyam and Nutting (Ref. 18)], or second-phase copolymer spheroids dispersed in a continuous first phase of polystyrene [Meisner (Ref. 19)]) in some aggregate;
  (ii) $s(y)$ denotes the size distribution of the circular sections of the particles on random plane sections through the aggregate;
  (iii) a denotes an upper bound on the maximum size of the spherical particles;
  (iv) m denotes the average radius of the spheres.

In contrast to the geometric probability application discussed above, one is now interested in estimating $u(x)$ from realizations

$$d_i = s(y_i) + \varepsilon_i, \qquad i = 0, 1, 2, \ldots, n, \qquad \varepsilon_i = \text{random errors} \qquad (16)$$

of $s(y)$ on random plane sections taken through a given sample. What we observe are independent realizations of a random variable with probability density $s(y)$, and not $s(y)$ itself. Obviously, the approximate distribution which such data yield is very noisy. In such situations, (14)-(16) can only be solved to yield a reliable estimate of $u(x)$ when the number of available observations is sufficiently large ($n > 100$).

2.4. Second-Kind Abel Integral Equations in Stereology.
See Goldsmith (Ref. 20) and Jakeman (Ref. 21):

$$s(y) + Tu(y) = \frac{2y}{2m + T} \int_y^a \frac{u(x)}{(x^2 - y^2)^{1/2}} dx, \qquad 0 \le y \le x \le a < \infty, \quad (17)$$

$$m = \int_0^a u(y)\, dy = \pi/2H, \qquad H = \int_0^a \frac{s(y)}{y}\, dy, \qquad (18)$$

where
  (i) $u(x)$ denotes the size distribution of spherical particles in some aggregate;
  (ii) $s(y)$ denotes the size distribution of the circular sections of the particles observed using transmission electron microscopy on random thin sections of the aggregate;
  (iii) T denotes the thickness of the thin sections;
  (iv) a denotes an upper bound on the maximum size of the spherical particles;
  (v) m denotes the average radius of the spheres.
  One is now interested in estimating $u(x)$ from realizations

$$d_i = s(y_i) + \varepsilon_i, \qquad i = 0, 1, 2, \ldots, n, \qquad \varepsilon_i = \text{random errors} \quad (19)$$

of $s(y)$ on random thin sections taken through a sample. As in Section 2.3, we observe $\{d_i\}$ as independent realizations of a random variable with probability density $s(y)$. Thus, the approximate distribution which such data yield will be very noisy, and the need for a sufficiently large sample, to guarantee a reliable estimate of $u(x)$ from solving (17)-(19), will again apply.

2.5. The Abel Integral Equation of Seismology.
See Jeffreys (Ref. 22), Bateman (Ref. 23), Knott (Ref. 24), and Macelwane (Ref. 25).

From observations of the epicentral arc of travel $\Delta(p)$ on a spherically symmetric Earth and the corresponding time travel $T(p)$ for various earthquake waves with parameter p, the problem is to calculate the elastic wave velocity $v(r)$ as a function of depth (radius) r. The basic equation is

$$s(p) = p \int_p^{R/V} \frac{\partial \log r / \partial u}{(u^2 - p^2)^{1/2}}\, du,$$

with $p = dT/d\Delta$ the ray parameter, $s(p) = \Delta(p)/2$, r the radius at which ray with parameter p bottoms, $u = r/v(t)$, $V = v(R)$, and R the radius of the Earth.

The integral equations which relate epicentral arc of travel $\Delta(p)$, travel time $T(p)$, and velocity $v(r)$ as a function of radius r are nonlinear in $v(r)$, the unknown of interest. They are usually derived on the basis of Fermat's principle of least action which implies that the time of travel of any ray through the Earth must be stationary. Thus, equations for the path of travel, and hence $v(r)$, can be derived from the Euler-Lagrange equations for the functional which defines the time of travel. By redefining what the dependent and independent variables are (i.e., by introducing an appropriate change of variables), the basic nonlinear equation becomes a linear Abel-type equation [Jeffreys (Ref. 22, Section 2.05) and Macelwane (Ref. 25, p. 273)]. The computational methods of solution applied to real data are based on this redefinition of variables.

It is clear from these examples that, in applications, Abel integral equations have to be solved for various forms of data which range over the full spectrum of possibilities from analytic to very noisy observational.

We conclude this section with an example where Abel equations play an implicit role in the formulation process in that they are used to derive a mathematical model, the final form of which is not an Abel integral equation.

2.6. Abel Integral Equations in Tomography. Reconstruction of two-dimensional structure from projections, which reduces to the solution of the Radon transform, is a basic tool in nondestructive testing and diagnosis. One of the better known aspects is tomography, which is the name given to its use in medical diagnosis [Gordon *et al.* (Ref. 26)]. Without the formal mathematical tools which have been developed for the solution of the Radon transform, the design of sophisticated CAT scanners, now available for medical diagnostic purposes, would not have been possible [Scudder (Ref. 27)].

Among the various methods used to invert the Radon transform, one of the more popular is circular Fourier averaging, which leads naturally to nonstandard Abel integral equations that possess inversion formulas.

In fact, it is easily illustrated that Abel integral equations represent a natural framework in which to study reconstruction from projections. Consider the situation where we take axial projections of radius R of an object with variable cross-sectional density $\rho(x, y)$. Since the relationship between projections and density is linear, it follows that circular averaging of the projections to yield $s(r)$, $0 \leq r \leq R$, corresponds to circular averaging of the cross-sectional density $\rho(x, y)$ to yield $u(r)$, $0 \le r \le R$. A standard argument (Merzkirch, Ref. 17, Chapter 8, Section 3.A) shows that

$$s(y) = 2 \int_y^R \frac{ru(r)}{(r^2 - y^2)^{1/2}} \, dr, \qquad 0 \le y \le r \le R < \infty. \tag{20}$$

Since $s(R) = 0$, it follows from the known inversion formula [Eq. (29b) below] that

$$u(r) = \frac{1}{\pi} \int_r^R \frac{s'(y)}{(y^2 - r^2)^{1/2}} \, dy, \qquad s'(y) = ds(y)/dy, \tag{21}$$

which proves the uniqueness of the reconstruction from projections for an axial symmetric geometry. The interesting point about this result is that it is in full accord with Radon's original result that the reconstruction is uniquely determined by the continuum of projections, since, for an axial symmetric geometry, one projection defines the continuum.

In addition, it follows from (21), because the circular averaging leaves the density unchanged on the axis of rotation, that

$$\rho(0, 0) = u(0) = \frac{1}{\pi} \int_0^R \frac{s'(y)}{y} \, dy, \tag{22}$$

which corresponds to one of the inversion formulas often used in tomography [Vest (Ref. 28)], since a change of origin is all that is necessary to make it generally applicable.

If, instead of simply averaging the projections, circular Fourier averaging is applied to them, the relevant mathematical framework becomes

$$s_m(r) = 2 \int_r^R \frac{T_m(r/\tau)}{(\tau^2 - r^2)^{1/2}} u_m(\tau) \, d\tau, \tag{23}$$

where $s_m(r)$ and $u_m(r)$ denote, respectively, the mth circular Fourier averaging of the projections and density, and $T_m(z)$ denotes the Chebyshev polynomial of degree m. This is the basis of Cormack's (Refs. 29–30) work which has been used widely in applications such as TOKAMAK monitoring in plasma physics [Sauthoff and von Goeler (Ref. 31)]. In fact, Cormack was the first to show that (23) possesses the inversion formula

$$u_m(r) = \frac{1}{\pi} \frac{d}{dr} \int_r^R \frac{T_m(\tau/r) \tau s_m(\tau)}{r(\tau^2 - r^2)^{1/2}} \, dr. \tag{24}$$

3. The Numerical Analysis of Abel Integral Equations

As the list of applications of Section 2 indicates, a general analysis of Abel integral equations is beyond the scope of this chapter. We therefore limit attention in the sequel to first-kind equations of which the following canonical forms are representative:

(a) $\quad s(y) = 2 \int_y^a \frac{xu(x)}{(x^2 - y^2)^{1/2}} dx, \qquad 0 \le y \le x \le a < \infty;$ (25)

(b) $\quad s(y) = 2 \int_0^y \frac{xu(x)}{(y^2 - x^2)^{1/2}} dx, \qquad 0 \le x \le y < \infty;$ (26)

(c) $\quad s(y) = \int_y^a \frac{u(x)}{(x - y)^{1/2}} dx, \qquad 0 \le y \le x \le a < \infty;$ (27)

(d) $\quad s(y) = \int_0^y \frac{u(x)}{(y - x)^{1/2}} dx, \qquad 0 \le x \le y < \infty.$ (28)

Under mild regularity constraints [Kowalewski (Ref. 32, Section 1, pp. 80–82)], they possess uniquely defined solutions with known explicit inversion formulas:

(a) $\quad u(x) = -\frac{1}{\pi x} \frac{d}{dx} \int_x^a \frac{ys(y)}{(y^2 - x^2)^{1/2}} dy,$ (29a)

$\qquad = -\frac{1}{\pi} \int_x^a \frac{s'(y)}{(y^2 - x^2)^{1/2}} dy, \qquad s'(y) = \frac{ds(y)}{dy},$ (29b)

if $s(a) = 0$;

(b) $\quad u(x) = \frac{1}{\pi x} \frac{d}{dx} \int_0^x \frac{ys(y)}{(x^2 - y^2)^{1/2}} dy,$ (30a)

$\qquad = \frac{1}{\pi} \int_0^x \frac{s'(y)}{(x^2 - y^2)^{1/2}} dy, \qquad s'(y) = \frac{ds(y)}{dy},$ (30b)

if $s(0) = 0$;

(c) $\quad u(x) = -\frac{1}{\pi} \frac{d}{dx} \int_x^a \frac{s(y)}{(y - x)^{1/2}} dy,$ (31a)

$$= -\frac{1}{\pi} \int_x^a \frac{s'(y)}{(y-x)^{1/2}} dy, \qquad s'(y) = \frac{ds(y)}{dy}, \tag{31b}$$

if $s(a) = 0$;

(d) $$u(x) = \frac{1}{\pi} \frac{d}{dx} \int_0^x \frac{s(y)}{(x-y)^{1/2}} dy, \tag{32a}$$

$$= \frac{1}{\pi} \int_0^x \frac{s'(y)}{(x-y)^{1/2}} dy, \qquad s'(y) = \frac{ds(y)}{dy}, \tag{32b}$$

if $s(0) = 0$.

At this point, it might appear that it is only necessary to choose a grid

$$\mathbf{G} = \{y_i; 0 = y_0 < y_1 < \cdots < y_n = a, y_i = ih, h = a/n\}$$

and to evaluate the appropriate inversion formula using discrete methods. This can certainly be done when $s(y)$ is given analytically, but (as explained in the Introduction) the situation is more complex when $s(y)$ is only available as observational data $\{d_i\}$.

To illustrate, we examine one of the most popular discrete methods for the inversion of (25) via the evaluation of (29a), namely, that of Nestor and Olsen (Ref. 33). On the grid $\mathbf{G}$, their method reduces to the evaluation of

$$U_i = -\frac{1}{\pi h} \sum_{j=i}^n B_{ij} s_j, \qquad i = 0, 1, \ldots, n, \tag{33}$$

where U_i denotes the resulting approximation to $u(y_i)$, and

$$B_{ij} = \begin{cases} -A_{i,i}, & i = j, \\ A_{i,j-1} - A_{i,j}, & j \geq i+1, \end{cases}$$

with

$$A_{i,j} = 1/\{[(j+1)^2 - i^2]^{1/2} + [j^2 - i^2]^{1/2}\}.$$

If perturbed data $s_i + \delta s_i$ ($i = 0, 1, \ldots, n$) are used in (33), the resulting approximations are $U_i + \delta U_i$ where

$$\delta U_i = -\frac{2}{\pi h} \sum_{j=i}^n B_{ij} \delta s_i, \qquad i = 0, 1, \ldots, n.$$

In particular, the effect of a single perturbation δs_k is such that

$$\delta U_i = \begin{cases} 0, & i = k+1, \ldots, n, \\ -2B_{ik}\delta s_k/(\pi h), & i = 0, 1, \ldots, k. \end{cases} \quad (34)$$

For $k = i$, (34) yields

$$\delta U_i = -\delta s_i / \pi [h(2y_i + h)]^{1/2}. \quad (35)$$

Thus, an error in $s(y)$ is amplified locally by a factor of the order of $h^{-1/2}$. On the other hand, it is well known that, for standard finite-difference approximations to the derivative of a function, the corresponding amplification factor is of the order of h^{-1}. This leads naturally to the conclusion that, from a numerical point of view, the inversion of first-kind Abel-type integral equations of the form (25)–(28) is as badly posed as a "half-differentiation." This connection to differentiation can be made more precise, if we recall that [Sneddon (Ref. 34) and Erdélyi (Ref. 5)]

$$s(y) = \frac{1}{\Gamma(1+\alpha)} \int_0^y u(x)/(y-x)^\alpha \, dx, \quad 0 < \alpha < 1, \quad (36)$$

can be used formally to define fractional differentiation of order α.

The correspondence with fractional differentiation represents the reason why first-kind Abel equations are classified as improperly posed even though inversion formulas exist. Such equations are however improperly posed in terms of the sensitivity of the solution $u(x)$ to small perturbations in the data $s(y)$. The connection with fractional differentiation indicates that, if order of differentiation is used to quantify the degree of such sensitivity, then first-kind Abel-type equations are only *weak* improperly posed.

Remark 3.1. The essential improperly posed nature of Abel equation inversion appears to have been overlooked by a number of authors such as Chan and Lu (Ref. 35) and Deutsch and Beniaminy (Ref. 36). They first observe that there exist inversion formulas (corresponding to those listed above) which do not involve the explicit differentiation of $s(y)$. Next, they derive finite-difference methods for such inversion formulas and conclude (from numerical experimentation) that they are stable. The flaw in such arguments has been explained in some detail by Anderssen and de Hoog (Ref. 37) who observe that

"... computational difficulties associated with the inversion of improperly posed problems such as Abel integral equations cannot be removed by simply manipulating the problem mathematically into an alternative form. Stabilization can only be obtained through the introduction of additional structure."

The above comment about an amplification factor of $h^{-1/2}$ now becomes crucial. Numerically, it implies that, if standard finite-difference methods are used to evaluate inversion formulas, there is a minimum grid spacing below which h cannot go if reliable approximations are to be guaranteed. This will of course lead to a loss of resolution. On the other hand, since the data are observational, it is important statistically to use more rather than less observations. Such tradeoffs are typical of situations which arise when simple numerical methods are used to solve improperly posed problems.

Remark 3.2. We note in passing that the stable finite-difference formulas developed by Anderssen and de Hoog (Ref. 38) for numerical differentiation could be generalized to cover fractional differentiation and hence Abel-type integral equations. However, this aspect is not pursued here since such formulas are only applicable to situations when there is an abundance of data on a fine grid.

Though this leads to the conclusion that specialized methods rather than simple ones are required when the data are observational, the choice is not so straightforward. In fact, the choice must be based on the accuracy and number of data available. If the data are limited in number (such as occurs in the interferometry application discussed above in Section 2.2), then only simple methods should be used, but are unlikely to yield reliable results unless the data are sufficiently accurate. If the number of data are such that specialized methods can be applied, the use of simple methods will invariably yield inferior results. Clearly, more sophisticated methods should be used in such situations.

However, the utility of simple methods, in situations where limited but accurate data are available, should not be used as justification for excessive grouping or pruning of observational data when the data are quite noisy.

There is, of course, a great deal of technology, such as regularization, which has been developed to deal with improperly posed problems which could be applied to obtain numerical solutions of Abel equations. In fact, regularization has been proposed by Baev and Glasko (Ref. 11). However, since such equations are very tractable analytically, we take the view here that features, such as the availability of inversion formulas, etc., should be fully exploited in any numerical schemes.

The three classes of methods we shall consider can be classified as:

(a) *Pseudoanalytic Methods.* To illustrate what we mean by such methods, we examine (25) under the assumption that the data take the form (8), and let $\{\varphi_k\}$ denote a system of functions for which the inversion of (25) can be performed analytically to yield $\{\psi_k\}$. The essence of the method is to approximate $s(y)$ by

$$\hat{s}_m(y) = \sum_{k=1}^{m} a_k^{(m)} \varphi_k(y), \qquad a_k^{(m)} = \text{const}, \qquad m < n, \qquad (37)$$

for then the required approximation $\hat{u}_m(x)$ to $u(x)$ is automatically obtained by

$$\hat{u}_m(x) = \sum_{k=1}^{m} a_k^{(m)} \psi_k(x). \qquad (38)$$

Such methods are very popular because, computationally, they reduce to parameter estimation for which reliable algorithms are available; for example, determine the $a_k^{(m)}$ as the minimizer of

$$\min_{\mathbf{a}} \left\{ \sum_{j=0}^{n} \left[d_j - \sum_{k=1}^{m} a_k \varphi_k(y_j) \right]^2 \right\}, \qquad \mathbf{a} = (a_1, a_2, \ldots, a_m)^{\mathrm{T}}. \qquad (39)$$

They are often used when the data are limited, but, as will be shown in Section 4, the approximation they yield can be grossly inaccurate when they fail to cope with the weakly improperly posed nature of the Abel integral equation being solved.

(b) *Wiener Filtering.* The work of Anderssen and Bloomfield (Refs. 4, 39) on the use of Wiener filtering methods for the numerical differentiation of observational data extends naturally to the solution of Abel-type equations, since they correspond (after an appropriate change of variables when necessary) to fractional differentiation. Such methods are discussed in Section 5, though the full potential of this approach has yet to be exploited.

(c) *Stabilized Evaluation of Inversion Formulas.* To date, the most successful methods have been based on the stabilized evaluation of inversion formulas. The direct evaluation of inversion formulas will be discussed in Section 6. It is shown that best results are obtained if (a) and (b) are viewed as compatible methods and used to construct the stabilized evaluation so that their advantages complement each other.

In many applications, it is not the solution $u(x)$ of (25) which is used for inference purposes, but linear functionals (e.g., moments) defined on $u(x)$, namely,

$$m_\theta(u) = \int_0^a \theta(x) u(x) \, dx, \qquad \theta(x) \equiv \text{known}. \qquad (40)$$

Often, such functionals can be redefined as functionals on $s(y)$ (i.e., data functionals), namely,

$$m_\varphi(s) = \int_0^a \varphi(\theta; y) s(y)\, dy. \tag{41}$$

Thus, the estimation of the linear functionals required for inference purposes reduces to evaluating the corresponding data functionals (41) on the observations $\{d_i\}$. The need to solve an Abel equation in conjunction with observational data before estimating the required functionals (40) is thereby circumvented and replaced by the better-posed problem of evaluating (41) on the data. The utility of this approach will be discussed in Section 7.

4. Pseudoanalytic Methods

A specific property of Abel integral equations on which such methods are based is the existence of complete sets of basis functions for which the inversion of (25) is known analytically. We initially define and examine such methods for a general densely defined operator equation

$$\mathcal{L}u = s, \qquad \mathcal{L}: S_1 \to S_2, \tag{42}$$

where S_1 and S_2 are Hilbert spaces with norms denoted by $\|\cdot\|_1$ and $\|\cdot\|_2$, respectively. We assume that the inverse of $\mathcal{L}$, namely $\mathcal{L}^{-1}$, formally exists, and let $\{\psi_k\}$ denote an explicitly-known coordinate system which spans the domain of $\mathcal{L}$, $D(\mathcal{L})$, such that the corresponding system $\{\varphi_k = \mathcal{L}\psi_k\}$ is also known explicitly. If $\mathcal{L}$ is bounded, then $\{\varphi_k\}$ spans the range of $\mathcal{L}$, $R(\mathcal{L})$. For the canonical Abel equation (25), it is only necessary to evaluate the integrals (25) with u replaced by ψ_k to obtain the corresponding φ_k.

Remark 4.1. Because these norms will be used to characterize the extent to which pseudoanalytic methods will fail to stabilize the half-differentiation associated with the inversion of the Abel equation, it is assumed that they are unweighted norms from Hilbert spaces in which half-differentiation remains unbounded.

For such systems, if

$$s = \sum_{k=1}^m a_k^{(m)} \varphi_k, \qquad m < \infty,$$

Table 1. Invertible Coordinate Systems for the Abel Equation (25)

1.	$\phi_0(y) = (a^2 - y^2)^{1/2},$ $\phi_k(y) = (a^2 - y^2)^{k-1}, \quad k \geq 1.$	$\psi_0(x) = 1/\pi$ $\psi_k(x) = \Gamma(k + \tfrac{1}{2}(a^2 - x^2)^{k-1/2}/\Gamma(k)\Gamma(\tfrac{1}{2}), \quad k \geq 1$	Minnerbo and Levy (Ref. 41)
2.	$\phi_k(y) = t^\beta T_k(1 - 2t), \quad k \geq 0,$ $t = a^2 - y^2,$ $T_k(\cdot)$ Chebyshev polynomial of degree k, $\beta = \text{const} > -\tfrac{1}{2}.$	$\psi_k(x) = (t^{\mu+\beta-1}\Gamma(1+\beta))/(\Gamma(\mu+\beta)\Gamma(1-\mu))$ $\times {}_3F_2(-k, k, \beta + 1; \tfrac{1}{2}, \mu + \beta; y),$ $t = a^2 - x^2$	Piessens and Verbaeten (Ref. 42)
3.	$(y - y_j)^m_+,$ $z^m_+ = \begin{cases} z^m, & z \geq 0, \\ 0, & z < 0, \end{cases}$ $y_j = $ fixed point.	$\left\{ \dfrac{m}{x} \int_0^x \dfrac{(y - y_j)}{(y^2 - x^2)^{1/2}} \, dy \right.$	Smarzewski and Malinowski (Ref. 67)

then the inversion of (42) can be performed analytically to yield

$$u = \sum_{k=1}^{m} a_k^{(m)} \psi_k, \qquad m < \infty.$$

Because of the obvious connection with variational and projection methods, we shall refer to the dual system $\{\psi_k, \varphi_k\}$, $\varphi_k = \mathscr{L}\psi_k$, as an $\mathscr{L}$-*invertible coordinate system*. Some examples of $\mathscr{L}$-invertible coordinate systems for the canonical Abel integral equation (25) are listed in Table 1.

In order to cope with the observational nature of the data $\{d_i\}$, the existence of an $\mathscr{L}$-invertible coordinate system $\{\psi_k, \varphi_k\}$ is utilized in the following way. The data $\{d_i\}$ are smoothed using the model

$$s_m(y) = \sum_{k=1}^{m} a_k^{(m)} \varphi_k(y), \qquad m \ll n, \qquad (43)$$

and an appropriate statistical fitting strategy, such as least squares, is applied to yield an estimate of $s(y)$, viz.

$$\hat{s}_m(y) = \sum_{k=1}^{m} \hat{a}_k^{(m)} \varphi_k(y), \qquad (44)$$

where the $\hat{a}_k^{(m)}$, $k = 1, 2, \ldots, m$, denote the estimates of the $a_k^{(m)}$, $k = 1, 2, \ldots, m$. This estimate of $s(y)$ is then inverted analytically to yield *the pseudoanalytic approximation to* $u(x)$ defined by

$$\hat{u}_m(x) = \sum_{k=1}^{m} \hat{a}_k^{(m)} \psi_k(x). \qquad (45)$$

The *tacit assumption* of such methods is: *if*

$$\left\| s(y) - \sum_{k=1}^{m} \hat{a}_k^{(m)} \varphi_k(y) \right\|_2$$

is small, then

$$\left\| u(x) - \sum_{k=1}^{m} \hat{a}_k^{(m)} \psi_k(x) \right\|_1$$

is also small. Though this is true when $\mathscr{L}$ is properly posed, it is not necessarily the case when $\mathscr{L}$ is (weakly) improperly posed.

The difficulty can be formally identified in the following way. For a general s, there exists an r, orthogonal to $\text{span}(\varphi_1, \ldots, \varphi_m)$, such that

$$s = s_m + r, \qquad s_m = \sum_{j=1}^{m} a_j \varphi_j(x). \tag{46}$$

Assuming for the moment that s_m is estimated exactly, it follows as an immediate consequence of the pseudoanalytic method that the corresponding approximation u_m satisfies

$$\mathscr{L} u_m = s_m, \tag{47}$$

and, since $\mathscr{L}^{-1}$ exists, that

$$u - u_m = \mathscr{L}^{-1}(s - s_m) = \mathscr{L}^{-1} r. \tag{48}$$

This result clearly indicates that, even if $\|r\|_2$ and other appropriate measures are small, $\|u - u_m\|_1$ and the corresponding measures can be large when $\mathscr{L}^{-1}$ is unbounded.

Numerically, because s is actually observed as

$$d_j = s(x_j) + \varepsilon_j, \qquad j = 0, 1, 2, \ldots, n,$$

where the ε_j denote random observational errors, the a_j of (46) are only estimated approximately as $\hat{a}_j^{(m)}$. Thus,

$$s - \hat{s}_m = \sum_{j=1}^{m} (a_j - \hat{a}_j^{(m)}) \varphi_j(x) + r(x), \tag{49}$$

and hence

$$u - \hat{u}_m = \sum_{j=1}^{m} (a_j - \hat{a}_j^{(m)}) \psi_j(x) + \mathscr{L}^{-1} r(x). \tag{50}$$

This indicates that terms of the form $(a_j - \hat{a}_j^{(m)}) \psi_j(x)$ also contribute to the error in $u - \hat{u}_m$.

In order to justify the extra effort involved in replacing pseudoanalytic methods by more sophisticated ones, it was shown by Anderssen (Ref. 9) that there exists data for which: (a) some pseudoanalytic models are inappropriate numerically; and (b) some pseudoanalytic approximations differ globally from the exact solution.

The data used were obtained from a study of steady supersonic flow past a cone [Landensberg and Bershader (Ref. 40)]. A feature of these data, as shown in Fig. 1, is the near discontinuity in $s(t)$ in the neighborhood of

Abel Integral Equations

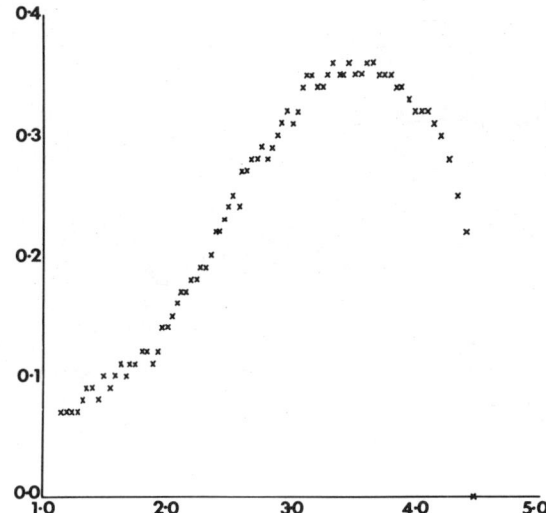

Fig. 1. Abel equation data from shock tube experiments.

$t = a$ ($a = 4.47$ in Fig. 1). In order to study such methods, the following synthetic data were used by Anderssen (Ref. 9) to model the above supersonic data:

$$s(y) = 2 \int_y^a xu(x)/(x^2 - y^2)^{1/2} \, dx \tag{51}$$

with $a = 1$ and

$$u(x) = \begin{cases} K + b(\tfrac{1}{2} - x)^p \sin(c\pi x - \pi/2), & 0 \le x \le \tfrac{1}{2}, \\ K + q(\tfrac{1}{2} - x)^p, & \tfrac{1}{2} \le x \le 1, \end{cases} \tag{52}$$

where K, b, c, p, and q are positive constants. A particular case is illustrated in Fig. 2.

For different values of K, b, c, p, and q, the coefficients $a_j^{(m)}$ of (43) were estimated using least squares. Both the Minerbo and Levy (Ref. 41) model (Table 1) and the model of Piessens and Verbaeten (Ref. 42) (Table 1) were used. For both these models, the values of $a_i^{(m)}$, for fixed i, failed to converge as m increased. In addition, the values of $a_i^{(m)}$ for the Minerbo and Levy (Ref. 41) model tended to grow quite rapidly for fixed i and increasing m. It follows that, for the data derived from (51)–(52), the Minerbo and Levy (Ref. 41) model for $s(y)$ is inappropriate.

Since there is a nonzero difference between the fitted and actual structure in the data (even when it is exact), it follows from the improperly

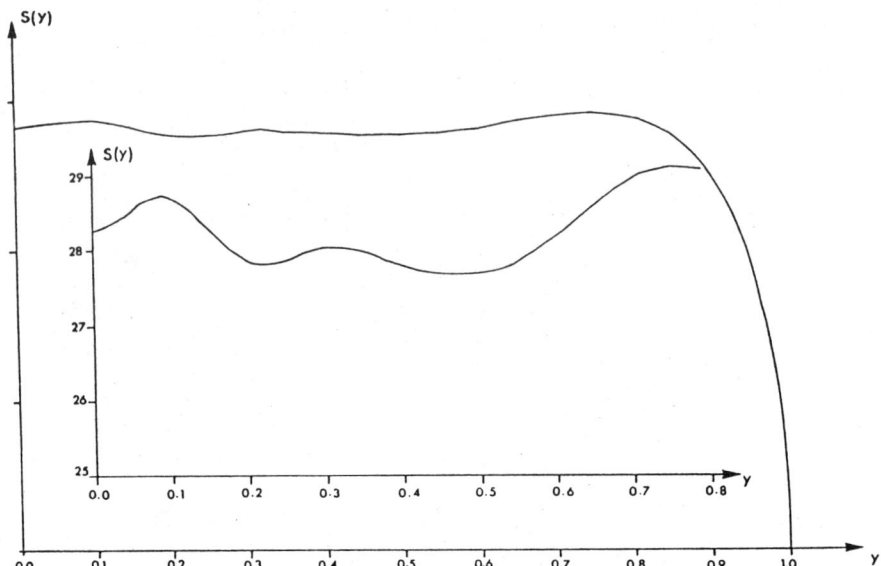

Fig. 2. Synthetic data for shock tube experiments.

posed nature of Abel's equation that the pseudoanalytic approximation to the required solution may differ radically from the exact. That this is in fact the case is illustrated in Fig. 3 for the application of the Minerbo and Levy (Ref. 41) method to the data shown in Fig. 2.

The above discussion and exemplification pinpoints qualitatively the defect associated with the use of a pseudoanalytic method for the solution of Abel integral equations. To quantify it rigorously, it is necessary to establish the sense in which the direct least-squares solution defined by

$$\bar{u}_m = \sum_{j=1}^{m} \hat{b}_j^{(m)} \psi_j, \tag{53}$$

with the $\hat{b}_j^{(m)}$, $j = 1, 2, \ldots, m$, the unique minimizer of

$$\min_{\mathbf{b}} \left\| u - \sum_{j=1}^{m} b_j \psi_j \right\|_1^2, \qquad \mathbf{b} = (b_1, b_2, \ldots, b_m)^{\mathrm{T}},$$

is equivalent to the corresponding pseudoanalytic solution $\hat{u}_m(x)$ of (45).

In fact, they are equivalent, when the systems $\{\varphi_j\}$ and $\{\psi_j\}$ are such that there exist inner products $(\cdot, \cdot)_2$ and $(\cdot, \cdot)_1$, respectively, with respect to which the $\{\varphi_j\}$ and $\{\psi_j\}$ are orthogonal. It follows from Fourier analysis

Abel Integral Equations

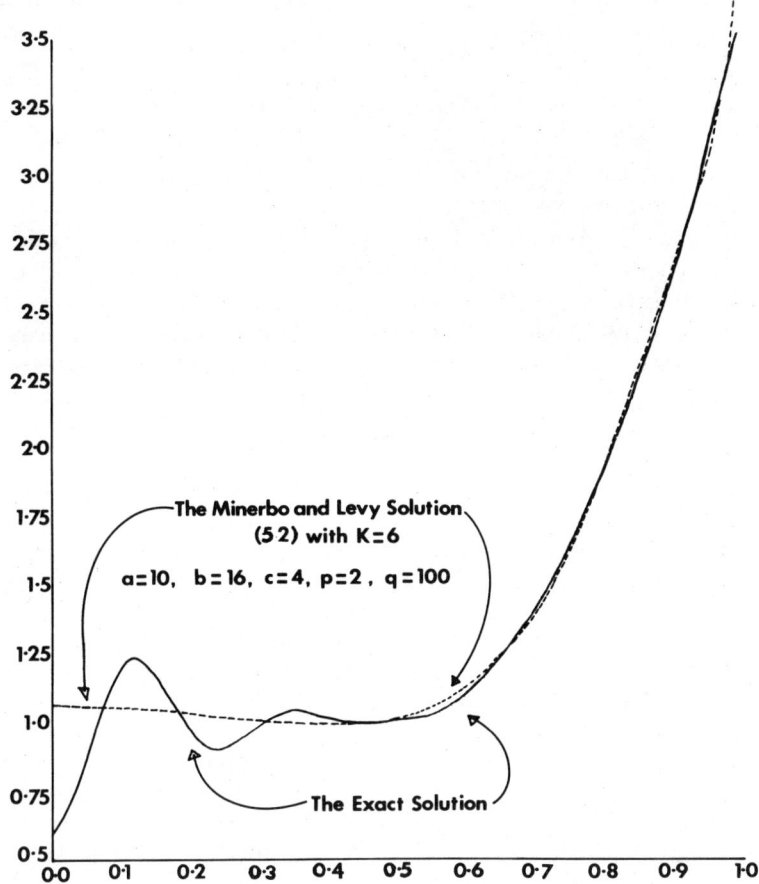

Fig. 3. Minerbo–Levy solution for synthetic data.

that

$$u = \sum_{j=1}^{\infty} \frac{(u, \psi_j)_1}{(\psi_j, \psi_j)_1} \psi_j. \tag{54}$$

In addition

$$s = \sum_{j=1}^{\infty} \frac{(s, \phi_j)_2}{(\phi_j, \phi_j)_2} \phi_j,$$

and hence

$$u = \mathscr{L}^{-1} s = \sum_{j=1}^{\infty} \frac{(s, \phi_j)_2}{(\phi_j, \phi_j)_2} \psi_j.$$

Since the coefficients in the Fourier expansion (54) are uniquely defined, it follows that

$$b_j^{(m)} = \frac{(u, \psi_j)_1}{(\psi_j, \psi_j)_1} = \frac{(s, \phi_j)_2}{(\phi_j, \phi_j)_2} = a_j^{(m)}.$$

Hence, in this case, the pseudoanalytic solution is equivalent to the least-squares solution (with respect to $\|\cdot\|_1$).

This result indicates that pseudoanalytic approximations can be quite inferior when the $\{\varphi_j\}$ are chosen arbitrarily. Ideally, we would like to choose an $\mathscr{L}$-invertible system which is orthogonal. However, even if this is possible, it should be noted that the data are not given continuously, and hence, that the estimate for $\hat{s}$ must be calculated using discrete least squares. Thus, the corresponding pseudoanalytic solution may not be optimal.

5. Wiener Filtering

We recall that an operator $\mathscr{L}$ is a linear (Fourier) filter if there exists a function $l(\omega)$, independent of t, such that

$$\mathscr{L} \exp(i\omega t) = l(\omega) \exp(i\omega t). \tag{55}$$

Clearly, when $\mathscr{L}^{-1}$ exists, we have

$$\mathscr{L}^{-1} \exp(i\omega t) = l^{-1}(\omega) \exp(i\omega t).$$

Examples of such filters are differentiation, for which $l(\omega) = i\omega$, and the Abel transform defined by [Bracewell, Ref. 43, p. 263; Erdélyi, Ref. 5, pp. 8 and 64]

$$\mathscr{A}u = \int_t^\infty \frac{u(\tau)}{\sqrt{\tau - t}} d\tau, \tag{56}$$

for which $a(\omega) = i\pi^{1/2}(i\omega)^{-1/2}$. Clearly, the inverse $\mathscr{L} = \mathscr{A}^{-1}$ of the Abel transform, when written in the canonical form of (56), is half-differentiation for which $l(\omega) = -i\pi^{-1/2}(i\omega)^{1/2}$.

Consider the set of observations

$$\{v_k; k = 0, 1, 2, \ldots, n\} \tag{57}$$

of some phenomenon at the evenly spaced time points $t_k = k\Delta$ ($k = 0, 1, 2, \ldots, n$) where Δ denotes a constant steplength. We say that a time series is statistical (as opposed to deterministic), if its future values can be described only in terms of a probability distribution. Further, if a statistical phenomenon evolves in time according to probability laws, then it is referred to as a *stochastic process*. Thus, we can regard a time series as a realization of a stochastic process, the properties of which we wish to investigate on the basis of information contained in the time series.

The concept of a *stationary (stochastic) process* is based on the assumption that the process is in a particular state of *statistical equilibrium*. In fact, we say that a stochastic process is *strictly stationary* if its properties are unaffected by a change of time origin. Thus, for a discrete process to be strictly stationary, the joint distribution of any set of observations must be unaffected by shifts in the time origin of length $j\Delta$ (j an integer). It follows from this result that the mean and variance of a stationary stochastic process are constant.

Let P_k denote the *probability density function* of the random variable v_k. Since P_k is the same for all k, its shape can be inferred from the histogram of the observations $v_0, v_1, v_2, \ldots, v_n$. This is the fundamental result on which the analysis of a stationary stochastic process hinges: *The analysis of the properties of a stationary stochastic process which generated the given time series can be derived from the statistical properties of this time series.* This has led to the development of a very powerful theory for the analysis of time series. For the breadth and depth of the subject, the reader is referred to Box and Jenkins (Ref. 44) for an introduction to time series and the autoregressive approach, to Hamming (Refs. 45-46) for an introduction to the frequency approach in numerical analysis, to Doob (Ref. 47) for the basic analysis of stochastic processes, to Hannan (Refs. 48-49), Jenkins and Watts (Ref. 50), and Grenander and Rosenblatt (Ref. 51) for the spectral approach to time series, and to Bloomfield (Ref. 52) and Tukey (Ref. 53) for the use of the fast Fourier transform to compute the numerical spectrum.

In this section, we show how the results from this general theory can be used to estimate from observational data the effect of applying a linear filter to the underlying signal. The central assumption is the stationarity of the time series. Though this will not be true for most given time series, it is often easy to introduce some transformation of the data [for example, the removal of a (linear) trend] which reduces it to near stationarity. Any transformation is valid as long as it does not prevent us from filtering the data or does not enhance the sensitivity of the method with respect to observational errors.

For the transformed data

$$\{d_k; k = 0, 1, \ldots, n\},$$

we assume that $s(t)$ defines the underlying signal to which we wish to apply the linear filter $\mathscr{L}$. In addition, we take $s(t)$ to be a stationary stochastic process with a continuous time parameter t, that is, a random function. Without loss of generality, we shall also assume that $s(t)$ is observed, or sampled, at the times $t_k = k\Delta$, $k = 0, 1, \ldots, n = 1/\Delta$, and that the measurement error in the kth observation is ε_k. We assume further that $\{\varepsilon_k\}$ is a stationary stochastic process with the discrete parameter k, that is, a random sequence. Thus

$$d_k = s(t_k) + \varepsilon_k. \tag{58}$$

Without further loss of generality, we take both $s(t)$ and $\{\varepsilon_k\}$ to have zero means. The Wiener–Khintchine theory of generalized harmonic analysis [Wiener (Ref. 54) and Khintchine (Ref. 55)] now implies that $s(t)$ and $\{\varepsilon_k\}$ may be represented as

$$s(t) = \int_{-\infty}^{\infty} \exp(i\omega t/\Delta)\, dZ_s(\omega), \tag{59a}$$

$$\varepsilon_k = \int_{-\pi}^{\pi} \exp(i\omega k)\, dZ_\varepsilon(\omega), \tag{59b}$$

where the terms $Z_s(\omega)$ and $Z_\varepsilon(\omega)$ are referred to as random spectral measures [Koopmans, Ref. 56, Chapter 2]. The term $\exp(i\omega t/\Delta)$ is not standard, but is chosen here since we have a natural time unit Δ.

From the point of view of the present discussion, it is not the actual definitions of $Z_s(\omega)$ and $Z_\varepsilon(\omega)$ which are important, but the mathematical properties of the representation (59). If $\mathscr{L}$ is the linear filter we wish to apply to $s(t)$, then it follows from the properties of the above representation for $s(t)$ that

$$\mathscr{L}s(t) = \int_{-\infty}^{\infty} l(\omega/\Delta) \exp(i\omega t/\Delta)\, dZ_s(\omega). \tag{60}$$

The *basic strategy* is the construction of a discrete linear filter of the form

$$\sum_{r=-\infty}^{\infty} \hat{l}_r d_{k-r}, \qquad \hat{l}_{(\omega)} = \sum_{r=-\infty}^{\infty} \hat{l}_r \exp(i\omega r), \tag{61}$$

which estimates $\mathscr{L}s(t_k)$ so that the difference

$$\mathscr{L}s(t_k) - \sum_{r=-\infty}^{\infty} \hat{l}_r d_{k-r} \tag{62}$$

Abel Integral Equations

is suitably small with respect to some appropriately chosen criterion such as variance.

On using the property that $Z_d(\omega) = Z_s(\omega) + Z_\varepsilon(\omega)$, the value of the difference follows naturally from (59) to be

$$\mathscr{L}s(t_k) - \sum_{r=-\infty}^{\infty} \hat{l}_r d_{k-r} = \int_{-\infty}^{\infty} \exp(i\omega k)[l(\omega/\Delta) - \hat{l}(\omega)] \, dZ_s(\omega)$$

$$- \int_{-\pi}^{\pi} \exp(i\omega k) \hat{l}(\omega) \, dZ_\varepsilon(\omega). \qquad (63)$$

To compute the variance of this difference, we use the fact that the variance of an integral of the form

$$\int_I \theta(\omega) \, dZ_s(\omega)$$

is given by

$$\int_I |\theta(\omega)|^2 \, dG_s(\omega),$$

where $G_s(\omega)$ is the nondecreasing function called the *spectral distribution function* of s. An analogous result holds for $Z_\varepsilon(\omega)$. We shall assume that neither the $s(t)$ nor the $\{\varepsilon_k\}$ process contains any purely oscillatory component, for then there exist *spectral density functions* $g_s(\omega)$ [and $g_\varepsilon(\omega)$] such that the last integral becomes

$$\int_I |\theta(\omega)|^2 g_s(\omega) \, d\omega.$$

Thus, it follows from (63) that

$$\operatorname{var}\left[\mathscr{L}s(t_k) - \sum_{r=-\infty}^{\infty} \hat{l}_r d_{k-r} \right] = \int_{-\infty}^{\infty} |l(\omega/\Delta) - l(\omega)|^2 g_s(\omega) \, d\omega$$

$$+ \int_{-\pi}^{\pi} |\hat{l}(\omega)|^2 g_\varepsilon(\omega) \, d\omega. \qquad (64)$$

Since it is essential in guaranteeing that Wiener filtering works, we now assume that Δ has been taken to be sufficiently small so that little or no detail of $s(t)$ has been lost in the sampling. This implies that $g_s(\omega)$ is either small or zero outside the interval $[-\pi, \pi]$ [Hannan (Ref. 49, Section III.6)], so we can approximate (3.10) by

$$\int_{-\pi}^{\pi} [|l(\omega/\Delta) - \hat{l}(\omega)|^2 g_s(\omega) + |\hat{l}(\omega)|^2 g_\varepsilon(\omega)] \, d\omega. \tag{65}$$

From the point of view of the subsequent analysis, it is more appropriate to work with $l_*(\omega) = \hat{l}(\omega)/l(\omega/\Delta)$. Using this notation, (65) becomes

$$\int_{-\pi}^{\pi} l^2(\omega/\Delta)[|1 - l_*(\omega)|^2 g_s(\omega) + |l_*(\omega)|^2 g_\varepsilon(\omega)] \, d\omega, \tag{66}$$

which we denote by $L(l_*(\omega))$.

The key point to which we have been working is:

Theorem 5.1. Set

$$W(\omega) = g_s(\omega)/[g_s(\omega) + g_\varepsilon(\omega)]. \tag{67}$$

Then

$$L(l_*) = L(W) + \|W - l_*\|_\alpha^2, \tag{68}$$

where $\|\cdot\|_\alpha$ denotes the L_2-norm with weight function

$$l^2(\omega/\Delta)[g_s(\omega) + g_\varepsilon(\omega)].$$

Proof. Because

$$L(l_*) = L(W + l_* - W)$$

$$= \int_{-\pi}^{\pi} l^2(\omega/\Delta)[|1 - W(\omega) - l_*(\omega) + W(\omega)|^2 g_s(\omega)$$

$$\qquad + |W(\omega) + l_*(\omega) - W(\omega)|^2 g_\varepsilon(\omega)] \, d\omega$$

$$= L(W) + \|W - l_*\|_\alpha^2$$

$$\quad - 2 \operatorname{Re} \int_{-\pi}^{\pi} l^2(\omega/\Delta)\{[1 - W(\omega)][l_*(\omega) - W(\omega)]g_s(\omega)$$

$$\qquad - W(\omega)[l_*(\omega) - W(\omega)]g_\varepsilon(\omega)\} \, d\omega,$$

Abel Integral Equations

the required result follows since it is easily verified that this last integral vanishes. □

As an immediate consequence, we obtain

Corollary 5.1. The filter (61) which minimizes the variance (64) is defined by

$$\hat{l}(\omega) = l(\omega/\Delta) W(\omega). \tag{69}$$

Mathematically, this can be interpreted in the following way. Let $d(t)$, $-\infty < t < \infty$, represent a "continuous realization" of the signal and noise processes which generated $\{d_k\}$. In addition, let $\hat{\mathscr{L}}$ denote the filter defined by $\hat{l}(\omega)$ and let $\mathscr{F}$ denote the Fourier transform. Then, formally, (69) yields

$$\hat{\mathscr{L}} d(t) = \mathscr{F}^{-1} l(\omega) \mathscr{F} d(t) = \mathscr{F}^{-1} l(\omega/\Delta) W(\omega) \mathscr{F} d(t)$$

$$= \mathscr{F}^{-1} l(\omega/\Delta) \mathscr{F} s(t) = \mathscr{L} s(t). \tag{70}$$

That is, the filter $\hat{\mathscr{L}}$, when applied to the "observed" $d(t)$, is such that it gives the required $\mathscr{L}s(t)$.

Because of the central role it plays, $W(\omega)$ is called the *Wiener window*. As the above discussion indicated, its role is essentially that of windowing the data in the frequency domain to remove the observational noise $\{\varepsilon_k\}$ from the signal $s(t)$, so that $\mathscr{L}$ can be applied directly to $s(t)$ to yield the desired $\mathscr{L}s(t)$. In reality, the situation is much more complex, since

(i) we must work with the $\{d_k\}$, not $d(t)$, $-\infty < t < \infty$;
(ii) though the structure of $W(\omega)$ is known formally, it is necessary to invoke additional assumptions about the form of $s(t)$ and $\{\varepsilon_k\}$ before it can be estimated using the $\{d_k\}$.

In fact, if $\mathscr{F}_*\{d_j\}$ denotes the discrete Fourier transform of $\{d_j\}$, then in essence we compute

$$\mathscr{F}_*^{-1}\{l(\omega_k/\Delta) \hat{W}(\omega_k) \mathscr{F}_*\{d_j\}\} = [\mathscr{L}s(t)]_{t=t_k}^*, \tag{71}$$

as our estimate of $\mathscr{L}s(t_k)$, where $\omega_k = 2\pi k/n$ and $\hat{W}(\omega)$ denotes an appropriate estimate of $W(\omega)$.

Except for some remarks about the role and estimation of $W(\omega)$, the actual details of implementation will not be pursued further in this paper, since the details for numerical differentiation, which we shall utilize below, can be found in Anderssen and Bloomfield (Refs. 4, 39).

From a practical point of view, because its role is that of removing the higher-frequency components of $\mathscr{F}_*\{d_j\}$ (which are associated with the observational errors $\{\varepsilon_k\}$) and retaining only the lower components [associated with $s(t)$], the estimates obtained using (71) tend to be stabilized with respect to small perturbations in $\{d_j\}$. In addition, these estimates yield good approximations to $\mathscr{L}s(t_k)$, especially when $s(t)$ is a smooth process with power in its spectrum concentrated at low frequencies, and the power in the spectrum of $\{\varepsilon_k\}$ is flat and much smaller than that of $s(t)$.

As a result, Wiener filtering can be viewed as a form of regularization. The implicit nature of this connection has been discussed by Wahba (Ref. 57). The explicit nature was first established by Anderssen and Bloomfield (Ref. 39) for numerical differentiation. In fact, they showed that Wiener filtering and regularization are equivalent in the following sense: *if $\acute{s}(t_k; \alpha, \|\cdot\|_S)$ denotes the regularized derivative of $s(t)$ constructed using Cullum's (Ref. 58) procedure with a Sobolev norm $\|\cdot\|_S$ as the smoothing norm and α as the regularization parameter, and $\acute{s}(t_k; \hat{W}(\omega))$ the corresponding spectral (Wiener) derivative, then there exists a function $\lambda(\omega; \alpha, \|\cdot\|_S)$ such that*

$$\acute{s}(t_k; \hat{W}(\omega)) = \acute{s}(t_k; \alpha, \|\cdot\|_S), \tag{72}$$

if and only if

$$\hat{W}(\omega) = \lambda(\omega; \alpha, \|\cdot\|_S). \tag{73}$$

The indeterminacy in regularization (involving the choice of $\|\cdot\|_S$ and α) is also present in Wiener filtering, since $W(\omega)$ can only be estimated using the data $\{d_j\}$.

This result extends naturally to any filtering process $\mathscr{L}$. The details are not given here, because they do not relate directly to the subject matter of this paper; namely, Abel integral equations. We therefore only pause to note that not only does this result characterize the sense in which Wiener filtering and regularization are equivalent, but it can also be used to construct maximum likelihood estimates for the regularization parameter α directly from the data $\{d_j\}$ [Anderssen and Bloomfield (Refs. 4, 39)].

6. Stabilized Evaluation of Inversion Formulas

On the basis of the discussion of Section 5, an obvious strategy for the solution of Abel equations is the use of Wiener filtering methods. Unfortunately, this is easier said than done. A number of important practical factors prevent their direct application. They are:

(i) In many applications, the data $\{d_j\}$ are not given on an even grid, and consequently the general theory of discrete Fourier transform analysis is not applicable [Bloomfield (Ref. 52)], and the use of fast Fourier transform algorithms for the evaluation of (71) is not possible.

(ii) Even when the data are given on an even grid, the form of the associated Abel transform may not correspond to that of half-differentiation, viz. (56). Thus, a transformation of variables is necessary to take the associated Abel transforms into the desired form (56), which automatically destroys the original grid of the data. The analytic form of the inverse filter corresponding to the given Abel transform could be constructed and used but, as indicated by Bloomfield (Ref. 59), this is not a trivial matter from both a theoretical and practical point of view.

This leads naturally to stabilizing the evaluation of the associated inversion formulas as the basis for solving Abel integral equations numerically. To illustrate, we examine (25) and its inversion formulas (29a) and (29b). The choice is between (a) the evaluation of the singular integrals of (29a) with respect to the given $\{d_j\}$ using product integration followed by the spectral differentiation of the resulting data, and (b) the spectral differentiation of the data $\{d_j\}$ followed by the evaluation of the singular integrals of (29b) using product integration. Except when the data are given on an even grid, the choice is in fact limited to (a) since it allows naturally for data given on a noneven grid to be integrated onto an even one [Anderssen (Ref. 9)], and thereby guarantees that spectral differentiation can be used in the stabilization.

The difficulty associated with the differentiation of data on a noneven grid, and hence the general use of (29b) rather than (29a), can be circumvented using smoothing splines [Wahba (Refs. 60-61)]. Though it represents a viable alternative it is not pursued here, partly because the choice of α via generalized cross-validation is based on the best smoothing of the data and not on the best smoothing for the derivative which is the case for spectral differentiation.

However, the stabilized evaluation of inversion formulas poses a major difficulty. The use of spectral differentiation requires that the data to be differentiated be stationary-like—a strong regularity constraint not usually satisfied by arbitrary given data. On the other hand, though they do not stabilize, pseudoanalytic methods can be viewed as simple strategies for generating first-order (low-parameterized) approximations to the required solution independently of the signal structure in the data $\{d_j\}$, but not more accurate approximations, while the stabilized evaluation of inversion formulas defines a sophisticated method for computing an accurate approximation to the solution but only when the signal structure in the data $\{d_j\}$ satisfies conditions compatible with the application of such a method (i.e., with the use of spectral differentiation).

Thus, the sensible strategy would appear to be to use them as compatible methods so that their advantages support each other at the expense of their disadvantages. In particular, for given data $\{d_j\}$, generate a first-order (low-parameterized) approximation to the required solution using a pseudoanalytic method in such a way that the residual data have a form compatible with the assumptions which underlie the use of the stabilized inversion formulas. This is the basis for the computational strategy proposed by Anderssen (Ref. 9) and studied in some detail in Jakeman (Ref. 21) and Anderssen and Jakeman (Ref. 62).

7. The Data Functional Strategy

As mentioned in the Introduction, it is often not the solution $u(x)$ of (25) which is used for inference purposes, but linear functionals (e.g., moments) defined on $u(x)$, namely,

$$m_\theta(u) = \int_0^a \theta(x)u(x)\,dx, \qquad \theta \equiv \text{known}. \tag{74}$$

An illustrative example has been discussed in some detail by Anderssen (Ref. 65). When such functionals can be redefined as functionals on $s(y)$, i.e., as *data functionals*

$$m_\varphi(x) = \int_0^a \varphi(\theta;y)s(y)\,dy, \tag{75}$$

the possibility exists of circumventing the difficulties associated with the evaluation of $u(x)$.

Such a strategy is not limited to Abel equations. In fact, if we consider the general problem

$$\mathscr{L}u = s, \qquad \mathscr{L}: \mathsf{D}(\mathscr{L}) \to \mathsf{R}(\mathscr{L}), \tag{76}$$

where $\mathsf{D}(\mathscr{L})$ and $\mathsf{R}(\mathscr{L})$ denote respectively the *domain* and *range* of $\mathscr{L}$, then the applicability of the data functional strategy can be characterized as follows:

Theorem 7.1. Let $\mathscr{L}^*$ denote the adjoint of $\mathscr{L}$ with respect to the L_2-inner product

$$(u, v) = \int_0^a uv\,dx. \tag{77}$$

If the known θ which defines $m_\theta(u)$ is contained in $R(\mathscr{L}^*)$, then the required φ which defines $m_\varphi(s)$ is given by

$$\mathscr{L}^*\varphi = \theta. \tag{78}$$

Proof.

$$m_\theta(u) = (\theta, u) = (\mathscr{L}^*\varphi, u) = (\varphi, \mathscr{L}u) = (\varphi, s) = m_\varphi(s). \quad \square$$

It follows that, once we have solved (78) for a given θ, the evaluation of (74) reduces to the estimation of (75) with respect to given observational data $\{d_j\}$. The need to use product integration methods has been examined in some detail by Anderssen and Jakeman (Ref. 63).

Golberg (Ref. 64) has used this result to explain the use of the data functional strategy for Abel integral equations, and reverse flow theorems, but has not explored its more general applicability. This is limited by the requirement that θ be contained in $R(\mathscr{L}^*)$.

Conditions under which given classes of θ are contained in $R(\mathscr{L}^*)$ when $\mathscr{L}$ defines a Volterra or a Fredholm operator have been examined by Anderssen (Ref. 65). In that paper, numerical experimentation with real and synthetic data is used to confirm the viability of this strategy in size distribution analysis, which is an important applications area for Abel integral equations.

In conclusion, we note that if $\mathscr{L}$ is improperly posed, then so is $\mathscr{L}^*$. The advantage of solving $\mathscr{L}^*\varphi = \theta$ over $\mathscr{L}u = s$ is that the data θ of the former are analytic so that, if $\mathscr{L}^*\varphi = \theta$ cannot be solved explicitly, solving $\mathscr{L}^*\varphi = \theta$ numerically is more tractable than $\mathscr{L}u = s$ for which the data are only observational. A more detailed discussion of this and other aspects associated with the linear functional strategy can be found in Anderssen (Ref. 66).

8. Choice of Algorithm

Computationally, when solving improperly posed problems, the decision making is more involved than for properly posed problems. It is not simply a matter of being able to rely on standard procedure, such as Gaussian elimination for (nonsingular) linear algebraic equations. For example, even though the use of singular value decomposition has been popularized as the standard procedure for the solution of discretizations of improperly posed problems, its actual use in such situations is not straightforward. In order to ensure that such procedures produce reliable information, the user must have a clear understanding of the degree of improperly posedness of the actual problem being solved and, within the framework of that knowledge, decide how the singular values should be filtered before the actual singular-value solution is constructed. Although this can be automated to a certain extent, it adds a complexity which does not exist for standard procedures such as Gaussian elimination.

Thus, just as the above analysis of Abel's equation depended crucially on the form of the data $s(x)$, so the choice of algorithm to use in a particular situation must be based on similar considerations.

Thus, for analytic data $s(x)$, the first step is to examine whether the relevant Abel equation can be solved exactly (using, for example, tabulated fractional integrals, etc.). Once it is apparent that this is either impossible or not very likely, the next step is not to immediately replace $s(x)$ by a discrete (numerical) counterpart. Since the solution of the relevant Abel equation for discrete (numerical) data only yields an approximate solution, the aim should be to exploit this availability of analytic data $s(x)$ so as to minimize the (approximation) error introduced. One therefore checks whether a decomposition (not necessarily unique)

$$s(x) = s_e(x) + r(x)$$

can be found such that

(i) the relevant Abel equation can be solved exactly for $s_e(x)$ to yield $u_e(y)$;

(ii) the residual $r(x)$ is such that the corresponding solution $u_r(y)$ (of the relevant Abel equation) is suitably small when compared with $u_e(y)$, e.g.,

$$\max_y |u_r(y)| \ll \max_y |u_e(y)|.$$

Clearly, there is a tradeoff between these two constraints which must be handled with some care. For example, it would be best not to invoke such a decomposition, if it were unlikely that $u_r(y)$ only made a small contribution to the solution $u(y)$ when compared with $u_e(y)$.

The biggest challenge arises when the data are discrete (numerical). In theory, such problems can often be solved with high accuracy, using simple algorithms. However, such accuracy can only be guaranteed if the types of tradeoff, which must be implemented for the numerical differentiation of discrete (numerical) data using simple finite-difference formulas, are rigorously respected, in particular, the need to control the step length between data points (impose a lower bound) with respect to the accuracy with which the data has been evaluated and will be manipulated. However, such discipline is really a form of intuitive (implicit) stabilization (regularization), and therefore implicitly acknowledges the overriding conclusion of this chapter that some form of stabilization is always necessary.

For obvious reasons one would like, where possible, to use simple algorithms. Consequently, explicit stabilization (using, for example, regularization techniques, or constrained optimization) does not in general represent a viable alternative, because of the inherent (computational) complexity of such procedures. The choice is further complicated by the fact that discrete (numerical) data are like highly accurate observational data so that, for the Abel equation, only some mild form of stabilization is required. As a general guide, implement simple methods in situations where the user has explicit interactive control over the solution process, and apply some form of explicit stabilization when the relevant Abel equation will be solved automatically within a larger context with little or no user interaction.

Whatever the choice, at all cost avoid the algorithms (discussed in Remark 3.1 above) for which claims are made that they bypass the inherent improperly posedness of the Abel equation. Though the claim may have limited validity, the scope of that validity will not be apparent. The fact is that "computational difficulties associated with the inversion of improperly posed problems such as Abel's equation cannot be removed by simply manipulating the problem mathematically into an alternative form" [Anderssen and de Hoog (Ref. 37)]. Thus, the required stabilization can only be achieved through the introduction of additional structure (constraints) into the computational representation used to solve the improperly posed problem. As already shown above, such constraints can even take on an intuitive (implicit) form, as long as they ensure that, computationally, the underlying problem is properly posed.

When the data are observational, there is no choice but to implement some form of explicit stabilization. The choice again depends on the form of the data as the following guidelines indicate:

(i) For highly accurate data on a fine grid, which recovers accurately the structure of $s(x)$, the above discussion for discrete (numerical data) applies.

(ii) For observational data on a fine grid, which recovers graphically the basic detail of $s(x)$, mild stabilization is all that is required such as regularization with a small regularization parameter and a second derivative regularizer.

(iii) For highly inaccurate data on a fine grid, which tends to smear out the basic detail of $s(x)$, strong stabilization is required such as regularization with a large regularization parameter and a second or higher derivative regularizer.

(iv) For observational data on a medium grid which only allows a partial picture of the form of $s(x)$ to be recovered, a strong form of regularization is required. Now, however, it is not simply a matter of stabilizing the computational process with respect to the observational errors. It is also necessary, where possible, to choose a regularizer which is consistent with the missing information (structure) (i.e., choose a regularizer which reflects some property of the solution, such as minimum energy).

(v) For inaccurate data on a medium grid, or observational (or inaccurate) data on a coarse grid, which only recover a very poor (or inadequate) picture of the structure of $s(x)$, the role of the stabilization reduces to extracting from the available data as much information as possible about the specific properties of $u(y)$ required for decision-making purposes, when those properties can be determined as properly posed operations to be performed on the data. The linear functional strategy plays a crucial role in such situations.

Though the choice of the actual form of and manner of implementation of the stabilization is at the user's discretion, it is crucial not to lose sight of the key fact that, for best results, the algorithms used should be tailored to and exploit the mathematical structure of the problem being solved. Consequently, a general-purpose regularization algorithm will not perform as well for the Abel equation as one tailored to its underlying half-differentiation (such as a Wiener filtering method). The use of the linear functional strategy to only extract the information about $u(y)$ which will be used for decision-making purposes is applicable to all the situations discussed above.

References

1. ABEL, N. H., *Auflosung einer Mechanischen Aufgabe*, Journal fur Reine Angewandte Mathematik, Vol. 1, pp. 153-157, 1826.
2. LONSETH, A. T., *Sources and Applications of Integral Equations*, SIAM Review, Vol. 19, pp. 241-278, 1977.
3. LINZ, P., *Analytical and Numerical Methods for Volterra Equations*, SIAM, Philadelphia, Pennsylvania, 1985.
4. ANDERSSEN, R. S., AND BLOOMFIELD, P., *Numerical Differentiation Procedures for Non-exact Data*, Numerische Mathematik, Vol. 22, pp. 157-182, 1974.
5. ERDÉLYI, A., *Bateman Manuscript Project: Tables of Integral Transforms*, McGraw-Hill, New York, New York, 1959.
6. MORAN, P. A. P., *The Probabilistic Basis of Stereology*, Advances in Applied Probability, Vol. 4, pp. 69-91, 1972.
7. SANTALÓ, L. A., *Sobre la Distribucion de los Tamaños os de Corpusclos Contenidos en un Cuerpo a Partir de la Distribucion en Sus Secciones o Proyecciones*, Trabados de Estadica, Vol. 6, pp. 181-196, 1955.
8. BRUNNER, H., *A Survey of Recent Advances in the Numerical Treatment of Volterra Integral Equations*, Journal of Computational and Applied Mathematics, Vol. 8, pp. 213-228, 1982.

9. ANDERSSEN, R. S., *Stable Procedures for the Inversion of Abel's Equation*, Journal of the Institute of Mathematics and its Applications, Vol. 17, pp. 329–342, 1976.
10. LUKAS, M. A., *Regularization, The Application and Numerical Solution of Integral Equations*, Edited by R. S. Anderssen, F. R. de Hoog, and M. A. Lukas, Sitjhoff and Noordhoff, The Netherlands, 1980.
11. BAEV, A. V., AND GLASKO, V. B., *On the Solution of the Converse Kinematic Problem of Seismics by Means of a Regularizing Algorithm*, USSR Computational Mathematics and Mathematical Physics, Vol. 16, pp. 96–106, 1976.
12. ANDERSSEN, R. S., AND GUSTAFSON, S., *Linear Programming Methods for the Inversion of Data*, Computational Techniques and Applications, CTAC-83, Edited by J. Noye and C. Fletcher, Elsevier Science Publishers, Amsterdam, Holland, 1984.
13. ANDERSSEN, R. S., *Application and Numerical Solution of Abel-type Integral Equations*, University of Wisconsin, Madison, Mathematics Research Center, Technical Summary Report, No. 1787, 1977.
14. JAKEMAN, A. J., AND ANDERSSEN, R. S., *Abel-Type Integral Equations in Stereology I. General Discussion*, Journal of Microscopy, Vol. 105, pp. 121–133, 1975.
15. WICKSELL, S. D., *The Corpuscle Problem*, Biometrika, Vol. 17, pp. 84–89, 1925.
16. WICKSELL, S. D., *The Corpuscle Problem II*, Biometrika, Vol. 18, pp. 151–172, 1926.
17. MERZKIRCH, W., *Flow Visualization*, Academic Press, New York, New York, 1974.
18. HYAM, E. D., AND NUTTING, J., *The Tempering of Plain Carbon Steels*, Journal of the Iron and Steel Institute, Vol. 148, pp. 148–165, 1956.
19. MEISNER, J., *Estimation of the Distribution of Diameters of Spherical Particles from a Given Grouped Distribution of Diameters of Observed Circles Formed by a Plane Section*, Statistica Neerlandica, Vol. 21, pp. 11–30, 1967.
20. GOLDSMITH, P. L., *The Calculation of True Particle Size Distributions from the Sizes Observed in a Thin Slice*, British Journal of Applied Physics, Vol. 18, pp. 813–830, 1967.
21. JAKEMAN, A. J., *Numerical Inversion of Abel Type Equations in Stereology*, Australian National University, PhD Thesis, 1975.
22. JEFFREYS, H., *The Earth*, Fourth Edition, Cambridge University Press, Cambridge, England, 1962.
23. BATEMAN, H., *The Solution of the Integral Equation which Connects the Velocity of Propagation of an Earthquake Wave in the Interior of the Earth with Times which the Disturbance Takes to Travel to Different Stations of the Earth's Surface*, Philosophical Magazine, Vol. 19, pp. 576–587, 1910.
24. KNOTT, C. C., *The Propagation of Earthquake Waves Through the Earth and Connected Problems*, Royal Society of Edinburgh Proceedings, Vol. 39, pp. 157–208, 1919.
25. MACELWANE, J. B., *Evidence on the Interior of the Earth Derived From Seismic Sources*, Internal Constitution of the Earth, Edited by B. Gutenberg, Dover, New York, New York, 1951.

26. GORDON, R., HERMAN, G. T., AND JOHNSON, S. A., *Image Reconstruction from Projections*, Scientific American, Vol. 233, pp. 56-68, 1975.
27. SCUDDER, H. J., *Introduction to Computer Aided Tomography*, Proceedings IEEE, Vol. 66, pp. 628-637, 1978.
28. VEST, C. M., *Formation of Images from Projections: Radon and Abel Transforms*, Journal of the Optical Society of America, Vol. 64, pp. 1215-1218, 1974.
29. CORMACK, A. M., *Representation of a Function by its Line Integrals with Some Radiological Applications*, Journal of Applied Physics, Vol. 34, pp. 2722-2727, 1963.
30. CORMACK, A. M., *Representation of a Function by its Line Integrals, with Some Radiological Applications*, Journal of Applied Physics, Vol. 34, pp. 2908-2913, 1964.
31. SAUTHOFF, N. R., AND VON GOELER, S., *Techniques for the Reconstruction of Two-Dimensional Images from Projections*, IEEE Transactions on Plasma Physics, Vol. PS-7, pp. 141-147, 1979.
32. KOWALEWSKI, G., *Integralgleichungen*, Walter De Gruyter and Company, Leipzig, Germany, 1930.
33. NESTOR, O. H., AND OLSEN, H. H., *Numerical Methods for Reducing Line and Surface Probe Data*, SIAM Review, Vol. 2, pp. 200-207, 1960.
34. SNEDDON, I. N., *Mixed Boundary Value Problems in Potential Theory*, North-Holland Publishers, Amsterdam, Holland, 1966.
35. CHAN, C. K., AND LU, P., *On the Stability of the Solution of Abel's Integral Equation*, Journal of Physics A, pp. 575-578, 1981.
36. DEUTSCH, M., AND BENIAMINY, I., *Derivative-free Inversion of Abel's Integral Equation*, Applied Physics Letters, Vol. 41, pp. 27-28, 1982.
37. ANDERSSEN, R. S., AND DE HOOG, F. R., *On the Methods of Chan and Lu for Abel's Integral Equation*, Journal of Physics A, Vol. 14, pp. 3117-3121, 1981.
38. ANDERSSEN, R. S., AND DE HOOG, F. R., *Finte Difference Methods for the Numerical Differentiation of Non-exact Data*, Computing, Vol. 33, pp. 259-267, 1984.
39. ANDERSSEN, R. S., AND BLOOMFIELD, P., *A Time Series Approach to Numerical Differentiation*, Technometrics, Vol. 16, pp. 69-75, 1974.
40. LANDENSBERG, R. W., AND BERSHADER, D., *Physical Measurements in Gas Dynamics and Combustion*, High Speed Aerodynamics and Jet Propulsion, Vol. 9, Oxford University Press, Oxford, England, 1957.
41. MINNERBO, G. N., AND LEVY, M. E., *Inversion of Abel's Integral Equation by Means of Orthogonal Polynomials*, SIAM Journal on Numerical Analysis, Vol. 6, pp. 598-616, 1969.
42. PIESSENS, R., AND VERBAETEN, P., *Numerical Solution of the Abel Integral Equation*, BIT, Vol. 13, pp. 451-457, 1973.
43. BRACEWELL, R., *The Fourier Transform and its Application*, McGraw-Hill, New York, New York, 1965.
44. BOX, G. E. P., AND JENKINS, G. M., *Time Series Analysis*, Forecasting and Control, Holden-Day, San Francisco, California, 1970.
45. HAMMING, R. W., *Numerical Methods for Scientists and Engineers*, McGraw-Hill, New York, New York, 1963.

46. HAMMING, R. W., *The Frequency Approach to Numerical Analysis*, Studies in Numerical Analysis: Papers Presented to Cornelius Lanczos, Academic Press, New York, New York, 1973.
47. DOOB, J. L., *Stochastic Processes*, John Wiley, New York, New York, 1953.
48. HANNAN, E. J., *Time Series Analysis*, Methuen, London, England, 1960.
49. HANNAN, E. J., *Multiple Time Series*, John Wiley, New York, New York, 1970.
50. JENKINS, G. M., AND WATTS, D. G., *Spectral Analysis and its Application*, Holden-Day, San Francisco, California, 1968.
51. GRENANDER, U., AND ROSENBLATT, M., *Statistical Analysis of Stationary Time Series*, John Wiley, New York, New York, 1957.
52. BLOOMFIELD, P., *Fourier Analysis of Time Series: An Introduction*, John Wiley, New York, New York, 1976.
53. TUKEY, J. W., *An Introduction to the Calculations of Numerical Spectrum Analysis*, The Spectral Analysis of Time Series, Edited by B. Harris, John Wiley, New York, New York, 1967.
54. WIENER, N., *Generalized Harmonic Analysis*, Acta Mathematica, Vol. 55, pp. 117-258, 1930.
55. KHINTCHINE, A., YA., *Korrelations theorie der Stationaren Stochastischen Prozesse*, Mathematische Annallen, Vol. 109, pp. 604-616, 1934.
56. KOOPMANS, L. H., *The Spectral Analysis of Time Series*, Academic Press, New York, New York, 1974.
57. WAHBA, G., *On the Approximate Solution of Fredholm Equations of the First Kind*, University of Wisconsin, Madison, Mathematics Research Center, Technical Summary Report No. 990, 1969.
58. CULLUM, J., *Numerical Differentiation and Regularization*, SIAM Journal on Numerical Analysis, Vol. 8, pp. 254-265, 1971.
59. BLOOMFIELD, P., *The Fourier Transform Solution of Abel's Equation in Stereology*, unpublished manuscript.
60. WAHBA, G., *A Survey of Some Smoothing Problems and the Method of Generalized Cross-validation for Solving Them*, University of Wisconsin, Madison, Statistics Technical Report No. 457, 1976.
61. WAHBA, G., *Practical Approximate Solutions to Linear Operator Equations When the Data are Noisy*, SIAM Journal on Numerical Analysis, Vol. 14, pp. 651-667, 1977.
62. ANDERSSEN, R. S., AND JAKEMAN, A. J., *Abel Type Integral Equations in Stereology II. Computational Methods of Solution and the Random Spheres Approximation*, Journal of Microscopy, Vol. 105, pp. 135-153, 1975.
63. ANDERSSEN, R. S., AND JAKEMAN, A. J., *Product Integration for Functionals of Particle Size Distributions*, Utilitas Mathematica, Vol. 8, pp. 111-126, 1975.
64. GOLBERG, M. A., *A Method of Adjoints for Solving Some Ill-posed Equations of the First Kind*, Applied Mathematics and Computation, Vol. 5, pp. 123-130, 1979.
65. ANDERSSEN, R. S., *On the Use of Linear Functionals for Abel-Type Integral Equations in Applications*, The Application and Numerical Solution of Integral Equations, Edited by R. S. Anderssen, F. R. de Hoog, and M. A. Lukas, Sitjhoff and Noordhoff, The Netherlands, 1980.

66. ANDERSSEN, R. S., *The Linear Functional Strategy for Improperly Posed Problems*, Inverse Problems, Edited by J. R. Cannon and V. Hornung, Birkhauser Verlag, Basel, Switzerland, 1986.
67. SMARZEWSKI, R., AND MALINOWSKI, H., *Numerical Solution of A Class of Abel Integral Equations*, Journal of the Institute of Mathematics and its Applications, Vol. 22, pp. 159–170, 1978.

Index

Abel integral equations
  algorithm choice in computation, 403–406
  complications related to, 375–376
  data functional strategy, 402–403
  first-kind
    in geometric probability, 377
    in interferometry, 377
    in stereology, 378
  numerical analysis of, 382–387
  original formulation of, 373–374
  pseudoanalytic methods, 386, 387–394
  second-kind, in stereology, 379
  in seismology, 379–380
  stabilized evaluation of inversion formulas, 386–387, 400–402
  in tomography, 380–381
  Weiner filtering, 386, 394–400
Accelerated projection methods, 89–90
Affine operator, 142, 143
Airfoil equation: *see* Generalized airfoil equation
Approximate methods, computer use and, 323
Aspherical gravitational potential, in numerical method in astrodynamics, 147–148
Astrodynamics: *see* Numerical methods in astrodynamics
Atkinson–Bogomolny direct analysis, 110

Banach contraction theorem, 153
Banach space, 73–74, 76
Boundary element methods, 11, 14–21
  collocation methods, 19–20, 21
  strongly elliptical operator framework, 14–15
  theorems in, 15–19, 20–21
Boundary integral equations
  boundary element methods, 11, 14–21
    collocation methods, 19–20, 21

Boundary integral equations (*cont.*)
  boundary elements methods (*cont.*)
    strongly elliptical operator framework, 14–15
    theorems in, 15–19, 20–21
  iterative methods, 25–29
    two-grid method, 26–29
  Laplace equation reformulations
    direct BIE methods, 4–5, 7
    indirect BIE methods, 5–7
    piecewise smooth boundaries, 8–9
    smooth boundary case, 7–8
  numerical integration, 22–25
    boundary element integrations, 22–25
  numerical methods for, 10–14
    collocation methods, 11
    Galerkin methods, 10, 12, 15
    global methods, 12–14
    problems with, 1
  superapproximation, 37
Boundary value problem: *see* Planing surface problem

Canonical function, 315
Cauchy–Schwarz inequality, 216
Cauchy singular integral equations, 72, 73, 89
  applications of, 183–184
  collocation method, 220–229
    continuous data, 221–222
    polynomial collocation, 228–229
  convergence, 272–296
    of collocation method, 284–287
    of discrete Galerkin method, 287–293
    of Gaussian quadrature method, 293–296
    mean square convergence of Galerkin method, 273–281

Cauchy singular integral equations (*cont.*)
  Sloan iterate, 283–284
  uniform convergence of Galerkin method, 281–283
  conversion to logarithmic equation, 250–254
  degenerate kernel methods, 207–209
  determining unknown constant c, 190–193
    Kutta condition, 191–193
  direct methods, 197–198
  Galerkin method, 209–220
  generalized airfoil equation
    analytical solution of, 185–190
    mapping properties of airfoil operator, 198–204
    numerical methods for, 193–194
    operator formulation, 204–207
  indirect methods, 197–198
  Kalandiya method, 247-B249
  Kantorovich regularization, 241–246
  planing surface problem in, 364–365
  polynomial approximation methods, 258–272
    Chawla and Kumar method, 270–272
    Cohen method, 269–270
    collocation method, 263–264
    Galerkin method, 262–263
    Gaussian quadrature method, 264–266
    Hashmi and Delves method, 272
    Lobatto quadrature, 266–268
    piecewise polynomial methods, 272
    quadrature method, 264
  product quadrature, 246–247
  quadrature methods
    Gaussian quadrature method, 233–236
    Lobatto quadrature, 236–241
    quadrature rules for principle value integrals, 230–233
  of second kind, 157–168
  superconvergence and, 37
Chawla and Kumar method, Cauchy singular integral equations, 270–272
Chebyshev polynomial of degree, 77
Chebyshev polynomials, 198, 199, 218–219, 251, 257
Cohen method, Cauchy singular integral equations, 269–270
Collocation method
  boundary integral equation reformulations, 11, 19–20, 21

Collocation method (*cont.*)
  Cauchy singular integral equations, 220–229, 263–264
    continuous data, 221–222
    polynomial collocation, 228–229
    convergence, 284–287
  direct analysis of discrete method, 113–117
  discrete collocation method, 222
  and multidimensional problems, 79
  singular integral equations, 328
  superconvergence
    collocation at Gauss points, 59–63
    discrete collocation method, 64
    iterated collocation method, 56–63
    versus iterated Galerkin method, 63–64
    for piecewise polynomials, 58–59
    for Volterra equations, 117–120
  zeros as collocation points, 229
Compound integration rule, 117
Conjugate gradient method, 26
Consistency conditions, 317
Convergence
  Cauchy singular integral equations, 272–296
    of collocation method, 284–287
    of discrete Galerkin method, 287–293
    of Gaussian quadrature method, 293–296
    mean square convergence of Galerkin method, 273–281
    Sloan iterate, 283–284
    uniform convergence of Galerkin method, 281–283
  Fromme-Goldberg rule and, 224–225, 227
  Gaussian quadrature method, 235–236, 293–296
Convergence analysis, of projection methods, 81–84
Convergence theorems and singular integral equations
  collocation method, 328, 349–356
  discrete Galerkin method, 356–359
  Galerkin method, 323–328
  Galerkin-Petrov method, 342–348
Correction step, 27

Data functional strategy, Abel integral equations, 402–403
Degenerate kernel methods, Cauchy singular integral equations, 207–209

Index

Degenerate kernel operator, 207
Dellwo-Friedman method, perturbed
  Dellwo-Friedman method, 95
Direct boundary integral equations, Laplace
  equation reformulation, 4-5, 7
Dirichlet problem, 4, 8, 19, 28
Discrete collocation method, 222
  superconvergence, 64
Discrete Galerkin method, 213, 219, 327
  convergence, 287-293
  superconvergence, 55
Discretization error, 327

Earth satellite orbits
  equation for, 148-149
  in numerical method in astrodynamics,
    148-149
Eigenvalue problem, 37
Erdös-Turan theorem, 286

Fourier transform techniques, 77
Frechet derivative, 142, 143
Frechet differentiable, 141
Fredholm equation, 92, 96, 140, 143, 163,
  166, 168, 198, 230, 241
Fredholm operator, 140, 319
Fromme-Goldberg quadrature rule, 224-225
Fundamental function Z, 314-315
Fundamental Theorem of Calculus, 136

Galerkin method, 77-80, 95-106
  boundary integral equation
    reformulations, 10, 12, 15
  Cauchy singular integral equations,
    209-220, 262-263
  convergence
    of discrete Galerkin method, 287-293
    mean square convergence, 273-281
    uniform convergence, 281-283
  direct analysis with quadrature errors,
    106-110
  discrete Galerkin method, 213, 219, 327
    convergence of, 287-293
  iterated Galerkin method, 327, 348
  in numerical method in astrodynamics,
    158-161, 164-165
  perturbed Galerkin method, 141,
    158-161, 164-165
    in numerical method in astrodynamics,
      158-161, 164-165

Galerkin method (cont.)
  for positive-definite dominant part
    equations, 120-126
  singular integral equations, 323-328
  superconvergence, 36, 38-40
    discrete Galerkin method, 55
    versus iterated collocation method,
      63-64
    iterated Galerkin method, 51-53, 55
    linear functions of Galerkin
      approximation, 48-51
Galerkin-Petrov method, 211-212
  singular integral equations, 342-348
Gaussian elimination, 403-404
Gaussian gravitational constant, 146
Gaussian quadrature formula, 24-25, 37
Gaussian quadrature method
  Cauchy singular integral equations,
    236-241, 264-266
  convergence, 235-236, 293-296
Gaussian rules, 231-232, 263
Gauss-Legendre quadrature rule, 133
Gauss points
  collocation at, 59-60
  quadrature points, 79
Generalized airfoil equation
  analytical solution of, 185-190
  mapping properties of airfoil operator,
    198-204
  numerical methods for, 193-194
  operator formulation, 204-207
Geometric probability, Abel integral
  equations, 377
Gibbs-type phenomenon, 214
Global methods, boundary integral equation
  reformulations, 12-14
Global positioning system
  in numerical method in astrodynamics,
    149
  perturbative forces on, 149
Global superconvergence, 36
Gravitational constants
  of central celestial body, 146
  Gaussian, 146
  of Newton, 145
Green's function, 132

Hammerstein equation, 65, 141, 143, 153
  Green's function, 132
Hammerstein operator, 141, 144, 154, 157
Harmonic coefficients, 148

Hashmi and Delves' method, Cauchy singular integral equations, 272
Hilbert space, in numerical method in astrodynamics, 138-140
HMP method, superconvergence, 53-55
Hodograph technique, 369
Hohmann orbit, 175

Indirect boundary integral equations, Laplace equation reformulation, 5-7
Inner-product space, 74
Interferometry, Abel integral equations, 377
Iterated collocation method
  superconvergence, 56-58
    versus iterated Galerkin method, 63-64
    for piecewise-polynomial spaces, 58-63
Iterated Galerkin method, 327, 348
  superconvergence, 51-53, 55
Iterated Kantorovich method, superconvergence, 45-48
Iterated Kantorovich regularization, as variant of projection method, 88-89
Iterated projection method, superconvergence, 40-45
Iterative methods
  boundary integral equations, 25-29
  two-grid methods, 26-29

Jacobian of the transformation, 17

Kalandiya's method, Cauchy singular integral equations, 247-249
Kantorovich method
  iterated method, 45-48
  perturbed method, 93-94
Kantorovich regularization, 45
  Cauchy singular integral equations, 241-246
  iterated regularization, 88-89
  as variant of projection method, 87-88
Kantorovich theorem, 153, 157
Keplerian motion
  Lambert's problem and, 147
  in numerical method in astrodynamics, 147, 155-156
  perturbed Keplerian motion, 155-156
  two-body problem, 147
Khevedelidze theorem, 317
Kutta condition, 191-193, 195, 252, 366

Lagrangian interpolation polynomial of degree, 352, 355
Lambert problem, 147, 156
Laplace equation reformulations
  integral equation reformulations
    direct BIE methods, 4-5
    indirect BIE methods, 5-7
    piecewise smooth boundaries, 8-9
    Poisson equation, 9
    smooth boundary case, 7-8
Legendre polynomials, 77
  in numerical method in astrodynamics, 136-138
Leibnitz formula, 135
Lifting line theory, 370
Linear functions of Galerkin approximation, superconvergence, 48-51
L-invertible coordinate system, 389, 394
Lobatto quadrature, Cauchy singular integral equations, 266-268
Lobatto rules, 117

Mean square convergence
  Galerkin's method, 273-281
    direct method, 276-278
    indirect method, 273-276
    $\nu = 1$, 278-281
Motion, equations of
  astronomical units in, 146
  earth satellite orbits, 148-149
  Gaussian gravitational constant, 146
  in geocentric coordinate system, 145
  gravitational constant of the central celestial body, 146
  in heliocentric coordinate system, 145
  Keplerian motion, 147
  Newton universal gravitational constant in, 145
  perturbative forces on global positioning system, 149
Multidimensional problems, collocation methods as preference in, 79

Neumann-Kelvin boundary value problem, 363-364
Neumann problem, 4, 25
Newton-Kantorovich method
  astrodynamics numerical technique, 133, 141-144
  iteration, 157, 165-167

Index

Newton universal gravitational constant, 145
Noether operators, 319
Nonlinear integral equations, superconvergence, 64–66
Numerical method in astrodynamics
 equation of motion, 144–149
 Green's function, 132
 mathematical preliminaries, 134–135
  Hilbert space, 138–140
  integral equation and, 135–136
  Legendre equation and, 136–138
  Newton–Kantorovich method, 141–144
  operator equations, 140–141
 numerical examples
  Earth–Mars trajectories, 174–177
  trajectory optimization, 177–178
 parallel algorithms, 165–174
  constant vector setup in, 168–169
  Legendre polynomial expansion, 171–172, 173
  matrix setups in, 167–168
  solution of linear system, 170–171
 perturbed Galerkin method, 158–161, 164–165
  analytic principle in, 163–164
  convergence result, 164–165
  equivalence in, 161–163
 well-posedness issue, 149–150
 method of patched conics, 151–152
 Newton–Kantorovich iteration, 157, 165–167
 perturbed Keplerian motion, 155–156
 Picard iteration, 154–155
 Voyager II example, 152–153
 See also Motion, equations of
Nyström method, 77, 110, 131, 133, 158, 161, 162–163, 165

Panel method, 21
Parallel processing, 150
 See also Numerical method in astrodynamics
Parseval theorem, 312–313, 343
Patched conics, method of, 151–152
Perturbation theory, 72
Perturbed Galerkin method, in numerical method in astrodynamics, 158–161, 164–165
Perturbed Keplerian motion, in numerical method in astrodynamics, 155–156
Perturbed projection methods, 91–95, 287–293
 perturbed Dellwo–Friedman method, 95
 perturbed Kantorovich iterate, 94
 perturbed Kantorovich method, 93–94
 perturbed Sloan iterate, 92–93
Picard iteration, in numerical method in astrodynamics, 154–155
Piecewise polynomials
 Cauchy singular integral equations, 272
 iterated collocation for, 58–59
Piecewise smooth boundaries, Laplace equation reformulation, 8–9
Planing surface problem, 363–371
 generalizations, 368–371
  surface tension, 369
  two-dimensional flow assumptions, 368–369
 planing equation, 364–368
  airfoil equation analogy, 365–366
  Cauchy singular integral equation, 364–365
  Kutta condition, 366
  premature detachment in, 368
Poincare–Bertrand formula, 312, 313, 317, 322, 343
Polynomial approximation methods
 Cauchy singular integral equations, 258–272
 Chawla and Kumar's method, 270–272
 Cohen method, 269–270
 collocation method, 263–264
 Galerkin method, 262–263
 Gaussian quadrature method, 264–266
 Hashmi and Delves method, 272
 Lobatto quadrature, 266–268
 piecewise polynomial methods, 272
 quadrature method, 264
Polynomial basis functions, superconvergence, 36
Positive definiteness property, Galerkins method and, 120–126
Premature detachment, planing surface problem in, 368
Probability density function, 395
Product quadrature, Cauchy singular integral equations, 246–247
Projection methods
 collocation methods, 110–120
  direct analysis of discrete method, 113–117

Projection methods (*cont.*)
  collocation methods (*cont.*)
    superconvergence for Volterra equations, 117–120
    convergence analysis of, 81–84
    definition of projection method, 75
    Galerkin method, 77–80, 95–106
      direct analysis with quadrature errors, 106–110
    for positive-definite dominant part equations, 120–126
    perturbed projection methods, 91–95
      perturbed Dellwo-Friedman method, 95
      perturbed Kantorovich iterate, 94
      perturbed Kantorovich method, 93–94
      perturbed Sloan iterate, 92–93
    variants of
      accelerated projection methods, 89–90
      iterated Kantorovich regularization, 88–89
      Kantorovich regularization, 87–88
      Sloan iterate, 84–86
Prolongation operator, 27, 324
Pseudoanalytic methods, Abel integral equations, 386, 387–394
Pseudodifferential operators, 124
Pseudoinverse, 318

Quadrature errors
  collocation method with, 110–120
  Galerkin method with, 106–110
Quadrature methods, Cauchy singular integral equations
  Gaussian quadrature method, 233–236
  Lobatto quadrature, 236–241
  quadrature rules for principle value integrals, 230–233
Quasi-linearization. *See* Newton-Kantorovich method

Radon transform, 380–381
Restriction operator, 27, 325

Seismology, Abel integral equations, 379–380
Ship hydrodynamics
  hovercraft, 364
  See also Planing surface problem

Singular integral equations
  direct methods, 328–359
    collocation method, 349–356
    discrete Galerkin method, 356–359
    Galerkin-Petrov method, 342–348
  indirect methods, 323–328
    collocation method, 328
    Galerkin method, 323–328
  theory for, 310–322
Singular value decomposition, 330
Sloan iterate, 105, 115
  convergence, 283–284
  perturbed Sloan iterate, 92–93
  as variant of projection method, 84–86
Smooth boundary case, Laplace equation reformulation, 7–8
Smoothing step, 27
Sohngen inversion formula, 191, 192, 197
Sokhotski-Plemelj formulas, 314
Spectral density function, 397
Spectral distribution function, 397
Sphere of influence, of planet, 151
Stability, 334–335
Stabilized evaluation of inversion formulas, Abel integral equations, 386–397, 400–402
Stereology, Abel integral equations, 378, 379
Stochastic process, stationary/strictly stationary process, 395
Strongly elliptic operators, boundary integral equation reformulations, 7, 14–15
Superapproximation, 37
Superconvergence
  Cauchy singular integral equations and, 37
  of collocation for Volterra equations, 117–120
  discrete collocation method, 64
  discrete Galerkin methods, 55
  Galerkin method, 36, 38–40
  global superconvergence, 36
  HMP method, 53–55
  iterated collocation method, 56–58
    versus iterated Galerkin method, 63–64
  for piecewise-polynomial spaces, 58–63
  iterated Galerkin method, 51–53, 55
  iterated Kantorovich method, 45–48

Superconvergence (*cont.*)
  iterated projection method, 40–45
  linear functions of Galerkin approximation, 48–51
  nonlinear integral equations, 64–66
  use of term, 35–36
  Volterra integral equations and, 37

Taylor expansion, 143
Tomography, Abel integral equations, 380–381
Tuck method, 252–254
Two-body problem, 147
Two-grid method, boundary integral equations, 26–29

Two-point boundary value problem
  computational aspects, 133–134
  equivalence to Hammerstein integral equation, 132
  See also Numerical method in astrodynamics

Uniform convergence, Galerkin method, 281–283

Volterra equations, 116
  superconvergence and, 37
Voyager 2, 152–153

Weiner filtering, Abel integral equations, 386, 394–400